Wearable Technology in Medicine and Health Care

Wearable Technology in Medicine and Health Care

Raymond Kai-Yu Tong

*Department of Biomedical Engineering, Faculty of Engineering,
The Chinese University of Hong Kong, Hong Kong*

ACADEMIC PRESS

An imprint of Elsevier

Academic Press is an imprint of Elsevier
125 London Wall, London EC2Y 5AS, United Kingdom
525 B Street, Suite 1650, San Diego, CA 92101, United States
50 Hampshire Street, 5th Floor, Cambridge, MA 02139, United States
The Boulevard, Langford Lane, Kidlington, Oxford OX5 1GB, United Kingdom

Notices
Knowledge and best practice in this field are constantly changing. As new research and experience broaden our understanding, changes in research methods, professional practices, or medical treatment may become necessary.

Practitioners and researchers must always rely on their own experience and knowledge in evaluating and using any information, methods, compounds, or experiments described herein. In using such information or methods they should be mindful of their own safety and the safety of others, including parties for whom they have a professional responsibility.

To the fullest extent of the law, neither the Publisher nor the authors, contributors, or editors, assume any liability for any injury and/or damage to persons or property as a matter of products liability, negligence or otherwise, or from any use or operation of any methods, products, instructions, or ideas contained in the material herein.

British Library Cataloguing-in-Publication Data
A catalogue record for this book is available from the British Library

Library of Congress Cataloging-in-Publication Data
A catalog record for this book is available from the Library of Congress

ISBN: 978-0-12-811810-8

For Information on all Academic Press publications
visit our website at https://www.elsevier.com/books-and-journals

Working together
to grow libraries in
developing countries

www.elsevier.com • www.bookaid.org

Publisher: Mara Conner
Acquisition Editor: Chris Katsaropoulos
Editorial Project Manager: Joshua Mearns
Production Project Manager: R. Vijay Bharath
Cover Designer: Christian J Bilbow

Typeset by MPS Limited, Chennai, India

Contents

Chapter 4: Upper Limb Wearable Exoskeleton Systems for Rehabilitation: State of the Art Review and a Case Study of the EXO-UL8—Dual-Arm Exoskeleton System....... 71

Yang Shen, Peter Walker Ferguson, Ji Ma and Jacob Rosen

Chapter 5: Lower Limb Exoskeleton Robot to Facilitate the Gait of Stroke Patients....... 91

Ling-Fung Yeung and Raymond Kai-Yu Tong

Chapter 9: WearUp: Wearable Smart Textiles for Telemedicine Intervention of Movement Disorders

*Mohammadreza Abtahi, Nicholas P. Constant, Joshua V. Gyllinsky,
Brandon Paesang, Susan E. D'Andrea, Umer Akbar and Kunal Mankodiya*

List of Contributors

Mohammadreza Abtahi   Electrical, Computer, and Biomedical Engineering, University of Rhode Island, Kingston, RI, United States

Talha Agcayazi   Electrical and Computer Engineering, North Carolina State University, Raleigh, NC, United States

Umer Akbar   Department of Neurology, Rhode Island Hospital, Providence, RI, United States

C.W Antuvan   PhD Student, School of Mechanical and Aerospace Engineering, Nanyang Technological University, Singapore

Thomas Boillat   Stanford University School of Medicine, Stanford, CA, United States

Alper Bozkurt   Electrical and Computer Engineering, North Carolina State University, Raleigh, NC, United States

Serhat Burmaoglu   Department of Health Management, Izmir Katip Celebi University, Izmir, Turkey

L. Cappello   Postdoctoral Fellow, Biorobotics Institute, Scuola Superiore Sant'Anna, Pisa, Italy

Brian Caulfield   Insight Centre for Data Analytics, University College Dublin, Dublin, Ireland

Nicholas P. Constant   Electrical, Computer, and Biomedical Engineering, University of Rhode Island, Kingston, RI, United States

Fani Deligianni   Hamlyn Centre, Institute of Global Health Innovation, Imperial College, London, United Kingdom

K.B. Dhinh   PhD Student, School of Mechanical and Aerospace Engineering, Nanyang Technological University, Singapore

Susan E. D'Andrea   Providence Veteran Affairs Medical Center, Providence, RI, United States

Peter Walker Ferguson   Department of Mechanical and Aerospace Engineering, University of California Los Angeles (UCLA), Los Angeles, CA, United States

Daniel Freer   Hamlyn Centre, Institute of Global Health Innovation, Imperial College, London, United Kingdom

Tushar Ghosh   College of Textiles, North Carolina State University, Raleigh, NC, United States

Joshua V. Gyllinsky   Computer Science and Statistics, University of Rhode Island, Kingston, RI, United States

X. Jiang   Menrva Research Group, Schools of Mechatronic Systems and Engineering Science, Simon Fraser University, Metro Vancouver, BC, Canada

Melisa Junata   Department of Biomedical Engineering, Faculty of Engineering, The Chinese University of Hong Kong, Hong Kong

John Kedzierski   University of Toronto Mississauga, Department of Biology, Mississauga, Canada

Chwee Teck Lim   Department of Biomedical Engineering, National University of Singapore, Singapore; Mechanobiology Institute, National University of Singapore, Singapore; Biomedical Institute for Global Health Research and Technology, National University of Singapore, Singapore

Jindong Liu   Hamlyn Centre, Institute of Global Health Innovation, Imperial College, London, United Kingdom

John H.T. Luong   Sektion Experimentelle Anaesthesiologie, University Hospital Ulm, Ulm, Germany

Ji Ma   Department of Mechanical and Aerospace Engineering, University of California Los Angeles (UCLA), Los Angeles, CA, United States

Kunal Mankodiya   Electrical, Computer, and Biomedical Engineering, University of Rhode Island, Kingston, RI, United States

Sana Maqbool   University of Toronto Mississauga, Communication, Culture, Information & Technology, Mississauga, Canada

L. Masia   Assistant Professor, School of Mechanical & Aerospace Engineering, Nanyang Technological University, Singapore

Michael McKnight   Electrical and Computer Engineering, North Carolina State University, Raleigh, NC, United States

C. Menon   Menrva Research Group, Schools of Mechatronic Systems and Engineering Science, Simon Fraser University, Metro Vancouver, BC, Canada

Qasim Muhammad   University of Toronto Mississauga, Department of Biology, Mississauga, Canada

Brandon Paesang   Textiles, Fashion Merchandising, and Design, University of Rhode Island, Kingston, RI, United States

Jayson L. Parker   University of Toronto Mississauga, Department of Biology, Mississauga, Canada

Homero Rivas   Division of Bariatric and Minimally Invasive Surgery, Stanford Medical Center, Stanford, CA, United States

Jacob Rosen   Department of Mechanical and Aerospace Engineering, University of California Los Angeles (UCLA), Los Angeles, CA, United States

Yang Shen   Department of Mechanical and Aerospace Engineering, University of California Los Angeles (UCLA), Los Angeles, CA, United States

Jingjing Shi   Northeastern University, Shenyang, China

Patrick Slevin   Insight Centre for Data Analytics, University College Dublin, Dublin, Ireland

Raymond Kai-Yu Tong   Department of Biomedical Engineering, Faculty of Engineering, The Chinese University of Hong Kong, Hong Kong

Vladimir Trajkovik   Faculty of Computer Science and Engineering, Ss Cyril and Methodius University, Skopje, FYR Macedonia

Tatjana Loncar Tutukalo   Faculty of Technical Sciences, University of Novi Sad, Novi Sad, Serbia

Rejin John Varghese   Hamlyn Centre, Institute of Global Health Innovation, Imperial College, London, United Kingdom

Sandeep Kumar Vashist   Immunodiagnostic Systems, Liege, Belgium

M.N. Victorino   Menrva Research Group, Schools of Mechatronic Systems and Engineering Science, Simon Fraser University, Metro Vancouver, BC, Canada

Jianqing Wang   Nagoya Institute of Technology, Nagoya, Japan

M. Xiloyannis   PhD Student, Program for Research in Future Healthcare in the Interdisciplinary Graduate School (IGS), Nanyang Technological University, Singapore

Haydar Yalcin   Department of Information Management, Izmir Katip Celebi University, Izmir, Turkey

Guang-Zhong Yang   Hamlyn Centre, Institute of Global Health Innovation, Imperial College, London, United Kingdom

Joo Chuan Yeo   Department of Biomedical Engineering, National University of Singapore, Singapore

Ling-Fung Yeung   Department of Biomedical Engineering, Faculty of Engineering, The Chinese University of Hong Kong, Hong Kong

Preface

Wearable Technology in Healthcare and Medicine is written for biomedical, and other clinical and technical professionals. Since wearable technology advances at an ever-increasing pace, we have included new and exciting topics, which reflect the modern technology for the clinical and biomedical applications. To successfully and smoothly apply wearable products, it requires to work with various key stakeholders in governments, manufacturers, hospitals, and medical doctors. Throughout the book, we present the main applications and challenges in the biomedical implementation of wearable devices. Government bodies will find this book useful to understand the state of the art of wearable technology to support their policy to improve the modern healthcare system. The medical device industry can use it to better understand and access this emerging market. Academics and students will find this book very important for their careers in biomedical engineering and wearable device—related fields.

This book would not have been possible without contributions from outstanding experts in various topics discussed in it. I wish to express my gratitude to all of them for their precious efforts and strong support. I would like to thank my research student, Melisa Junata, for her assistance.

Finally, many thanks to my family (parents, Wai-chuen Tong, and Lai-lin Tsui, and daughter and sons, Lok-ching, Lok-tin, and Lok-ting), for their support, encouragement, and patience. They have been my driving forces.

Raymond Kai-Yu Tong
Professor and Chairman, Department of Biomedical Engineering (BME)
Chinese University of Hong Kong (CUHK)

CHAPTER 1

Wearable Technology in Medicine and Health Care: Introduction

Melisa Junata and Raymond Kai-Yu Tong

Department of Biomedical Engineering, Faculty of Engineering, The Chinese University of Hong Kong, Hong Kong

In this era of continuous innovations, the health care world that we are living in will look different in the future. We heard all about new technology innovations in the health-care world. They all promise different outcome and leap from what we have discovered before. A hearing aid is the earliest form of wearable technology that people have ever created [1]. But with the developments in the computing world and the rise in ubiquitous computing technology that gets smaller by size, wearable technology will be a definite part of our future.

It was unimaginable for people in the early forms of computer years that now we can have a smartwatch that will detect our steps and our heartbeat as we walk. It was a significant breakthrough for people to have portable computers wherever and anywhere we go. As time goes on, they are no longer restricted to workplace and home. Wearable technology enables us to wear these devices. We got used to the recent wearable technology that focuses on fitness and health. However, these are not the only ones coming. More advanced devices have a broader use and more sophisticated than before. Rather than consumer-centered, the development of the device is trying to cater health-care professionals as well. This book is comprised of chapters of wearable technology that goes into different facets of technology and impacts health care like never before.

Wearable technology itself is evolving and proliferating in both commercial and research worlds. From Fig. 1.1, we can see how the projected size of the global market for wearable devices in health care grows incrementally from 2015 to 2021. With this trend, we can be confident that the wearable technology will be a big part of health care and medicine sector in the future and there is a significant market for them.

From Fig. 1.2 we can see that the term wearable technology in the medical world arises from 1946 with two studies published that year. Then the studies start to grow steeply from

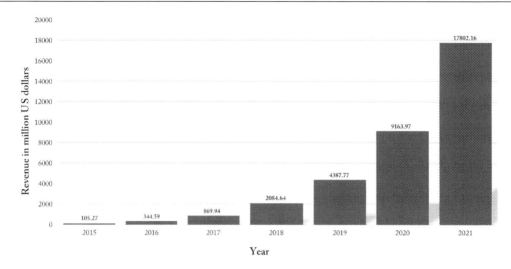

Figure 1.1

Projected size of the global market for wearable devices in the health-care sector from 2015 to 2021 [2].

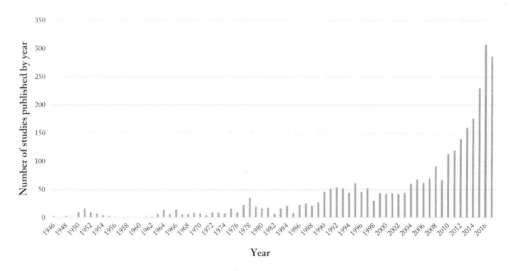

Figure 1.2

Number of studies resulted from a PubMed search with "wearable technology AND (health care OR medicine)" as the keyword [3].

the year 2010 onward. The exponential growth observed from 2010 onward is a sign that more scientists are developing as well as assessing wearable technology that is going to be deployed in the health-care world.

With increasing number of global aging population, paired with home-based health care and rehab, as well as personalized medicine, wearable technology seems like an integral part of the future medicine [4]. The chapters that follow contribute toward the different technological development that has recently turned into media buzzword heard daily: smart eyewear, robotics, Internet of things (IoT), and big data. Some chapters also talk about the intangible parts of technology: patent and regulatory. But they are not only buzzwords, they are also different growing developments in which any technology are going after.

Smart eyewear is a wearable computer glasses that add information alongside or to what the wearer sees [5]. Boillat and Rivas from Stanford University School of Medicine (United States) demonstrate how the use of smart eyewear in the surgery room will enable surgeons and other health professionals to perform things that were impossible or challenging to be done before. This concept is enabled by commercial smart eyewear such as Google Glass and it has benefited time efficiency and live collaboration.

Robotics is defined *as* a field of engineering that deals with machines that manage a task by mimicking human behavior [6]. Contributions for this book spans widely on wearable robotics. Varghese et al. from Imperial College London (United Kingdom) provide us a review of upper-limb wearable robotics for rehabilitation. The all-in-one review encompasses the design and mechanism of the upper-limb robots, as well as the challenges we may face to put this promising device out there for clinical applications. Shen et al. from the University of California, Los Angeles (United States), discuss the recent progress of upper-limb wearables for rehabilitation and also talk about their research team's "EXO-UL8" bilateral upper arm rehabilitation wearable. The next robotic wearable review covers the lower-limb wearable for rehabilitation in stroke patients by Yeung and Tong from the Chinese University of Hong Kong (HKSAR, China). They cover the mechanical, electronic, and control design of the lower-limb wearable that is relevant for its clinical application.

The *IoT* refers to a type of network to connect anything with the Internet through sensors and other monitoring and transmitting equipment to interact and communicate in order to reach common goals such as smart recognition, positioning, tracking, monitoring, and administration [7]. With IoT, health-care devices can be more interconnected than ever and this allows health care to be able to be administered out of conventional point-of-care such as hospitals and clinics to homes and elderly centers.

Sensors is one of the component that enables IoT to happen. Yeo and Lim from the National University of Singapore (Singapore) take us into a greater depth that many upper-limb wearable technologies are faced with, the sensors for upper-limb monitoring and they also discuss on how these technologies meet the requirements of upper-limb monitoring as well as the challenges involved. Victorino et al. from the Simon Fraser University (Canada) expanded the upper-limb sensor discussion by getting into details on the emerging force myography, which has shown capable of doing accurate continuous sensing of finger's movement.

Smart textile is another technology that we can adopt for use in wearable technology. McKnight et al. from North Carolina State University (United States) talk about the process and the challenges that is involved in creating sensing fabrics and how they can be integrated with the day-to-day textile that we normally wear, but still stay robust to perform its functions. Abtahi et al. from the University of Rhode Island in collaboration with Providence Veteran Affairs Medical Centre and Rhode Island Hospital (United States) describe the future potential of wearable smart textile to change the medical practice; they make their point by demonstrating their application of smart textile on their WearUP smart glove for Parkinson's disease patients. Xiloyannis et al. from Nanyang Technological University (Singapore) illustrate the possible switch that we may have from hard robots to more user-friendly made-from-fabric exosuit. They demonstrate the exosuit potential with their application for assisting elbow movements.

It is well understood that with more sensing technologies, we also created an even bigger pool of data to be monitored and makes sense of. *Big data* is a term that describes large volumes of high velocity, complex, and variable data that require advanced techniques and technologies to enable the capture, storage, distribution, management, and analysis of the information [2]. Shi and Wang from Northeastern University (China) and Nagoya Institute of Technology (Japan), respectively, discuss about human body communication-based wearable technology that involves sensing and transmitting vital signs. Vashist and Luong from Immunodiagnostic Systems (Belgium), and University College Cork (Ireland), respectively, illustrate an overview of the wearable technologies that are available commercially for personalized mobile health-care monitoring and management. Slevin and Caulfield from University College Dublin (Ireland) explore how the benefits and uses of data generated by patients will change the future of health care.

With all of the things happening for wearable technology, patent and regulatory is something that we need to consider when we want to bring this technology out to the marketplace. Burmaoglu et al. from Izmir Katip Celebi University (Turkey), University of Novi Sad (Serbia), Ss. Cyril and Methodius University (FYR Macedonia), and University College Dublin (Ireland) talk about how wearable technology in health-care patents evolve over time. Parker et al. from University of Toronto Mississauga (Canada) follows by presenting the chapter on the interconnected truth between design and regulations. This is a very essential chapter noting that regulations are strongly tied with the overall design of the medical device and its delivery to the market.

All in all, we hope that this book and its contents can help widen the horizon of its readers on what we can do to improve medicine and health care with more advanced wearable devices for clinical professionals and patients.

References

[1] M. Mills, Hearing aids and the history of electronics miniaturization, IEEE Annals of the History of Computing 33 (2) (2011) 24−44.

[2] TechAmerica Foundation's Federal Big Data Commission, Demystifying big data: a practical guide to transforming the business of Government.

[3] PubMed Search (accessed 29.01.18.).

[4] S. Majumder, T. Mondal, M.J. Deen, Wearable sensors for remote health monitoring, Sensors (Basel, Switzerland) 17 (1) (2017) 130.

[5] S. Mann, Vision 2.0, IEEE Spectrum 50 (3) (2013) 42−47.

[6] J. Simpson, E. Weiner, Oxford English Dictionary, 3rd ed., Clarendon, Oxford UK, 2013.

[7] G. Morabito, D. Giusto, A. Iera, L. Atzori, The Internet of Things, 1st Edition, Springer, Berlin, 2010.

CHAPTER 2

Empowering Medical Staff With Smart Glasses in Operating Rooms

Thomas Boillat[1] and Homero Rivas[2]

[1]*Stanford University School of Medicine, Stanford, CA, United States* [2]*Division of Bariatric and Minimally Invasive Surgery, Stanford Medical Center, Stanford, CA, United States*

2.1 Introduction

Health care is a knowledge-intensive field where education, experience, and rules, among many other factors affecting all stakeholders, intertwine into a complex ecosystem. Therefore, to guide medical staff's activities and manage knowledge, information technologies (ITs) have become key [1]. A place where ITs have a critical place and new devices are constantly evaluated is the operating room (OR). While the number of computers and screens exceeds the number of people in OR, accessing the right information at the right time remains a challenge. A promising technology that might change this story, though, is smart glasses. The main concept behind smart glasses is not new, but it has reached a higher level of fidelity and maturity. Before being "smart," these glasses were mainly known for their head-up mounted displays (HMDs).

Compared to traditional computers, smart glasses provide novel affordances that enable individuals to access computing resources in different ways [2]. In many fields, smart glasses primarily serve the purpose of improving efficiencies and security at work. When applied to an order picking scenario, smart glasses reduce pick error by 80% and completion time by 38% compared to traditional paper-based forms [3].

For more than two decades, HMDs have been experimented with in ORs. They are particularly appreciated for their capacity to display information such as patient electrocardiograms in a medical staff's field of view [4]. Thus, HMDs do not constrain any movements, keeping both hands free while providing information that would not be accessible otherwise. HMDs took another dimension in 2013 when Google introduced the Glass. In addition to the HMD, Google placed multiple sensors on the unit, including a GPS, accelerometers, and gyroscopes to sense the context in which the glasses are used.

Equipped with Wi-Fi and Bluetooth connections to access the Internet, the glasses can provide contextual and personalized services. All of this new hardware and software have transformed HMDs to what is presently referred to as smart glasses.

While the list of smart glasses providers grows nearly every month, it remains unclear how this relatively new technology can be leveraged in ORs and how it can enhance medical staff's capabilities. In this chapter, we intend to address this gap by analyzing the tasks executed by nurses, anesthesiologists (ANs), and surgeons (SUs) in ORs and the capacities of current smart glasses to support them. It will show the opportunities offered by the different smart glass types and conclude with our vision of how smart glasses will shape ORs in the future.

2.2 Smart Glasses

2.2.1 Definition

Smart glasses are a wearable technology that builds on spectacle frames to display contextualized information in a person's field of view [2]. The main piece of hardware is a head-mounted display that allows users to access texts, pictures, and videos. They are also equipped with a front-end camera, touchpad, microphone, and a series of sensors (e.g., accelerometer, gyroscope, GPS) [5]. Smart glasses are either connected to a mobile device (e.g., smartphone, tablet) or to a Wi-Fi network that enables access to the Internet or a company's information system. Compared to smartphones that require users to hold them in one hand, smart glasses offer new affordances [2]: they can be used hands-free via voice commands, and information is immediately displayed and accessible directly in their user's field of view. The front-end camera digitally reproduces, via pictures or videos, what their user sees. These affordances offer new opportunities to use computing capacities for supporting tasks that could not benefit from any other technologies (e.g., laptop computers, smartphones) due to physical constraints. For instance, in the field of inspection and maintenance, technicians and engineers can access checklists to guide them while their two hands are occupied [6]. Another field seen has a good candidate for the use of smart glasses is surgery for two reasons: SUs and some medical staff (e.g., scrub nurses (SNs)) handle themselves in a sterile environment and can therefore not touch any nonsterile objects (e.g., screens), while they need to focus their attention on patients rather than screens.

2.2.2 Overview of Available Smart Glasses

Google Glass has had a prominent place in the smart glass market. It was released mid-2013 via the Explorer Program to approximately 8000 partners, mostly developers. Given Google's notoriety, Glass has attracted a lot of interest and has been under the radar of many IT companies, analyst firms, universities, as well as industries. However, this popularity also led to many disillusions when Google suddenly stopped selling its Glass with unclear intentions of the future of its product. Though Google is not the inventor of smart glasses, it has shaped the market. Many other IT suppliers have benefited from Google Glass' user experience and use cases to design their own smart glasses.

Current smart glasses can be classified along three categories according to their materiality [7]: (1) smart eyewear (EW), which consists of a monocular lens that does not obstruct the field of view. It is mainly used to access information and capture pictures and videos, and is considered as extending smartphones' capacities, (2) augmented reality (AR) headsets are binocular lenses covering both eyes. They usually allow users to see through their glasses while the binocular lenses enable 3D capabilities, and (3) virtual reality (VR) headsets completely obstruct the field of view to immerse users in different worlds. Many of the EW and AR smart glasses are designed to support health care in an OR environment, while VR headsets by nature cannot be used in the OR. However, VR may have a few applications in simulation, management of phobias by desensitization and education, for instance.

Table 2.1 presents a nonexhaustive list of EW and AR headsets that are foreseen by analyst firms as the best candidates to be used in OR.

Among this list, two are EW headsets: Google Glass 2 and the Vuzix M300. The former is the new Google Glass headset that has been distributed to developers mid-2016 and currently available through Google's partner network. Aside from being able to fold the frame, the second version is very similar to the first one from a design perspective. The screen is still located on the right-hand side, while the headset is still very light, with a weight below 50 g. The Vuzix M300 is considered Google Glass's direct competitor. Its screen can be placed either on the left- or on the right-hand side and can be adjusted horizontally; however, it is slightly heavier than the Google Glass. The five other headsets are designed for AR and have two screens. The ODG R-7 is the closest AR headset to a traditional pair of glass (Fig. 2.1). It contains two screens with a resolution of 1280*720p, mounted behind 60%−80% transparent glasses. Optionally, traditional transparent glasses can be mounted. The Sony SmartEye and the Epson BT-300 are the most accessible AR headsets for consumers. They are controlled through an additional external touchpad linked to the glasses via a wire. The AiR Glasses by Atheer is characterized by its three cameras that enable users to navigate through and within applications via finger gestures. The glass is linked to an additional case in which the battery and part of the hardware are located. Lastly, the Microsoft Hololens uses holographs to increase user immersion and highly leverage finger gestures, too.

Through a smart glass's materiality, one can have a good idea of the ability of each headset to be worn by medical staff. This is referred to as the "wearability." The heavier the headset and the greater the requirement for hand interactions, the more impractical it becomes for medical staff, given that most of the time their hands are occupied or must remain sterile. Table 2.1 summarizes the characteristics of the selected smart glasses.

Table 2.1: Characteristics of the selected smart glasses

Characteristics		Google Glass 2[a]	Atheer AiR Glasses[b]	Microsoft Hololens[c]	Epson BT-300[d]	Vuzix M300[e]	Sony SmartEye[f]	ODG R-7[g]
	Brand	Google	Atheer	Microsoft	Epson	Vuzix	Sony	ODG
	Model	Glass 2[a]	AiR Glasses[b]	Hololens[c]	BT-300[d]	M300[e]	SmartEye[f]	R-7[g]
	Category	EW	AR	AR	AR	EW	AR	AR
	Availability	2016[h]	2016	2016	2016	2016	2016	2016
	Price	1500USD	3950USD	3000USD	750USD	1499USD	840USD	2750USD
	OR Wearability	High	Low	Low	Low	High	Low	Medium
	Memory	2 GB RAM, 16 GB storage	2 GB RAM, 128 GB storage	2 GB RAM, 16 GB storage	2 GB RAM, 16 GB storage	2 GB RAM, 16 GB storage	2 GB RAM, 16 GB storage	3 GB RAM, 64 GB storage
Inputs	Weight (g)	45	135 + 350 battery	579	59 + 130 controller	70	77 + 45 controller	170
	Touchpad	Yes	Yes	No	Yes	Yes	Yes	Yes
	Voice commands	Yes	Yes	No	No	Yes	Via the controller	Yes
	Gesture	No	Yes	Yes	No	No	No	Yes
Output	Camera	Photo 5Mp, Video 720p	Photo 4Mp, Video 720p	Photo 2Mp, Video 720p	Photo 5Mp, video 720p	Photo 13Mp, video 1080p	Photo 3Mp, video N/A	Photo 4Mp, video 1080p
	Screen	1 screen: 640 × 360	2 screens: 1280 × 720	2 screens: 1268 × 720	2 screens: 1280 × 720	1 screen: 640 × 360	2 screens: 419 × 138[i]	2 screens: 1280 × 720
	See-through	Yes	Yes	Partially	Yes	Partially	Yes	Yes
	Speaker	One ear: bone speaker	Built-in speaker	Built-in speaker	Additional earplugs	Built-in speaker	In the controller	Additional earplugs

[a]https://www.google.com/glass/start/.
[b]https://store.atheerair.com/air-glass.html.
[c]https://www.microsoft.com/en-us/hololens.
[d]https://epson.com/For-Home/Wearables/Smart-Glasses/c/h420.
[e]https://www.vuzix.com/products/m300-smart-glasses.
[f]https://developer.sony.com/develop/wearables/smarteyeglass-sdk/.
[g]https://shop.osterhoutgroup.com/products/r-7-glasses-system.
[h]Via Google Glass partners.
[i]Monochrome.

Figure 2.1
ODG R-7—the closest model to a traditional pair of glass.

2.3 Current Uses and Benefits of Smart Glasses in ORs

Much research has been conducted to analyze the extent to which smart glasses can support medical staff in ORs. In 1995, ANs were among the first to clinically evaluate smart glasses' capacities in ORs [4]. Their study focused on one of the main ANs' challenges: accessing a patient's vital signs while being away from their traditional screens. Over the years, many scholars have investigated the benefits of smart glasses in anesthesiology. Research shows that smart glasses reduce the number of times ANs have to shift their attention between patients and screens [8], in addition to helping them detect events when physically constrained [9,10]. Accessing vital signs can also be critical for SUs when an AN is not present in the OR (e.g., endoscopy) or when performing a risky surgery. Evaluated in a laboratory setting, SUs wearing smart glasses (Google Glass) to access vital signs spent 90% less time looking away from the procedural field to view traditional monitors during bronchoscopy, and recognized critical desaturation 8.8 seconds earlier than the group of SUs without smart glasses [11]. Another study shows that SUs can benefit from smart glasses in order to access patient information during surgeries, such as CT scans, X-rays, or notes from previous procedures [12,13].

Smart glasses are not only used in ORs to ease and speed the access to information; SUs have shown great interest in using them to get remote assistance while performing surgeries. Given the position of the camera adjacent to a SU's eye, smart glasses can capture what a SU really sees [5]. Traditionally, static cameras located at headlights or from a peripheral cameraman have been used. However, this has often been obstructive to the field of view of the SUs and surgical teams. Through Wi-Fi connections, smart glasses can also broadcast video to anyone who has an Internet access to communicate with SUs using the built-in microphone and speaker [14,15]. Remote assistance is particularly useful for junior physicians as well as critical cases when the advice of experts is required to ensure patients' safety [5]. Finally, studies report that smart glasses are also used to document cases. SUs can easily take pictures, videos, or notes during surgeries via voice commands. This shortens the documentation process and can be used for conducting further research [16].

Table 2.2: Selected smart glasses' capacities

	Display Short Content	Display Rich Content	Take Pictures/ Videos	Take Note/ Memo	Remote Assistance	Hands-Free Navigation	Not Disturb "Off Glass" Activities
Google Glass 2	+ + +	+	+ +	+ +	+ +	+ + +	+ + +
Atheer AiR Glasses	+ +	+ + +	+ +	+ + +	+ + +	+ + +	+
Microsoft Hololens	+ + +	+ + +	+ +	+ +	+ + +	+ + +	+
Epson BT-300	+ +	+ +	+ +	+ +	+ +	+	+
Vuzix M300	+ + +	+	+ + +	+ +	+ +	+ + +	+ +
Sony SmartEye	+ +	−	+	+	−	+	+ +
ODG R-7	+ + +	+ + +	+ +	+ + +	+ + +	+ + +	+ +

2.3.1 Capacities of Smart Glasses

Based on their materiality and the different uses taken from existing studies presented earlier, smart glasses can be evaluated along their capacities (Table 2.2). A "−" indicates an inability to support a distinct use, while "+ + +" indicates complete ability. For instance, the small screen of the Google Glass can perfectly display short content, such as a few lines of text or a simple picture. Thus, it receives "+ + +" for this ability. However, Google Glass cannot make long content easily readable, and for this reason it receives a " + " under this category. Capacity assessments were determined by the authors based on direct experience with the smart glasses, interpretation of other reviewers' assessments, or a combination of the two.

2.4 Activities Performed in ORs Among the Medical Staff

In order to have a comprehensive understanding of how smart glasses can support medical staff, we describe the main activities of circulating nurses (CNs), SNs, ANs, and SUs. For each of these activities, we also describe their underlying requirements as well as their constraints and challenges. We then show how the capacities of smart glasses can address these constraints and challenges to eventually increase the effectiveness and efficiencies of medical staff's activities along with patient safety.

2.4.1 Circulating Nurse

2.4.1.1 Activities and requirements

The roles of CNs vary among hospitals and clinics. In small institutions, CNs welcome, prepare (e.g., install an IV), and accompany patients to the OR. In large institutions, CNs do not leave the OR. Their activities start with picking up supplies from the inventories (Table 2.3—No. 1). Surgeries require 20—100 different supplies depending on their type (e.g., laparoscopic, robotic, open) and on the SU. Most of the time, CNs work with

Table 2.3: CNs' activities and opportunities to use smart glasses

	Activities			Smart Glasses	
No.	Description	Requirements	Constraints/Challenges	Opportunities	Candidates
1	Pick up supplies from the inventory	• Pick up the right supplies • Ensure all supplies are taken • Supplies must be documented in patient's medical record	• CNs often work on many different surgeries • Supplies are sometimes badly placed • Number of supplies can be as big as 100	• Indicate where supplies are located • Automatic counting and report of supplies	ODG R-7, Atheer AiR Glasses
2	Set up the room	• Equipment must be ready	• Unclear how some equipment works	• Access preparation checklist • Access documentation	
3	Install the patient	• Install the patient in the correct position • Install an IV	• The position varies depending on the surgery and the SUs	• Access documentation	
4	Safety checklist (sign-in)	• Relevant elements must be verified	• Checklist outcomes must be reported into patient's medical record	N/A	N/A
5	Count supply	• All the supplies must be counted • Supplies are recounted after staff shifts	• Sometimes unclear what the SNs or SUs have used	N/A	N/A

preference cards or checklists to ensure the correct supplies are gathered, where after they prepare equipment and machines such as laparoscopic cameras (Table 2.3—No. 2). Across the different surgical specializations, CNs may need to manipulate up to 50 different devices, some of them being very complex (e.g., surgical robots). After installing the patient on the table (Table 2.3—No. 3), the CNs will go through a safety checklist along with an AN to confirm patient identity, procedure, patient concerns, etc. (Table 2.3—No. 4). Finally, at the end of the surgery or when medical shifts occur, CNs count the supplies with an SN to ensure nothing has been forgotten in the patient's body (Table 2.3—No. 5).

2.4.1.2 Constraints and opportunities to use smart glasses

During CN's activities, multiple constraints and challenges can affect work quality and patient's safety, and for some of these, smart glasses offer new opportunities to increase

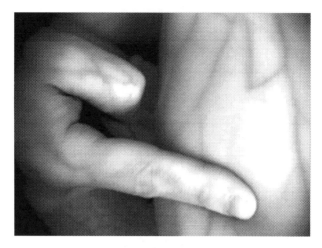

Figure 2.2
Use of an additional infrared module on smart glasses for viewing peripheral and femoral veins.

work effectiveness and efficiencies. While CNs often use paper based preference cards when picking up supplies from the inventory, it occupies one hand and provides little guidance to make sure the right supplies are taken. CNs would then benefit from smart glass applications already used in industrial settings. They directly display the supplies' list, indicate where supplies are located, and automatically scan their reference number to count and report them [6]. Smart glasses can also help set up the room, install the patient, and prepare the equipment by displaying the different tasks and the equipment's configuration. In case an IV has to be installed, infrared and ultrasound smart glasses allow for viewing peripheral veins and artery [17]. When it comes to executing the sign-in checklists or counting supplies, the needs for smart glasses are limited given that CNs can leverage other computing devices such as smartphones or tablets. However, they may support nurses by displaying the checklist's elements and the list of supplies to provide for more flexibility (Fig. 2.2).

2.4.1.3 Smart glass candidates

Potential uses of smart glasses for CNs include retrieving and reading large information content from pictures and text, therefore, a dual-lens smart glass is required. As CNs can touch nonsterile objects, smart glasses requiring touch navigation are possible; however, it is important for CNs to see through the glasses move freely. The battery life is not the first criterion given the little time the smart glasses are used. From the smart glasses we analyzed, the ODG R-7 and Atheer AiR Glasses are the best candidate. They are light, have a relatively small field of view, can be equipped with transparent lenses, and possess screen resolution suitable for large pictures and text contents.

Table 2.4: SNs' activities and opportunities to use smart glasses

| No. | Activities | | | Smart Glasses | |
	Description	Requirements	Constraints	Opportunities	Candidates
1	Setting up supplies	• Know how all supplies work	• Some SUs use specific supplies	• Access documentation (instructions)	Atheer AiR Glasses, Microsoft Hololens and the ODG R-7
2	Provide supplies to SU(s)	• Know the names of all supplies • Know how to prepare and give supplies	• Unclear what is on the table • Some supplies need specific preparation	• Access documentation	
3	Count supply	• All the supplies must be counted	• Supply can be forgotten in patient's body	• Keep track of what is used during the surgery	

2.4.2 Scrub Nurse

2.4.2.1 Activities and requirements

SN activities usually start with setting up supplies (Table 2.4—No. 1). This includes counting the supplies with the CN and preparing equipment (e.g., needles, knives). Once the surgery commences, the main role of SNs is to provide SUs with the required supplies, ready to be used (Table 2.4—No. 2). When a supply is meant to temporarily stay in the patient's body, SNs are often responsible for reminding the SUs before the end of the surgery. At the end of the surgery, the SN and the CN count the supplies once again (Table 2.4—No. 3).

2.4.2.2 Constraints and opportunities to use smart glasses

SNs work in a sterile environment, implying that they can only touch sterile objects. During the initialization of the supplies, it often happens, in large institutions, that SNs are not familiar with the supplies they must prepare given they are specific to the procedures and the SUs. While they can always be de-scrubbed or provided with external assistance, smart glasses can help by displaying instruction or documentation via voice commands. The same applies when preparing the supplies: SNs occasionally manage up to 50 different types of supplies and may not be familiar with all of them. It is not usual for the SN to take a couple of minutes to find the correct supply and prepare it. In such cases, smart glasses can help SN identify supplies based on the surgical steps and provide visual and textual information to guide the supplies' preparation. If a supply is temporarily in the patient's body, the SN can create a reminder via voice commands. It can facilitate the count of supplies and can automatically document what supplies were used in the patient's medical record.

2.4.2.3 Smart glass candidates

While SNs share the same potential uses for smart glasses as CNs, the fact they must remain sterile dictates that voice or gesture commands are required. Also, CNs need the smart glasses for a long time periods, which calls for a battery life of at least 2 hours, with the option to be extended. Additionally, the weight of the glass must be as small as possible. As candidates, we foresee the Microsoft Hololens (although they are the heaviest, the weight is well distributed), Atheer AiR Glasses, and the ODG R-7.

2.4.3 Anesthesiologists

2.4.3.1 Activities and requirements

Deciding the anesthetic procedure is among the first AN's activities (Table 2.5—No. 1). Together with the patient, ANs intend to find the best technique and medications suitable for the patient's health. Then, with the CN or SN, ANs conduct anesthesia or safety

Table 2.5: ANs' activities and opportunities to use smart glasses

		Activities		Smart Glasses	
No.	Description	Requirements	Constraints	Opportunities	Candidates
1	Choose the anesthesia procedure	• Know patient's information (e.g., allergies)	• Patient may forget information • Computers are not always available	N/A	N/A
2	Anesthesia/safety checklist	• Relevant elements must be verified	• Elements can potentially be forgotten	N/A	N/A
3	Intubate and anesthetize the patient	• Know the different techniques	• Access to documentation can be limited	N/A	N/A
4	Ensure patient's stability	• Keep track of the medications • Document medications in patient's medical record	• Consider patient's allergies	N/A	N/A
5	Monitor vital signs	• Must access the vital signs at all time	• Monitors are not always visible (when intervene on patient)	• Access patient's vital signs	Atheer AiR Glasses, Epson BT-300, ODG R-7
6	Awaken patient	• Know the different techniques	• Monitors are not always visible (when intervene on patient)	• Access patient's vital signs	

checklists (Table 2.5—No. 2) before commencing the anesthetic procedure (Table 2.5—No. 3). Once the patient is unconscious, ANs must ensure the patient's stability (Table 2.5—No. 3), which involves (1) the continuation of the anesthesia process and (2) reacting to unplanned events (e.g., intolerance to medicine). It then requires the AN to observe the patient as well as his or her vital signs (Table 2.5—No. 5). When the surgery concludes, ANs must terminate the anesthetic procedure (Table 2.5—No. 6).

2.4.3.2 Constraints and opportunities to use smart glasses

Given that ANs are not in a sterile area and usually have access to a computer, opportunities for smart glasses are limited. The main benefit from using them occurs when shifting attention between patients and monitors to monitor the patient's vital signs (Table 2.5—No. 5 and 6). As revealed by studies [9], the use of smart glasses in these situations allow ANs to focus more on patients without the risk of missing any signs from the monitors. It also gives more freedom to ANs who are sometimes stocked behind screens for multiple hours.

2.4.3.3 Smart glass candidates

ANs are not sterile and can therefore touch a controller if required. Additionally, once the vital signs are loaded on the HMD the only expected interactions are things such as zooming and selecting one specific vital sign (e.g., CO_2). Battery life, though, is an important criterion. Therefore, as potential candidates, we foresee the Atheer AiR Glasses, Epson BT-300, or the ODG R-7. While Google Glass has been used in many experiments, its small screen is often criticized for its inability to clearly display the history of multiple vital signs.

2.4.4 Surgeon

2.4.4.1 Activities and requirements

Prior to or when entering the OR, SUs consult the patient's latest updates (e.g., medical tests, given medication) and help prepare the patient for the surgery (Table 2.6—No. 1). Before the surgery starts, the SU in chief goes through the safety surgery checklist (often called "time-out") to ensure the medical staff has a common understanding of the patient's health (e.g., current state, allergies), that the required equipment is available and that the surgery's critical steps and concerns are known (Table 2.6—No. 2). Then, SUs proceed to the surgery (Table 2.6—No. 3). Either in open, endoscopic, laparoscopic, or robotic surgeries, SUs apply different techniques that vary based on their training, patient characteristics, and disease to be treated. Sometimes, SUs have to access additional patient information such as X-rays or vital signs. Once the surgery is finished, SUs debrief with the

Table 2.6: SUs' activities and opportunities to use smart glasses

| No. | Description | Activities | | Smart Glasses | |
		Requirements	Constraints	Opportunities	Candidates
1	Check patient last updates	• Access patient's medical record	• Limited access to computers	• Information access while on the move	Google Glass, Vuzix M300
2	Safety surgery checklist (time-out)	• Relevant elements must be verified	• Limited access to documentation	• Access checklist • Customize checklist • Document checklist	
3	Proceed to the surgery	• All steps must be performed with a high level of precision and quality • Access last X-rays, CT scans, MRI • Access vital signs • Ensure no supply is inside the patient's body	• Surgical steps can be forgotten (junior SUs) • Unplanned steps or cases • Little access to information	• Access case's information • Access patient's vital signs • Contact experts is needed and share videos/ pictures • Video record specific steps • Record notes	Google Glass, Vuzix M300, ODG R-7 for reading X-rays, CT scans
4	Debriefing/ sign out checklist	• Critical event must be documented	• Limited access to computers • Limited amount of time	• Record notes while on the move	Google Glass, Vuzix M300

medical staff and come back on any complications that might have happened during the surgery (Table 2.6—No. 4).

2.4.4.2 Constraints and opportunities to use smart glasses

SUs are very restricted given that they cannot touch any nonsterile equipment and must stay focused on patients. Smart glasses can support SUs when they need to access information and do not have access to a computer. During checklist execution, smart glasses can also increase the likelihood that SUs go through all checklist elements, as demonstrated in laboratory settings by our senior author [18] (Fig. 2.3). During the surgery, smart glasses can display critical surgical steps to ensure (junior) SUs do not miss any of them. In some cases, such as in surgical oncology, smart glasses enable SUs to access X-rays or CT scans and navigate through them via voice commands. As seen in existing studies, smart glasses can also be used to take notes, pictures, and videos to document the surgery in case of any specific concerns or complications. It facilitates the debriefing and make sure that critical information was documented.

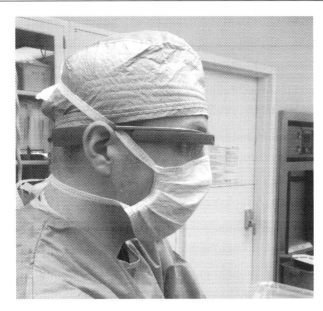

Figure 2.3
Dr. Rivas executes the safety checklist on smart glasses.

2.4.4.3 Smart glass candidates

The case of SUs is the most complicated given SUs require different types of support along their surgeries. Google Glass or Vuzix M300 are very convenient for displaying small amount of text or images. Their small screens and navigation via voice commands make them the perfect candidates for accessing small amounts of information (e.g., latest patient's update) or for executing checklists and increasing the likelihood that elements are all verified. Similarly, EW headsets can guide (junior) SUs during surgeries. They are also lightweight and do not obstruct the field of view. They can also easily be used to take pictures, videos, or getting remote assistance. When accessing richer content such as X-rays or CT scans, the ODG R-7 would be more suitable, but given the heavyweight and risk of obstructing the field of view, this may not be an appropriate trade-off.

2.5 Are ORs the Future of Smart Glasses?

Smart glasses have created not only excitement when the first models were announced, but also disillusions when little information was given about the capacities and availabilities of this technology. Almost 4 years after the prerelease of Google Glass, the capabilities of smart glasses are only slightly clearer. In addition, among the 50 different models on the market, only few of them are available as products.

Many use cases that show the benefits of using smart glasses in several fields, among them being health care. But are ORs a real playground for smart glasses? Scientific research partially answers this question via multiple studies that demonstrate the benefits of smart glasses. For instance, smart glasses decrease the time to detect critical desaturation as well as leverage knowledge that would be difficult to access otherwise (e.g., via videoconferences). Beyond these examples, smart glasses have the capacity to increase work efficiency and compliance, which is one of the main causes of complications and deaths in ORs. To reach this level, smart glass providers, software developers, and hospitals must work hard to produce the necessary changes, with the first being to provide and support smart glasses at a large scale. Additionally, smart glass reliability and battery life must be addressed given they remain insufficient for intensive uses as expressed by several studies [19,20]. Smart glasses must also be able to provide easy-to-use and compliant applications (e.g., Health Insurance Portability and Accountability Act, for the United States), while processes to manage these devices in ORs must be implemented. To achieve these vital milestones, it implies the collaboration of the aforementioned actors, during the design, implementation, and use of smart glasses.

In this book chapter, we described that not all smart glasses serve the same purpose. For instance, SUs require a lightweight, unobtrusive, and voice controlled headset, limiting the list to few models. On the other hand, CNs can afford to manipulate smart glasses with their hands, but battery life and big screens are required to read large contents. We also observed that some smart glasses are not necessarily the best candidates for ORs, namely the Sony SmartEyeWear, given its limited capacities. During our evaluation phase, almost all smart glass providers openly expressed awareness of their models' current limitations and are working on new versions. The conclusion is therefore that smart glasses undoubtedly have a future in ORs; however, the best is yet to come.

Ackowledgment

We would like to thank Peter Grantcharov for his help in collecting data related to the different smart glasses as well as the Swiss National Science Foundation and the Foundation of the University of Lausanne (Switzerland) for funding (P1LAP2−168407) part of this study.

References

[1]   J. Adler-Milstein, et al., Electronic health record adoption in US hospitals: Progress Continues, But Challenges Persist, Health Aff. (Millwood) 34 (12) (2015) 2174−2180.

[2]   T. Starner, Fundamentals of wearable computers and augmented reality, Wearable Computing: Meeting the Challenge, second ed., CRC Press, Boca Raton, FL, 2015.

[3]   A. Guo et al., A comparison of order picking assisted by head-up display (HUD), cart-mounted display (CMD), light, and paper pick list, in: Proceedings of the 2014 ACM International Symposium on Wearable Computers—ISWC '14, New York, 2014, pp. 71−78.

[4]   F.E. Block Jr., D.O. Yablok, J.S. McDonald, Clinical evaluation of the 'head-up' display of anesthesia data, Int. J. Clin. Monit. Comput. 12 (1) (1995) 21−24.

[5]   S. Vallurupalli, H. Paydak, S.K. Agarwal, M. Agarwal, C. Assad-Kottner, Wearable technology to improve education and patient outcomes in a cardiology fellowship program—a feasibility study, Health Technol. 3 (4) (2013) 267−270.

[6]   S. Elder, A. Vakaloudis, A technical evaluation of devices for smart glasses applications, in: 2015 Internet Technologies and Applications, ITA 2015—Proceedings of the 6th International Conference, 2015, pp. 98−103.

[7]   B. Kress, E. Saeedi, V. Brac-de-la-Pierriere, The segmentation of the HMD market: optics for smart glasses, smart eyewear, AR and VR headsets, Int. Soc. Opt. Eng. (2014) 92020D.

[8]   D.F. Ormerod, B. Ross, A. Naluai-Cecchini, Use of a see-through head-worn display of patient monitoring data to enhance anesthesiologists' response to abnormal clinical events, in: Wearable Computers—Sixth International Symposium on Wearable Computers, Seattle, WA, 2002, pp. 131−132.

[9]   D. Liu, S.A. Jenkins, P.M. Sanderson, Clinical implementation of a head-mounted display of patient vital signs, in: Proceedings—International Symposium on Wearable Computers, ISWC, Linz, Austria, 2009, pp. 47−54.

[10]   D. Liu, et al., Monitoring with head-mounted displays: performance and safety in a full-scale simulator and part-task trainer, Anesth. Analg. 109 (4) (2009) 1135−1146.

[11]   C.A. Liebert, M.A. Zayed, O. Aalami, J. Tran, J.N. Lau, Novel use of Google glass for procedural wireless vital sign monitoring, Surg. Innov. 23 (4) (2016) 366−373.

[12]   O.J. Muensterer, M. Lacher, C. Zoeller, M. Bronstein, J. Kübler, Google Glass in pediatric surgery: an exploratory study, Int. J. Surg. 12 (4) (2014) 281−289.

[13]   A. Scheck, Special report: seeing the (Google) Glass as half full, Emerg. Med. News vol. 36 (no. 2) (2014) 20−21.

[14]   A. Collman, First ever surgery conducted by doctor wearing Google glass, Daily Mail (2013).

[15]   P. Guillen, First surgery transmitted live via Google Glass, Clinica Cemtro, 2013. <http://blog.clinicacemtro.com/tag/google-glass/?lang = en>.

[16]   D.G. Armstrong, T.M. Rankin, N.A. Giovinco, J.L. Mills, Y. Matsuoka, A heads-up display for diabetic limb salvage surgery: a view through the google looking glass, J. Diabetes Sci. Technol. 8 (5) (2014) 951−956.

[17]   M. Kilroy, V. Christopherson, Evena medical eyes-on glasses 3.0 named one of the 'best medical technologies of 2015' by Medgadget|business wire, BusinessWire, 2016.

[18]   H. Rivas, Droiders Stanford glass surgical safety checklist, YouTube, 2014.

[19]   U.-V. Albrecht, et al., Google Glass for documentation of medical findings: evaluation in forensic medicine, J. Med. Internet Res. 16 (2) (2014).

[20]   C.R. Davis, L.K. Rosenfield, Looking at plastic surgery through Google Glass: Part 1. Systematic review of Google Glass evidence and the first plastic surgical procedures, Plast. Reconstr. Surg. 135 (3) (2015) 918−928.

CHAPTER 3

Wearable Robotics for Upper-Limb Rehabilitation and Assistance: A Review of the State-of-the-Art, Challenges, and Future Research

Rejin John Varghese*, Daniel Freer*, Fani Deligianni, Jindong Liu and Guang-Zhong Yang

Hamlyn Centre, Institute of Global Health Innovation, Imperial College, London, United Kingdom

3.1 Introduction

Several neuromuscular disorders and cerebrovascular diseases affect motor function with symptoms ranging from distractive involuntary movements (e.g., tremor) to inhibited movements (e.g., bradykinesia and permanent paralysis). Stroke, for example, affects more than 15 million people every year and is the second leading cause of disability after dementia [1]. The nearly 9 million survivors of stroke usually face significant motor dysfunction, with the type and severity of the dysfunction depending on the location and extent of the original brain damage. In stroke rehabilitation, the sensorimotor experience from the prompt application of intensive and continuous therapeutic exercise has shown to greatly influence recovery [2].

On the other hand, progressive neuromuscular disorders such as Parkinson's disease, amyotrophic lateral sclerosis, muscular dystrophy, Friedreich ataxia, and multiple sclerosis (MS) can affect the brain as well as the neuromuscular junctions and/or muscles and thus lead to significant sensorimotor disability, which intensifies over time. Pharmacological and surgical interventions have varying potency and induce significant side effects. Therefore, these individuals need intensified physical rehabilitation in order to manage their symptoms and slow down the progression of the disease in addition to assistance with everyday tasks. Furthermore, the increasingly aging population across the world will inevitably lead to an

---

*Co-first authors.

Wearable Technology in Medicine and Health Care.
DOI: https://doi.org/10.1016/B978-0-12-811810-8.00003-8

increased number of individuals with movement disorders and reduced neuromuscular function that need care and assistance [3].

Traditionally, physical therapy involves trained professionals working with the affected individual to perform repetitive movements and exercises. However, this process can be labor intensive, expensive, and psychologically challenging [4,5]. The need for repetitive, precise assistive movements makes this an ideal application for robotics. In addition to rehabilitation, robotic systems can also aid activities of daily living (ADLs), or even take over functionality in cases where muscular function has been totally lost. Many robotic systems with multiple degrees of freedom (DOF) have been developed over the last two decades and have demonstrated the ability to enhance rehabilitation and/or provide partial/complete assistance (Fig. 3.1). A number of systems have also been commercialized, but

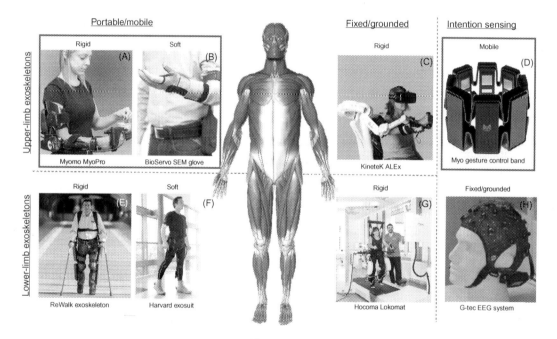

Figure 3.1

An overview of exoskeleton systems and HRI via intention sensing. The focus of this review (highlighted, in colour) is on portable and wearable robotic solutions for upper-limb rehabilitation and assistance, and wearable intention-sensing technologies. *Musculoskeletal system courtesy: Biodigital Inc., United States, www.biodigital.com. (A) Copyright Myomo Inc., United States, www.myomo.com. (B) Copyright Bioservo Technologies AB, Sweden, www.bioservo.com. (C) Copyright Wearable Robotics Srl (Kinetek Division), Italy, www.wearable-robotics.com. (D) Copyright Thalmic Labs Inc., United States, www.thalmic.com. (E) ReWalk Robotics, Inc., United States, www.rewalk.com. (F) Copyright Wyss Institute at Harvard University, United States, https://wyss.harvard.edu/darpa-contract-to-further-develop-soft-exosuit. (G) Copyright Hocoma AG, Switzerland, www.hocoma.com. (H) G-tec EEG system picture (courtesy Hamlyn Centre, Imperial College London).*

these are typically expensive, large, require experienced oversight, and are only available for in-clinic use [6]. An ideal system for rehabilitation and assistance should strive to be an extension of the affected limb, naturally sensing and assisting the intent of the wearer, without requiring extensive cognitive effort from the user. As the inspiration for exoskeletons originated in military applications which required strong materials and a rigid design architecture, a large volume of the robotic rehabilitation and assistance systems also employed the same design archetype [5,7]. Over the last few years, the increased practical challenges faced in translating lab-based rigid-bodied systems for real-world applications has resulted in a paradigm shift in the design philosophy of rehabilitative/assistive robotics [7,8]. This new design paradigm exploits the advancements and breakthroughs achieved in another domain of robotics—soft robotics. Soft robotics could possibly be a key driver in making wearable exoskeletons ubiquitous.

Comprehensive reviews [4,5,9] of upper-limb exoskeleton technologies have been published recently. The review by Gopura and Kiguchi [9] focuses on the mechanical design of upper-limb exoskeletons, the review by Heo et al. [4] focuses on hand exoskeletons, and the review by Maciejasz et al. [5] is an extensive compendium of fixed and wearable upper-limb exoskeleton technologies for both rehabilitation and assistance. Although these reviews encompass an exhaustive list of conventionally designed rehabilitation/assistance systems, there have been no reviews focused toward wearable and portable systems. The distinction between wearable and portable systems has been highlighted in this chapter as many wearable systems (like tethered wearable pneumatic and electric exoskeletons) lack portability, limiting their applications to physical therapy and rehabilitation. This chapter will focus on wearable systems that are portable or could potentially achieve full portability. The review emphasizes systems developed with soft robotics design principles, and different intention-sensing and control strategies that are crucial for making stand-alone portable assistive robots.

The organization of this chapter is briefly described here. Section 3.2 introduces the wearable device design problem from a clinical standpoint with a discussion of the biomechanics of the upper limb and the neuroscience associated with neuromuscular disease, rehabilitation, and assistance. Section 3.3 presents systems that have been developed for assistance and rehabilitation with the potential to achieve complete wearability and portability. This section discusses the systems first along the lines of the physical human—robot interaction (HRI) (i.e., the design paradigm and actuation principle) and then the cognitive HRI (intention-sensing, feedback, and control strategies). This section is followed by a brief discussion of the challenges currently faced, and the potential breakthroughs that would facilitate affordable, comfortable, wearable, and fully portable suits for rehabilitation and assistance.

3.2 Wearable Robotics Design Requirements From a Clinical Perspective

Exoskeletons usually have both a physical and a cognitive HRI component. Understanding the biomechanics as well as the neuroscience behind neuromuscular disease and rehabilitation, therefore, becomes crucial. A thorough understanding of the anatomy, biomechanics, and bio-inspiration is especially essential while designing soft-robotics-based systems, as the user's body is used as the frame for the robot [6,9−11]. Additionally, compared to the lower limb where the loading is largely consistent, in the upper limb, there is an inevitable uncertainty and variability due to the large number of DOF and the variety of tasks that are performed [12]. Understanding the neuroscience of disease and rehabilitation is crucial in designing intention-sensing algorithms and control systems that can adapt to neurological phenomenon like neuroplasticity, as well as being robust to the diversity introduced by the numerous real-world scenarios encountered [2].

3.2.1 Biomechanics of the Upper Limb

The upper limb consists of the shoulder complex, elbow, and hand. The shoulder complex consists of three bones and four articulations with the thorax providing the stable base [13,14]. The shoulder complex as a whole is modeled as a ball-and-socket joint, with the proximal part of the humerus (humeral head) and the concave part of the scapula (glenoid cavity) [14]. The shoulder complex has a total of 3 rotational DOF which allows yaw, roll, and pitch motions. An important caveat while modeling the shoulder joint as a conventional ball-and-socket joint is that the center of rotation is constantly changing, thereby making it necessary for it to be modeled as even a 5 or 6 DOF system [12,15]. The motions facilitated include flexion/extension (Fig. 3.2B), abduction/adduction (Fig. 3.2C), and internal/external rotation (Fig. 3.2D). In the study conducted in Refs. [12,16], the shoulder flexion moment required for ADL is around 14.3 Nm. In a study of the three-dimensional (3D) kinematics during upper-extremity functional tasks [17], the following range of motion (ROM) results were obtained for the shoulder complex: humeral internal rotation was 60 ± 9 degrees, humeral external rotation was 89 ± 13 degrees, humeral elevation during anterior flexion was recorded at 138 ± 9 degrees, and maximum humeral elevation during abduction was very similar at 133 ± 9 degrees.

The elbow joint is a complex joint and consists of three bones and two joints [14,18]. The humeroradial joint is essentially a ball-and-socket joint but gets restricted to 2 DOF due to close association with the humeroulnar and superior radioulnar joint. The elbow-joint complex allows 2 DOF: flexion/extension (Fig. 3.2E) and pronation/supination (Fig. 3.2F). Certain texts also discuss the pronation/supination movement as a DOF associated with the wrist rather than movement associated with the elbow joint. According to the findings presented in Ref. [16], elbow-joint moments of 3.1 Nm are needed to hold an average forearm against gravity with a bend angle of 90 degrees. Additionally, based on Ref. [19],

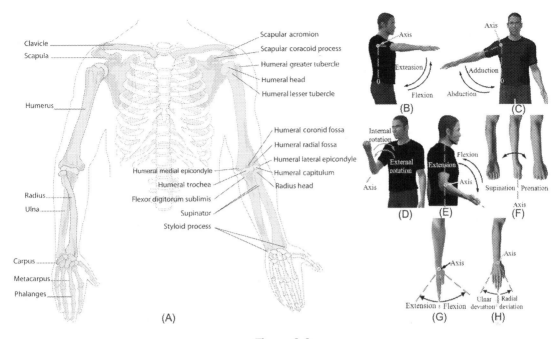

Figure 3.2

(A) Upper-limb anatomy (Wikimedia) and (B−H) upper-limb movements [9].

joint moment of around 5.8N m was needed to complete the assessed ADL tasks. In the study [17], forearm pronation ROM was 161 ± 16 degrees and the maximum elbow flexion was recorded at 143 ± 5 degrees.

The wrist joint is a deformable anatomic unit which is a collection of eight carpal bones and the surrounding soft structure. The wrist joint is traditionally modeled to contain 2 DOF: flexion/extension (Fig. 3.2G) and radial/ulnar deviation (Fig. 3.2H). Wrist motions are generated around an instantaneous center of rotation which is not fixed, but the path of the centroid is small and is ignored in traditional exoskeleton and biomechanics modeling. Neu et al. [20] established that the flexion, extension, radial and ulnar deviation axes are of four different axes. The offset of the rotational axes is approximately 5 mm [14,21]. In the study [17], the ROM for wrist palmar flexion was 66 ± 8 degrees and dorsal flexion was recorded to be -64 ± 6 degrees.

In addition to the wrist, the human hand is made up of digits—four fingers and a thumb. Each digit consists of two segments. The phalangeal segment is made up of three and two bones for the fingers and thumb, respectively. The metacarpophalangeal (MCP) joint allows for flexion/extension and abduction/adduction (2 DOF); the fingers also each have two interphalangeal (IP) joints which each allow flexion/extension (1 DOF). The thumb has one IP joint allowing flexion/extension (1 DOF), an MCP joint allowing flexion/extension, and abduction/adduction (2 DOF), and a carpometacarpal joint which again allows flexion/extension and abduction/adduction (2 DOF) making a total of 5 DOF for the thumb.

Studies conducted on grasping forces exerted by healthy individuals have shown maximum grip strength in the range of around 450 and 300 N for males and females, respectively [6,22,23]. However, in the case of individuals with impaired neuromuscular function, the ability could reduce to almost zero. A study by Matheus and Dollar in Ref. [24] to benchmark grasping and manipulating forces for ADL (like stirring a pot, grasping a glass of water or a fruit, picking up a wallet, etc.) showed that while observing a median coefficient of friction of 0.255, distal tip forces in the range of 10 N for each finger would be more than sufficient.

3.2.2 Neuroscience of the Motor System and Rehabilitation

Motor function in the human brain is mediated by system−level interactions between several brain regions that include the basal ganglia (BG), the cerebellum (CB), and motor-related cortical regions such as the primary motor cortex (M1) [25,26] (Fig. 3.3). Voluntary

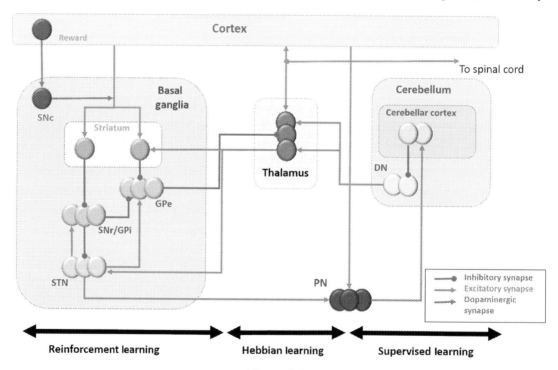

Figure 3.3

Schematic representation of the BG-thalamo-cortical loop and the cerebello-thalamo-cortical loop [25,27]. This is a simplified representation of the current consensus of several neuroanatomical and functional studies on the brain circuitry that underlies normal motor function [28]. Dentate Nucleus (DN), Pontine Nuclei (PN), Sub-Thalamic Nucleus (STN), internal/external segment of Globus Pallidus (GPi/GPe), Substantia pars reticulate/compacta (SNr/SNc).

movements are typically initiated by neural signals in the motor cortex and transmitted from the central nervous system to the peripheral nervous system which innervates muscles, causing a muscular contraction. The BG plays a key role in selecting motor responses by gating appropriate actions and suppressing others [27], whereas the motor-related functional role of the CB is essential for movement precision and accurate reaching performance [25].

Learning in the adult healthy brain takes place via changes in the response efficacy of neuronal synapses (synaptic strength), which is activity dependent. These changes occur mainly as a result of long-term potentiation or long-term depression, which cause enhancement or reduction of synaptic transmission, respectively. In other words, the more a neuron fires, the more its efficacy is enhanced and vice versa. Motor learning both in the healthy and diseased brain is mediated by three forms [25,29]:

1. *Reinforcement learning*: BG forms a complex network of a finely balanced excitatory and inhibitory connections modulated by a reward prediction signal. This has been related to a pattern classification step combined with a reward prediction step.
2. *Supervised learning*: The feedback loop between the CB and cerebral cortex via the thalamus is thought to have a functional refining role of movement precision similar to supervised learning.
3. *Unsupervised (or Hebbian) learning*: In the motor network, the cerebral cortex plays a similar role to unsupervised learning in artificial deep neural networks. It learns features and representations without external feedback on whether the movement was successful or not.

Neurological diseases that affect motor function, also called movement disorders, can be characterized by a reduction/slowness of movements such as bradykinesia and hypokinesia (Parkinsonism) or from an excess of abnormal involuntary movements, such as tremor, dystonia, chorea, ballism, athetosis, and myoclonus [26]. The neurological substrate of these disorders can be either neurodegenerative (primary movement disorders) or a consequence of a specific cause, such as brain injury and cerebrovascular diseases (secondary movement disorders).

Stroke, which is a cerebrovascular disease that causes 22% of secondary movement disorders, provides a representative example. When a stroke occurs the blood supply to a part of the brain is cut off, resulting in immediate and lasting brain damage [30]. The location in the brain and extent of strokes vary greatly, which makes it difficult to generalize the effects stroke may have, though many strokes occur because of thrombosis or an aneurysm in the mid-cerebral artery which has proximity to the sensory-motor cortex [31]. The motor function of the hand and arm is affected in around 85% of stroke survivors [32]. This severely affects the ability of patients to complete everyday tasks, which normally require coordination of both hands. During the early poststroke period, significant brain reorganization occurs as a result of the degeneration of the damaged neurons and biological processes that can be also activity dependent. These processes occur over several months and spread to various brain regions according to the connectivity substrate of the injured area. More active neural pathways would sprout new axons that influence the resulting pattern of brain connectivity and trigger spontaneous recovery.

Therefore, the timing of initiating therapeutic exercises and movements as well as the content of the exercises are of paramount importance. For example, it has been found in hemiparetic rodents that training the nonparetic forelimb before the onset of rehabilitative training of the paretic limb results in increased impairment and profound nonuse of the paretic limb lasting for several months [32,33]. On the contrary, in the absence of prior training of the healthy limb, training of the paretic limb improves its performance and also engages the peri-infarct motor cortex, which surrounds the damaged brain tissue. It has been well established that developing compensatory movements of the nonparetic side disrupts recovery in the paretic side with lasting suboptimal effects. This is not very problematic for lower-limb rehabilitation because most leg tasks such as walking require the use and coordination of both legs. In contrast, many simple daily activities of the upper limb can be achieved by a single arm, which makes it more likely that the affected arm will be ignored. To overcome the nonuse phenomenon, sensory feedback that encourages the movement of the affected limb can be provided to the patient during rehabilitation [34—38]. An alternative to the strict encouragement of the paretic limb, the healthy limb may also be inhibited through various constraints, which would force use of the affected limb [39—43]. Studies investigating such an approach (constraint-induced movement therapy) have indicated that the learned nonuse phenomenon is reversible by the mandatory use of the paretic limb [33].

In addition to stroke, many other neuromuscular diseases impede physical function, e.g., there are persistent yet nonprogressive diseases such as cerebral palsy. Cerebral palsy occurs after a lesion or maldevelopment in the brain and may have effects from early childhood carrying on into adulthood. However, without proper management, the challenges of a person with cerebral palsy may intensify over time, making physical therapy and assistance crucial to management. Many other diseases such as MS are progressive. In progressive neuromuscular diseases, symptomatic treatment is the norm. For example, tremor can be a symptom of several movement disorders including Parkinson's, lesions in the CB, the thalamus, and so on. However, response to medication varies considerably and additional motor rehabilitation is needed to slow down the progression of the disease. Furthermore, as symptoms of these diseases will only worsen, increased daily assistance over time should be the primary goal for management [44].

To conclude, there is a strong motivation to build computational neurorehabilitation models that learn to predict motor learning and recovery [29]. For example, modeling the reward motor adaptation system based on a consensus circuitry of the BG and a virtual biomechanical arm can explain behavioral results of motor adaptation [27]. Using such models as a reference could guide therapeutic intervention and help to determine the amount of assistance that should be provided by a robotic exoskeleton to optimally promote recovery. In addition, comparison to a generic model could more aptly point out individual differences in patients, allowing for the development of bespoke rehabilitation solutions. Finally, the understanding gained from the neuroscience and associated learning mechanisms could also lead to the design of wearable robots with patient-specific HRI interfaces and control software [27].

3.3 Upper-Limb Wearable Exoskeletons

In the previous section, the biomechanics and neuroscience associated with sensorimotor impairments and rehabilitation were discussed, the understanding of which is crucial to the development of upper-limb wearable orthoses. This section discusses the different developed and in-development robotic solutions based on their design principle, actuation, and sensing methodologies. The two main applications for robotic systems in this research domain are rehabilitation (physical therapy) and assistance (powered assistance or tremor suppression) to perform ADLs. While there is a need for systems that can assist individuals to accomplish basic ADL at home, technological and economic limitations are major hindrances to widespread adoption [5]. Only a few upper-limb systems have achieved complete wearability and portability. Wearability and portability are especially key for assistive applications. These attributes would further help fulfill the need for portable powered at-home rehabilitation devices, e.g., the portable version of the Gloreha (Idrogenet Srl, Lumezzane, Italy) rehabilitation system and research systems like the system proposed by Polygerinos et al. [6].

The primary classification in this review is based on the design philosophy—the traditional design paradigm of rigid systems versus the more recent design philosophy of using soft-robotics-based design principles. The soft- and rigid-bodied wearable systems are then discussed on the basis of actuation principle, control input/intention-sensing, and control and rehabilitation strategies. The features are discussed highlighting the differences in adopted design methodologies for rigid-structured versus soft-structured orthoses. Some lower-limb systems have been briefly mentioned as the principles used in their development can be translated to upper-limb systems as well. The scope of this review is limited to robotic systems that are wearable (both portable and potentially portable), and hence soft grounded systems (exoskeleton based or otherwise) are not discussed. Also, systems that substitute movements (like prosthetics, robotic wheelchairs, and autonomous robots) have also not been considered in this review. Table 3.1 tabulates the different soft wearable robotic devices and selectively tabulates the rigid wearable robotic systems.

Table 3.1: Wearable rigid and soft upper-limb rehabilitative/assistive robotic systems

System Name/ Reference	Year	Country	Type	Portable	DOF	Supported Movements	Actuation Scheme	User Input/ Sensing Scheme	Status/Clinical Study Details	Application
Exo-Glove Poly, Kang et al. [11]	2016	South Korea	Soft glove	Yes	2	Hand—grasping and pinching [fingers (FE), thumb (FE, AA)]	Electric; 2 × DC motors (antagonistic arrangement); Bowden cable	Handheld switch	C0; 1hs	Rehabilitation and assistance
Xiloyannis et al. [45]	2016	Singapore	Soft exosuit	Yes	2	Hand—functional grasping + elbow—F/E	Electric; 2 × DC motors; Bowden cable + tendon-driving unit to control index, middle and thumb with single motor	Flex sensor	C0	Rehabilitation and assistance
Polygerinos et al. [6]	2015	United States	Soft glove	Yes	5	Hand—grasping [fingers (FE), thumb (FE, AA)]	Hydraulic; 5 × fiber-reinforced multisegment elastomers	sEMG	C1; 1hs + 1 subject muscular dystrophy	Rehabilitation and assistance
Yap et al. [10,46]	2015	Singapore	Soft glove	Yes	5	Hand—grasping and pinching [fingers (FE), thumb (FE, AA)]	Pneumatic; 5 × Pneu-Net actuators	sEMG + RFID	C0; 1hs	Rehabilitation and assistance
Exo-Glove, In et al. [47]	2015	South Korea	Soft glove	Yes	2	Hand—grasping and pinching [fingers (FE), thumb (FE, AA)]	Electric; 4 × DC motors; Bowden cables: 1 tendon for index and middle finger + 1 tendon for thumb	Bend sensor on wrist	C1; 1hs and 1 with tetraplegia	Rehabilitation and assistance
Villoslada et al. [48]	2015	Spain	Soft glove	Potentially portable	1	Wrist (FE)	Electric; SMA; 1 × flexible Bowden cable SMA actuator	Flexible potentiometer	C0; 1hs	Rehabilitation and assistance
Delph et al. [49]	2013	United States	Soft glove	Yes	5	Hand [fingers and thumb (FE)]	Electric; 5 × DC motors; Bowden cable + stacked spool (antagonistic arrangement)	Switches; sEMG	C0; 1hs	Rehabilitation and assistance
MUNDUS, Pedrocchi et al. [50]	2013	Italy	Modularrigid orthosis; wheelchair mounted	Yes	[3] + [2] + 1	[Shoulder—FE, AA; Elbow—F/E]; [forearm—PS wrist—FE]; all fingers grasp assistance	Elastic elements or DC brakes, FES (optional), DC motor (optional hand orthosis)	sEMG, manual input, BCI; RFID—object label	C1; 3 spinal cord injury + 2 MS patients	Assistance

Reference	Year	Country	Type	Portable	DOF	Motion	Actuation	Sensors	CO	Rehabilitation and assistance
Galiana et al. [51]	2012	United States	Soft exosuit	Yes	1	Shoulder–A/A	Electric; 2 × DC motor + series elastic element (compliant brace and system); 2 × Bowden-cable-driven system for managing misalignments	Flex sensors and IMUs		Rehabilitation and assistance
Park et al. [52]	2012	United States	Soft sleeve	Potentially portable	3	Elbow/knee–active support and rigidity control	Pneumatic; 4 × (4 × PAMs)—Each set of 4 PAMs is one actuation unit	Strain sensors	Proof-of-concept	Active rigidity control and joint support
SEM Glove, Nilsson et al. [53]	2012	Sweden	Soft glove	Yes	3	Hand–grasping assistance;[2 × fingers (FE) + 1 for thumb opposition]	Electric; 3 × DC motors; Bowden cable	Force-sensing resistors; capacitive sensors; Hall effect sensors	Commercial product	Assistance
Kadowaki et al. [54]	2011	Japan	Soft glove	Potentially portable	6	Hand–grasping and pinching [fingers (FE)]; thumb (FE, AA)	Pneumatic; 5 × sheet-like curved rubber (CR) muscle (FE) + 1 × spiral CR muscle for thumb–A/A	Data glove; sEMG	C0; 1hs	Rehabilitation and assistance
Indrogenet Srl, Borboni et al. [55]	2011	Italy	Soft glove	Yes	5	Independent passive motion of each finger	Electric; 5 × motors	Finger positions	Commercial system	Rehabilitation
Hand of Hope, Ho et al. [56]	2011	Hong Kong	Rigid orthosis	Potentially portable	5	Each finger separately–FE	Electric; 5 × linear DC motors	sEMG	Commercial system	Rehabilitation
SUE, Allington et al. [57]	2011	United States	Rigid orthosis	No–extension for Armeo Spring	2	Forearm–PS, wrist–FE	Pneumatic; 2 × pneumatic chambers/pistons	Joint angles, forces	C1; 8 chronic stroke patients	Rehabilitation
Connelly et al. [58]	2010	United States	Soft glove	Potentially portable	5	Hand–assisted extension. unassisted flexion	Pneumatic; 5 × air bladders	Flex sensors	C2; 2 groups (1 control) and 7stroke patients	Rehabilitation/physical therapy
Tong et al. [59]	2010	Hong Kong	Rigid orthosis	Portable	5	Each finger–FE (×2); underactuated	Electric; 5 × linear electric motors + linkages	sEMG	C1; 2 chronic stroke patients	Rehabilitation
Pylatiuk et al. [15]	2009	Germany	Rigid orthosis + soft Actuator	No	1	Elbow (FE)	Hydraulic; 1 × FFA actuator	sEMG	Prototype	Rehabilitation

(Continued)

Table 3.1: (Continued)

System Name/Reference	Year	Country	Type	Portable	DOF	Supported Movements	Actuation Scheme	User Input/Sensing Scheme	Status/Clinical Study Details	Application
Vanderniepen et al. [60]	2009	Belgium	Rigid orthosis	No	1	Elbow (FE)	Electric; 2 × SEAs	Joint angles	Prototype	Rehabilitation
Ertas et al. [61]	2009	Turkey	Rigid orthosis	Potentially portable	1	Concurrent FE of 3 joints of a single finger	Electric; 1 × DC motor	Joint angles	C0; 4hs	Rehabilitation
Rosati et al. [62]	2009	Italy	Rigid orthosis	Potentially portable	1	4 fingers (no thumb)—FE together	Electric; 1 × SEA (DC motor)	N/A	Design	Rehabilitation
MR_CHIROD v.2, Khanicheh et al. [63]	2008	United States	Rigid orthosis	Potentially portable	[1]	Grasping—all fingers together (handgrip)	Electric; ERF brake	Finger position and torque	C0; 1hs	Rehabilitation; MR compatibility
Nathan et al. [64]	2008	United States	Soft glove	Potentially portable	1	Grasping—all fingers together	FES	Handheld switch	C1; 1hs + 2 stroke patients	Rehabilitation
Xing et al. [65]	2008	China	Rigid orthosis + soft actuator	No	2	Thumb—FE; other fingers together—FE	Pneumatic; 2 × PAMs + linkages	Position; force	C0; 3hs	Rehabilitation
RiceWrist, Gupta et al. [66]	2008	United States	Rigid orthosis	No—extension for MIME system	4	Forearm—PS, wrist—FE, AA	Frameless DC brushless motors	Joint angles, forces	Prototype	Force-feedback based rehabilitation
Ding et al. [67]	2008	Japan	Soft suit	No	4	Elbow—FE wrist—FE, AA forearm—PS	Pneumatic; 8 × pneumatic actuators	Joint angles; motion	C0; 6hs	Rehabilitation
Hasegawa et al. [68]	2008	Japan	Rigid orthosis	Potentially portable	11	Forearm—PSWrist—FE,AA Thumb(FE × 2), index finger (FE × 3), other fingers(FE × 3)	Electric; 11 × DC motors; fingers actuated by tendons and parallel link mechanism for wrist	sEMG	C0; 1hs	Grasp assistance
RUPERT IV, Balasubramanian et al. [69]	2008	United States	Rigid orthosis + soft actuator	Potentially portable	5	Shoulder—AA, RT; Elbow—FE; Forearm—PS; wrist—FE	Pneumatic; 5 × PAMs	Joint angles and actuator pressures	C1; 6 chronic stroke patients	Rehabilitation
Stein et al. [70]	2007	United States	Rigid orthosis	Yes	1	Elbow (FE)	Electric; 1 × DC motor	sEMG	Commercial product (FDA clearance);	Rehabilitation

Reference	Year	Country	Type	Portable	DOF	Joints/movement	Actuator	Sensors/feedback	Status	Application
Fuxiang [71]	2007	China	Rigid orthosis	No	4	Index finger (FE)—3 joints + AA	Electric; linear stepping motors	Joint positions and torques	C0; 3hs	Rehabilitation (continuous passive motion)
MRAGES, Winter and Bouzit [72]	2007	United States	Soft glove	Potentially portable	[5]	[All fingers—FE]	Electric; 5 × MRF brakes	Finger positions and torques	Prototype	Force-feedback based rehabilitation
Worsnopp et al. [73]	2007	United States	Rigid orthosis	No	3	Index finger—FE (× 3)	Electric; 6 × brushless DC servomotors + gear/cable driven	Joint angles	Prototype	Rehabilitation
WOTAS, Rocon et al. [74]	2007	Spain	Rigid orthosis	Potentially portable	[3]	[Elbow—FE forearm—PS wrist—FE]	Electric; 3 × DC motors	Angular velocity; torque	C1; study with different pathologies, mostly essential tremor	Tremor suppression
ArmeoPower, Hocoma AG, based on ARMin III, Nef et al. [75]	2007	Switzerland	Rigid orthosis	No	6 [+ 1]	Shoulder—FE, AA, RT; elbow—FE; forearm—PS; [fingers—grasping]	Electric; 6 × DC motors	Joint angles, grasp force	Commercial system	Rehabilitation
Li et al. [76]	2006	China	Rigid orthosis	No	5	Shoulder—FE, AA; elbow—FE; forearm—PS	Electric; 3 × AC motors + 2 × DC motors	sEMG signals from unaffected arm	Prototype	Rehabilitation
Kline et al. [77]	2005	United States	Soft glove	Potentially yes	1	Hand—all fingers simultaneous extension	Pneumatic; single bladder for full hand extension	Joint angles, sEMG	C1; 1 hs + 1 stroke patient	Rehabilitation/physical therapy
ASSIST, Sasaki et al. [78,79]	2005	Japan	Rigid orthosis + soft actuator	Potentially portable	1	Wrist (FE)	Pneumatic; 2 × [rotary-type soft actuator + McKibben PAMs]	Bend and tactile pressure sensors; sEMG	C0; 2hs	Rehabilitation and assistance (different prototypes)
Noritsugu [79]	2005	Japan	Rigid orthosis + soft actuator	Potentially portable	2	Elbow (FE) + wrist (FE)	Pneumatic; 2 × extended rotary-type rubber muscle	Not specified	C0; 1hs	Rehabilitation and assistance
Mavroidis et al. [80]	2005	United States	Rigid orthosis	Yes	1	Elbow (FE)	Electric; 1 × DC motors; direct drive	Force/torque sensors	Prototype	Rehabilitation
Loureiro et al. [81]	2005	United Kingdom and Spain	Rigid orthosis	No	1	Elbow (FE)	Electric; MRF brake	Hand motion (tremor)	C1; 1 essential tremor	Tremor suppression
Mulas et al. [82]	2005	Italy	Rigid orthosis	Potentially portable	2	Thumb—FE + other fingers—FE	Electric; 2 × servo motors	sEMG + pulley position	Prototype	Rehabilitation

(Continued)

Table 3.1: (Continued)

System Name/Reference	Year	Country	Type	Portable	DOF	Supported Movements	Actuation Scheme	User Input/Sensing Scheme	Status/Clinical Study Details	Application
TU Berlin Finger Exoskeleton, Wege and Hommel [83]	2005	Germany	Rigid orthosis	No	4	4 × fingers (no thumb)—FE (×3); underactuated	Electric; 4 × DC motors; Bowden cable + linkage driven	Joint angles	C0; 1 hs	Rehabilitation
Noritsugu [79,84]	2004	Japan	Soft glove	Potentially portable	6	Hand—grasping and inching [fingers (FE), thumb (FE, AA)]	Pneumatic; 5 × extended curved-type rubber muscle + 1 × Linear-type rubber muscle for thumb—A/A	Expiration switch; tactile pressure sensors; sEMG	C0; 1hs	Rehabilitation and assistance
Lucas et al. [85]	2004	United States	Rigid orthosis	Potentially portable	1	Index finger—flexion (passive extension)	Pneumatic; 2 × pneumatic chambers/pistons	sEMG	C1; 1 spinal cord injury	Rehabilitation
Kobayashi et al. [86]	2004	Japan	Rigid orthosis + soft actuator	No	4	Shoulder—FE, AA + elbow—FE	Pneumatic; 10 × PAMs	Joint angles	C0; 5hs	Rehabilitation
Hand Mentor, Kinematic Muscles Inc., Koeneman et al. [87]	2004	United States	Rigid orthosis + soft actuator	Portable	1	Wrist and 4 × fingers—FE together	Pneumatic; 1 × PAM	EMG bio-feedback; wrist angle and flexion torque	Commercial system	At-home rehabilitation
ArmeoSpring, Hocoma AG, based on: T-WREX, Sanchez et al. [88]	2004	United States	Rigid orthosis	No	[7]	{Shoulder—FE, AA, RT; elbow—FE; forearm—PS; fingers—grasping}	N/A (passive—springs)	Joint angles, grasp force	Commercial system	Rehabilitation
Salford Arm Rehabilitation Exoskeleton, Tsagarakis and Caldwell [89]	2003	United Kingdom	Rigid orthosis + soft actuator	No	7	Shoulder—FE, AA, RT; Elbow—FE Forearm—PS; wrist—FE, AA	Pneumatic; 14 × PAMs	Joint position and torques	Prototype	Rehabilitation
Rutgers Master II, Bouzit et al. [90]	2002	United States	Rigid orthosis	Potentially portable	4	Thumb, index, middle, ring finger—(FE)	Pneumatic; 4 × pneumatic chambers/pistons	Actuator translation and inclination	Research device	Rehabilitation

Öğce and Özyalçin [91]	2000	Turkey	Rigid orthosis	No	1	Elbow (FE)	Electric; 1 × DC motor	sEMG	C1; 2 brachial plexus injury	Rehabilitation
CyberGrasp, CyberGlove Systems LLC, Turner et al. [92]	1998	United States	Soft glove	No	[5]	[Resistive force for each finger]	Electric; 5 × DC motors	Joint angles	Commercial product	Force-feedback glove; interaction with VR
PowerGrip, Broadened Horizons, Inc. [93]	–	United States	Rigid orthosis	Portable	1	Grasping—thumb, index, and middle finger (FE)	Electric; 1 × DC motor	sEMG; handheld switch	Commercial system	Grasp assistance

AA, Abduction/adduction; *FE*, flexion/extension; *RT*, internal/external rotation.

[n], number of passive DOF in the robot.

Classification of Clinical Trials for Rehabilitation Studies (adapted from the guidelines provided by A.C. Lo, Clinical designs of recent robot rehabilitation trials, Am. J. Phys. Med. Rehabil. 91 (2012) S204–S216 [94] and P. Maciejasz, J. Eschweiler, G.-H. Kurt, J.-T. Arne, S. Leonhardt, K. Gerlach-Hahn, et al., A survey on robotic devices for upper limb rehabilitation, J. Neuroeng. Rehabil. 11 (1) (2014) 3:

1. Category 0 (C0)—Initial feasibility studies; trials performed with low number of healthy volunteers often using prototype of a device in order to evaluate its safety and clinical feasibility.
2. Category 1 (C1)—Pilot consideration-of-concept studies: Clinical trials aimed at testing device safety, clinical feasibility, and potential benefit. They are performed in a small population of subjects suffering from the target disease. There is either no control group in the trial or healthy subjects are used as control group.
3. Category 2 (C2)—Development-of-concept studies: Clinical studies aiming at verification of device efficacy. Includes a standardized description of the intervention, a control group, randomization, and blinded outcome assessment.
4. Category 3/4—Demonstration-of-concept studies/proof-of-concept studies: Further evaluation of the device efficacy. Similar to the C2, however, usually these are multicentered studies with a high number of participants.

3.3.1 Classification by Design Methodology

3.3.1.1 Rigid-bodied exoskeletons

3.3.1.1.1 Advantages of rigid-bodied exoskeletons

Research on exoskeletons began in the 1960s, with more significant growth happening since the 1990s with work by the Berkeley Robotics & Human Engineering Laboratory and other research groups worldwide [7,95]. Until recently, the dominant design paradigm has been an exoskeleton with a rigid frame [7,96]. Rigid exoskeletons are robust and provide stiff mechanical body support [97,98]. The rigid-body design also allows for the forces/torques to be transmitted without the anatomical equivalent (the user's limb) experiencing any load. Compared to soft-bodied exoskeletons, the rigid frame tolerates a simpler control system design due to simpler system dynamics [8,99]. This allows more complex manipulation scenarios to be achieved with relative ease.

The above advantages make rigid-bodied systems beneficial for applications with high-force/torque transmission requirements. Heavy industrial and some military requirements could be ideal applications. For rehabilitative purposes, individuals with low spasticity in their joints or higher force/torque requirements could also benefit from rigid-bodied exoskeleton systems. Lower-body rigid exoskeleton systems have been more widely researched and are also more prevalent commercially as opposed to upper-body systems. Examples of rigid-bodied upper-limb exoskeleton technologies for powered assistance include the HAL Single Joint Elbow (Cyberdyne Inc., Ibaraki Prefecture, Japan) [100,101] and the MyoPro Motion-G elbow—hand orthosis (Myomo Inc., MA, United States) [100] (Fig. 3.4). A rigid-bodied exoskeleton for tremor suppression is the PAULE (MedEXO

Figure 3.4

Commercially developed rigid wearable upper-limb exoskeletons: Myomo MyoPro elbow—hand orthosis [70]. *Copyright Myomo Inc., United States, www.myomo.com.*

Robotics, Kowloon, Hong Kong) system [100]. The portability of wearable rigid-bodied exoskeleton systems decreases with the increasing complexity of these systems. A few fully/partially portable systems assisting multiple joints have been developed by research groups, e.g., the fully portable 11 DOF hand exoskeleton proposed by Hasegawa et al. [68], the partially portable 9 DOF ESTEC exoskeleton [102], and the 20 DOF hand exoskeleton [103] developed at TU Berlin. Wearable exoskeletons have been developed for the elbow [101,104] and hand [68,102,103], but to our knowledge, no exoskeleton for the entire upper limb has been developed that has achieved complete wearability and portability.

3.3.1.1.2 Disadvantages of rigid-bodied exoskeletons

Rigid-bodied exoskeleton technologies have a few disadvantages inherently associated with the design methodology. Barring a few exceptions like the ones mentioned earlier, exoskeletons with rigid frames tend to be heavy, requiring them to be paired with high-torque actuators and subsequently ending up needing larger power sources [104,105]. Though this pairing could be apt for applications that entail supporting heavy loads across rough terrains (military) or where portability is not a necessity (industrial), this design archetype is suboptimal when the intended application is assisting patients recovering from stroke or some other neuromuscular disease. The size, weight, and inertia of these systems could impede individuals in accomplishing basic ADL. Some lightweight lower-body rigid exoskeletons, like those developed in Refs. [106−108], achieve wearability and portability by using expensive lightweight materials like carbon fiber [99].

The lack of compliance of the orthosis with the wearer's body also poses safety concerns to both the wearer and people interacting with the wearer [8,10,109]. Exoskeletons are typically designed to run parallel to their biological counterpart and subsequently augment the limb's functionality. The mechanics of most rigid exoskeleton systems are suboptimal when compared to the complex biomechanics of the human body, resulting in constrained and restricted movements [8,97,110]. Static misalignments lead to increased dynamic misalignments which hamper natural movements and could intensify discomfort, pain, and even injury [111]. This limitation increases the need for professional assistance, user-specific design, or the ability to adjust the design to allow universal fitting like in Ref. [112]. A few exoskeletons incorporate mechanisms in their design which self-align with the individual's joints like the finger and thumb wearable robot developed by Cempini et al. [113] and the system developed by Brokaw et al. [114], but most if not all self-aligning shoulder and elbow exoskeletons are fixed, e.g., the system developed by Ergin and Patoglu [115].

Rigid exoskeletons, by design, transfer forces/torques by bypassing the joints they are assisting [9]. This design would benefit an individual having an extremely low chance of recovering neuromuscular function and would enable him/her to perform basic ADLs. However, users having the potential of rehabilitating themselves to normalcy may benefit more from less rigid systems. The long-term use of such rigid exoskeletons has shown to lead to disuse atrophy and consequently makes the user more dependent on the orthosis [116,117]. Long-term use of the rigid device has also been shown to result in neural adaptation from the reduced muscle activity [118]. A potential solution to this problem could be an advanced control system that senses the capability of the user in real time and produces only assistance on an as-needed basis.

The high cost of current rigid-bodied systems is another factor hindering widespread adoption. Rigid-bodied systems would typically have a higher material cost and user-specific design cost in addition to the energy, service, maintenance, spare parts, training, and supervision costs [7,119,120]. The inherent shortcomings of the design paradigm are partly responsible for the high cost of current systems along with factors like low-volume manufacturing and the exoskeleton market still being in its infancy [119].

3.3.1.2 Soft-bodied exoskeletons

3.3.1.2.1 Soft robotics: a new approach to exoskeleton design

Robotic systems have become ubiquitous in the manufacturing sector due to their strength, speed, agility, precision, accuracy, repeatability, and lack of fatigue. The inordinate skew toward manufacturing as the primary application for robotics contributed to the use of stiff materials and rigid actuation systems in conventional mechanical systems design. This design architecture made human interaction with robots inherently unsafe, and it became common practice in the industrial setting to isolate human and robot workspaces [121]. As applications demanded increased HRI and robotics grew outside the realms of the manufacturing sector, soft robotics development gained impetus. Researchers realized the limitations of conventional robot mechanical design methodologies in applications involving complex unstructured environments and extensive HRI [122]. Soft robotics research originated with inspiration from nature to incorporate: (1) the inherent compliance of the soft tissues of animals and plants [122,123], and (2) *embodied intelligence* or *morphological computation*, the ability of natural organisms to embody intelligence physically in their body and to use it to control their movement [121,124].

The domain of soft robotics is still in its infancy and picked up momentum around a decade ago (Fig. 3.5A). A few seminal works between 2009 and 2012 laid the foundation for the current spurt in soft robotics research, with research ranging from actuation and sensing systems to system modeling and control [125]. Soft robotics is a highly interdisciplinary field with a community of researchers from biology and medicine, mathematics and modeling, material science and chemistry, to engineering and robotics. The 1000 publications mark in soft robotics publications was recently crossed (Fig. 3.5A).

Around 2010, two state-of-the-art full-body military exoskeleton projects, the HULC (Human Universal Load Carrier) by Ekso Bionics and Lockheed Martin, and the XOS/XOS 2 by

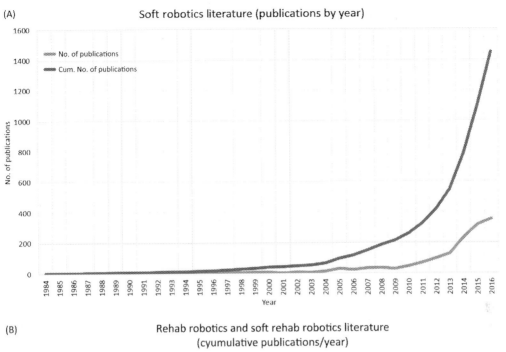

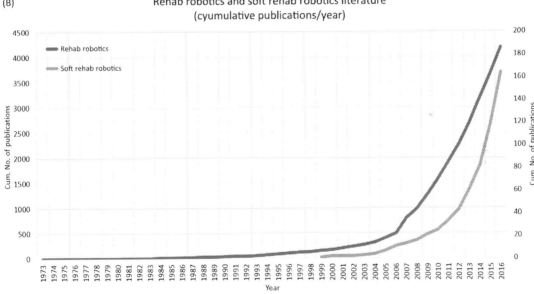

Figure 3.5

Soft robotics and soft wearable assistive/rehabilitation robotics research literature. (A) Data on soft robotics literature. (B) Comparison of rehabilitation/assistance robotics literature and soft rehabilitation/assistance robotics. *(A and B) From Scopus till December 2016.*

Sarcos/Raytheon were discontinued as the US military stopped expressing interest and support for these systems [8]. The designs ran into the vicious circle of having rigid heavy frames which needed powerful and heavy actuators, which in turn needed bigger power sources (for reasonable operation times), which in turn required heavier and stronger frames, and the loop continued [7,120]. The failure of these projects highlighted the limitations of rigid-bodied exoskeleton technology. The upswing in soft robotics research permeated into wearable (exoskeleton) robotics as it was a fitting application for this design paradigm, and had the ability to address many of the challenges currently faced by researchers.

It is only in the last decade that increased efforts and breakthroughs have been achieved in rehabilitation/assistance robotics [7]. Soft robotics applications in rehabilitation/assistance, by contrast, is an even more nascent development [121,125]. Though still at an early stage, the successful application of soft robotics in rehabilitation/assistance robotics is definitely seeing increased endorsement (Fig. 3.5B). Only a few upper-limb assistive exoskeleton systems have been commercialized, e.g., the Daiya Industry Power Assist Glove (Daiya Industry Co. Ltd., Okayama Prefecture, Japan) (Fig. 3.6A), and the Robotic SEM Glove (BioServo Technologies AB, Kista, Sweden) (Fig. 3.6B) [53] among others [8]. Some other soft exoskeleton research systems in an advanced development stage include the DARPA Warrior Web Program—Harvard Soft Robotic Exosuit (lower limb) [109], the Exo-Glove Poly [11], and other research systems. A completely fabric-based pneumatically actuated full-body exosuit concept is being developed by Roam Robotics (CA, United States), a spinout of Otherlab [126].

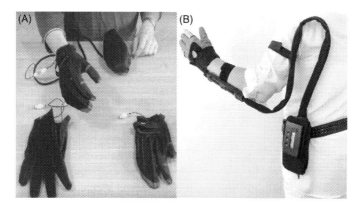

Figure 3.6
Commercially available soft wearable exoskeletons: (A) Daiya Industry Power Assist Glove. (B) Bioservo SEM Glove [53] *(A) Copyright Daiya Industry Co. Ltd., Japan, www.daiyak.co.jp. (B) Copyright Bioservo Technologies AB, Sweden, www.bioservo.com.*

3.3.1.2.2 Advantages of soft-bodied exoskeletons

When the rigid frame is removed from the exoskeleton, there are some significant advantages. Soft orthotic systems are lighter, less extensive, and have less inertia as compared to their rigid counterparts. This reduces the energy requirements and allows for longer running times compared to rigid-bodied equivalent systems using the same power source [99].

Soft exoskeletons are compliant and bend with the joints of the user's body. Custom fitting a soft-bodied exoskeleton also becomes simpler as they could potentially be discretized into the sizes of standard clothing. Soft-bodied systems are less restrictive to the user's joints and thereby reduce the possibility of discomfort and injury to the user [6,11]. The intrinsic compliance of soft wearable orthoses therefore greatly simplifies the mechanical design of these systems.

An intrinsic advantage of soft-bodied orthotics is the low profile and ease-of-concealment. When compared to rigid-bodied systems, this is a big advantage as the psychosocial aspect is crucial to the widespread adoption of exoskeleton technology. As soft-bodied systems could potentially be concealed underneath regular clothing, there is a high likelihood that these systems would enjoy greater acceptance. An example of a passive power-assist soft exosuit which can be worn with regular clothing is the Ski-Mojo (Ski-Mojo, Horsham, United Kingdom) [8].

The materials used to fabricate soft-bodied systems (like fabrics, elastomers, etc.) are much cheaper compared to metals or other materials used in rigid-bodied systems. The cost of manufacturing would also be comparable if not lower. Lower manufacturing and material costs would translate to lower costs for the end user and consequently facilitate greater affordability [8]. A lower price point makes the technology accessible to a significantly greater fraction of the population, bringing further cost benefits associated with manufacturing at scale. As soft-bodied systems could potentially be more energy efficient and made of cheaper materials, this could also help to bring down running costs, cost of spare parts, maintenance, etc. as well.

3.3.1.2.3 Disadvantages of soft-bodied exoskeletons

The soft-bodied exoskeleton's principal strength is also its biggest drawback—these systems have no rigid external frame. Soft-bodied systems essentially depend on the user's skeletal system to provide the rigid frame. This makes power transmission complicated and challenging. The entire power transfer happens through the user's body and therefore the extent of force/torque that can be applied is limited compared to rigid-bodied systems. The robotic system will supplement the user's muscles, but most if not all the force/torque will go through the user's body as there is no external frame through which the forces could be grounded [101,105,127].

As there is no rigid frame present, it becomes more difficult to mount sensors and motors. Direct-drive electric actuation systems become almost impossible in the case of soft-bodied systems. Pneumatic and hydraulic soft actuators are much lighter and have been used to directly actuate the hand, elbow, and shoulder [6,10,46]. In the case of completely soft systems actuated by motors, transferring the assistive forces and torques to the actual muscles is done by using tendon-based systems as implemented in Refs. [11,47,53]. Secondly, a limiting factor of actuators developed completely using soft robotics principles is their low force and/or displacement generation capability. Actuators based on different principles like flexible fluid actuators (FFAs, except pneumatic artificial muscles (PAMs)), shape memory materials (SMMs), dielectric actuators, etc. all have this limitation, which has been discussed in greater detail in the following subsection. Controller design for soft actuators also becomes much more challenging compared to rigid-bodied systems, as the user's biomechanics becomes an integral aspect of the system dynamics. The nonlinearities introduced by the soft materials also add to the complexity of the system dynamics and controller design [121,128,129].

A possible middle ground could overcome some of the limitations of completely soft-bodied systems. One possibility would involve having systems with both rigid and soft parts, like the systems proposed by Park et al. [129] and Sasaki et al. [78]. In this design strategy, the stiffer elements are placed alongside the bones, and the softer elements in parallel to the moving elements, i.e., the user's joints. Alternatively, solutions integrating materials and mechanisms that stiffen and soften on demand are also being investigated. Research works have explored robotic systems using materials like magnetorheological/electrorheological fluids (MRF/ERF) and elastomers, low melting point and glass transition-based materials, and SMMs as part of the robot's frame. Apart from materials, phenomena like layer- and granular-jamming among others are also being studied by researchers to produce on-demand variable stiffness mechanisms. A detailed review of different stiffening technologies has been compiled by Manti et al. [123].

3.3.2 Classification by Actuation Principle

The energy source in traditional rigid exoskeletons uses one of three forms: electric current, pneumatic pressure, or hydraulic fluid pressure [5]. The energy source primarily decides the type of actuator. Soft robotic exoskeleton systems have also predominantly made use of matured traditional actuation technologies that have achieved successful implementation. While in rigid exoskeletons the predominant mode of actuation is electric motors, in soft-bodied systems, other actuation principles have also been implemented extensively.

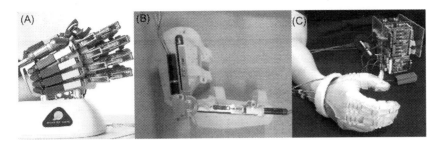

Figure 3.7

Electrically actuated upper-limb wearable orthosis: (A) rehab-robotics hand of hope [56] hand rehabilitation system. (B) MACCEPA by Vanderniepen et al. [60], SEA-based elbow orthosis. (C) Exo-Glove Poly [11], a polymer-body-based soft wearable orthosis for grasp and pinch assistance. *(A) Copyright Rehab-Robotics Co. Ltd., Hong Kong S.A.R., China, www.rehab-robotics.com.*

3.3.2.1 Electric actuation

Electric actuation in the form of electric motors is the most widely used mode of powering exoskeletons as they are easily available, reliable, and easy to install, operate, and control [4]. Electric actuators have a comparable power-to-weight ratio to pneumatic and hydraulic actuators if the weight of the power sources are accounted for, but have significantly lower power-to-weight ratio otherwise [5,130]. In this subsection, only electric motors are discussed. Actuation methods like electroactive polymers (EAPs), dielectric elastomers, etc. are discussed with other nonstandard actuation systems. Both AC and DC motors have been used, but all documented AC-motor-based rehabilitation devices are fixed rigid-bodied systems [4,5,9]. Systems using DC motors are more prevalent and some of the developed systems are both wearable and portable. Both rigid-bodied and soft-bodied exoskeleton systems have been developed using DC motors. Rigid-bodied portable wearable exoskeletons actuated by electric motors include the system developed by Mavroidis et al. [80], the MyoPro [57], the HAL Single Joint Elbow [101], and Hand of Hope (Rehab-Robotics Company Ltd., Hong Kong) [56]. Implementations include direct-drive systems like the system developed in Ref. [80], while others incorporate different transmission systems like tendons in the systems proposed by In et al. [47], linkages in the Hand of Hope system developed by Ho et al. [56] (Fig. 3.7A), chain-and-sprocket mechanism in the system developed by Looned et al. [131], etc.

A major limitation of electric motors compared to other actuation principles like pneumatics and hydraulics is the high impedance intrinsic in the actuator. This characteristic makes the entire orthosis very stiff especially when combined with rigid-bodied frames. The low compliance/high stiffness of the system makes it unsafe especially in the case of unexpected events like control system malfunction, power failure, or spasmodic events. Some exoskeletons have incorporated series elastic actuators (SEAs), which implements an elastic element in series with an electric motor to reduce the intrinsic impedance and inertia of the system, providing more compliant and safer force control [5,132]. SEAs mostly use springs or hydraulic pistons as the elastic element. An alternate approach has been to incorporate compliance in the software. A few wearable exoskeleton systems incorporating SEAs in their design include the systems developed by Vanderniepen et al. [60] and Rosati et al. [133] (Fig. 3.7B). Both systems use a spring as the elastic element. A disadvantage of using an SEA is the resulting narrow functional bandwidth, but this would not be a detriment for users with sensorimotor impairments [5].

In soft exoskeletons, electric actuators have mostly been used to manipulate tendon-driven systems as it is difficult to mount motors directly on the soft frame. Tendon-driven systems with soft tendon routing range from off-the-shelf fabric gloves coupled with tendons like the *Exo-Glove* by In et al. [47,134] (Fig. 3.7C), to glove designs with much more sophisticated soft bodies that adapt to varying hand sizes like the Exo-Glove Poly by Kang et al. [11,135]. A commercialized tendon-driven system is the Robotic SEM Glove (BioServo Technologies AB, Kista, Sweden) [8,53] for functional grasp assistance. A clear advantage of tendon-driven electric-motor-based systems over pneumatic and hydraulic actuators is the ease of setup, maintenance and installation, remote actuator placement, controllability, low profile (allowing concealment), and the ability to provide higher forces [11,46,136]. Most tendon-driven hand exoskeletons are underactuated and use only a single actuator for multiple joints [4]. The underactuation allows the hand to grasp objects with different profiles with relative ease. A disadvantage of tendon-based systems compared to actuators like FFAs is that the forces are localized at the point of attachment. This gives rise to the need to incorporate rigid elements in the design for better force distribution, facilitate higher force application, prevent damage/deformation of the soft structure, and discomfort to the user.

3.3.2.2 Pneumatic actuation

Pneumatic actuators are light, have low impedance compared to electric actuators, and have a significantly higher power-to-weight ratio if the pneumatic source (air compressor, air canister, etc.) is not considered. The high power-to-weight ratio allows for reasonable forces to be applied while maintaining a light frame. One of the implementations deriving inspiration from traditional pneumatic systems was developed by Lucas et al. [85] (Fig. 3.8A). This system used two pneumatic cylinders/pistons to move the proximal IP/ distal IP and MCP joints of the index finger for pinch assistance. A system proposed by Allington et al. [57] also used pneumatic cylinders coupled with mechanical linkages to assist wrist pronation/supination motion.

Another system that used pneumatic actuation with a mechanical linkage was developed by Xing et al. [65] (Fig. 3.8B), but rather than using pneumatic cylinders the system makes use of McKibben PAMs. PAMs have been widely used in rehabilitation/assistance

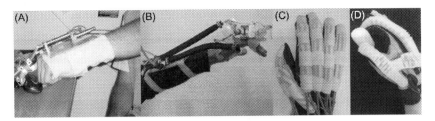

Figure 3.8
Pneumatically actuated wearable orthosis: (A) pneumatic-cylinder actuated pinch assist mechanism [85]. (B) PAM actuated finger and thumb orthosis [65]. (C) Soft power-assist glove [54]. (D) Pneu-Nets-based hand exoskeleton [46].

applications [5]. PAMs are a class of actuators that consist an internal bladder surrounded by a braided mesh shell with flexible, nonextensible threads. When the bladder is pressurized, the actuator expands its diameter and consequently shortens its length, thus applying tension [137]. PAMs have been used in both rigid-bodied systems like the commercialized Hand Mentor (Kinematic Muscle Inc., AZ, United States) [87], and soft-bodied wearable systems like the one proposed by Kobayashi et al. [86] and Park et al. [129]. PAMs have been used for actuation both directly [65] and indirectly [128] by using tendons and linkages. Some systems have modified the traditional McKibben PAMs with added reinforcements, etc. to achieve more complex manipulations [79].

Soft pneumatic actuators have garnered quite a bit of interest and a significant number of systems have been developed for rehabilitation and assistance. These systems range from simple single chamber/bladder glove for hand extension by Kline et al. [77] to more complex systems. Elastomer-based soft pneumatic actuators combined with reinforcements have been developed which allow the actuators to be programmed to achieve complex actuations. A series of mechanically programmed actuators developed by Noritsugu et al. [79,84], Sasaki et al. [78], and Kadowaki et al. [54] (Fig. 3.8C) from Okayama University, Japan, demonstrated extension, contraction, bending, and twisting movements. A curved rubber muscle developed by Sasaki et al.[78] was employed in an orthosis for wrist flexion/extension. Inspired by this research, many groups have developed mechanically programmed pneumatic and hydraulic actuators [10,138,139].

Another class of pneumatic actuators, known as Pneu-Nets (Pneumatic Networks) [140] is based on a series of channels and chambers inside an elastomer. This actuator combined with other elements like strain-limiting fabrics have been used in the systems developed by Polygerinos et al. [141], Mosadegh et al. [142], and Yap et al. [10,46,98] (Fig. 3.8D). Soft flexible actuators such as Pneu-Nets, PAMs, fiber-reinforced actuators, etc. belong to a class of actuators known as FFAs.

Pneumatic systems have some disadvantages as compared to electric actuators. One limitation is the significantly lower control bandwidth and slow response time. Another is the need to remain tethered to air compressors or tanks, thereby limiting portability. Some systems like those presented in Refs. [10,46] have employed portable air canisters to make these pneumatic systems portable. The low output forces and small actuation displacements are a big disadvantage and severely limit the applications of soft pneumatic actuators [130]. PAMs can generate reasonably high forces due to the higher operating pressure, but the actuation lengths are limited. These actuation and compression operations traditionally tend to also be very noisy but this drawback could be potentially improved with better design. Safety is also a big concern in both hydraulic and pneumatic pressure-based systems in the case of leaks, and the damage to the user and environment caused by the instantaneous release of pressurized fluid.

3.3.2.3 Hydraulic actuation

Very few rigid-bodied exoskeletons have been developed which use hydraulic actuation [5]. The actuators that have been developed are application specific and do not use traditional industrial actuators. Hydraulic actuation offers certain advantages over pneumatic actuation. They can exert more force/power compared to similarly sized pneumatic systems and have better efficiencies as pneumatic systems lose energy as heat during air compression. Hydraulics is more suitable for lifting heavy loads, but they have a few disadvantages as compared to pneumatic systems in exoskeleton systems. Nonetheless, hydraulic actuators are typically avoided because of their weight, slow response time, noise, and fluid leakages (and the associated environmental concerns) [130].

Most hydraulic systems developed so far are fixed-type exoskeletons. A wearable system was proposed by Pylatiuk et al. [15] (Fig. 3.9A) and uses an FFA for elbow-joint motion in conjunction with functional electrical stimulation (FES) for grasping. The actuation mechanism in spider legs is the inspiration for this FFA design [143]. This FFA lengthens when pressurized (and vice versa), the opposite behavior of PAMs. Hydraulics has also been used in conjunction with electric actuation mechanisms to make hydraulic SEAs. Hydraulic SEAs have been used in a number of fixed systems like the NeuroEXOS system developed by Lenzi et al. [144].

(A)   (B)

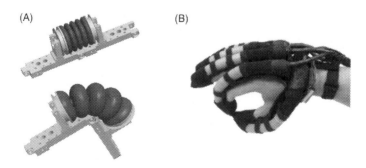

Figure 3.9
Hydraulically actuated wearable systems: (A) FFA actuator for elbow orthosis [15].
(B) Fiber-reinforced actuator-based hand rehabilitation glove [6].

An at-home rehabilitation glove developed by Polygerinos et al. [6,139] (Fig. 3.9B) uses fiber-reinforced elastomeric actuators. Fiber-reinforced actuators basically consist of an elastomeric chamber wrapped with inextensible reinforcements permitting only programmed actuation [145]. Connolly et al. [146] developed a solution to automatically derive the requisite configuration of reinforcements based on desired tip trajectories. The system developed in Ref. [6] used a portable waist pack which houses the hydraulic fluid, controllers, and other equipment. The system facilitates active flexion assistance for the fingers and opposition motion for the thumb but depends on the mechanical properties of the actuator to facilitate extension. Another limitation is that trajectory of the actuator is preprogrammed and cannot be changed on demand during operation. Additionally, the forces and the operating frequency though sufficient to allow rehabilitation exercises would be a limiting factor in assistive applications [6,139].

3.3.2.4 Other nascent actuation principles

A few nonstandard actuation methods have also been used for upper-limb exoskeleton systems. These have been incorporated in soft-, semirigid, and rigid-bodied wearable exoskeleton technologies. An actuation principle, FES, harnesses the natural actuators by electrically stimulating the user's own muscles. A few systems like the ones proposed by Pylatiuk et al. [15], Looned et al. [131], and Nathan et al. [71] use FES for functional grasping and releasing. These systems use FES in conjunction with other actuation methods used for assisting elbow motions. Another system developed by Gallego et al. [147] (Fig. 3.10A) uses FES as the principal actuation method for tremor suppression. The prototype uses a multichannel FES array and an inverse dynamics muscle model in conjunction with a multimodal brain—computer interface (BCI) (electroencephalography (EEG) + electromyography (EMG) + inertial measurement unit (IMU)) system. FES-based systems only comprise electrodes and a control unit. This allows the system to have a very low profile and be concealed under regular clothing. Additionally, FES systems allow patients to exercise unused/underused muscles, improve muscle bulk and strength, and prevent muscular atrophy [148]. However, the disadvantages of FES include poor

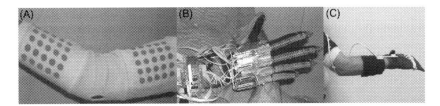

Figure 3.10
Wearable orthoses with other actuation systems: (A) soft wearable orthosis for tremor suppression with FES array [147]. (B) MRAGES force-feedback glove developed by Winter et al. [72]. (C) Flexible SMA actuator for wrist flexion/extension by Villoslada et al. [48].

controllability due to the nonlinear nature of the contracting muscles, muscle fatigue, discomfort, pain, and the erratic variation in resulting outputs [5,148]. Nevertheless, FES and other technologies using electrical stimulation still hold a lot of potential and are being investigated extensively (Fig. 3.10).

MRF and ERF are a class of fluids that change their rheological properties (viscosity) dynamically based on the applied magnetic and electric fields, respectively [123]. These fluids are being investigated by certain groups as means for achieving active variable energy dissipation. Groups have investigated using ERF and MRF as brakes and clutches to control the force output from electric motors in active systems. The systems that incorporate MRF and ERF are mostly fixed [5]. MRAGES, a glove developed by Winter and Bouzit [72] (Fig. 3.10B), used MRF to provide haptic force feedback. An orthosis for tremor suppression was developed by Loureiro et al. [81] which used an MRF-based controllable double viscous beam actuator. A portable elbow/knee orthosis which uses an ERF-based brake was proposed by Mavroidis et al. [80].

SMMs are comprised of alloys, polymers, and composites capable of recovering a predetermined geometric shape after a plastic deformation is induced through phase transformation [123]. The shape memory effect, exhibited by alloys like Ni–Ti and Cu–Al–Ni, is a result of the different mechanical properties that are associated with the different stable phases of these materials. A flexible shape memory alloy (SMA) actuator-based orthosis was developed by Villoslada et al. [48] (Fig. 3.10C) to assist wrist extension movements. The actuator used multiple runs of a single SMA wire wound around pulleys inside a Bowden cable to overcome the small strain limitation characteristic of SMAs and was also able to generate forces up to 35 N. The limitations of current SMA-based technologies are the limited ROM and forces, slow response time, safety issues associated with heating, and handling nonlinearities like hysteresis in the controller design. Another orthotic system, the SMART Wrist–Hand Orthosis developed by Makaran et al. [149] makes use of two SMA actuators for antagonistic grasping and releasing motion. A prosthetic system using SMA technology was proposed by Andrianesis and Tzes [150].

Robotics researchers, material scientists, and researchers in other disciplines are currently experimenting with new novel actuation methods like ionic and electronic EAPs [151], dielectric elastomers [152], twisted nylon coil artificial muscles [153], and other concepts. Though these actuation technologies are still in early stages of development, they could be potential actuation mechanisms in future soft-bodied exoskeleton systems [4].

3.3.3 Classification by Intention Sensing, User Feedback, and Control

One of the biggest challenges in the development of an assistive wearable robotic platform is the ability to sense the intention of the user. Because the main goal of an assistive robot

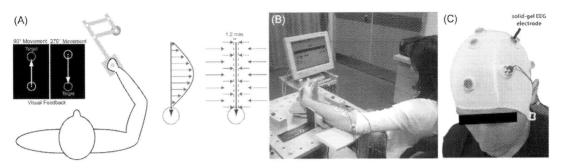

Figure 3.11

The main types of input to rehabilitative or assistive robots: (A) physical feedback, represented by a rehabilitation system regulated by impedance control [154]. (B) Muscular signals through EMG electrodes [155]. (C) A wearable EEG cap with solid-gel electrodes for control using a BCI [156].

is to aid in the completion of a task, it is first necessary to determine what the task is. For controlled situations, this may be easy to solve. For example, a common rehabilitation exercise used with the MIT-MANUS rehabilitation robot is a planar reaching task [157], for which the desired trajectory is predefined. In this situation, you can assume that the intended movement of the user is to move along this trajectory, and can, therefore, make modifications to the movement as needed. However, in daily life situations, determining the intention of the user is much more complicated, as the arm and hand are consistently interacting with different objects in complex ways. Therefore, even with the research developments described in this section, context awareness gained through other sensing modalities like cameras may be a crucial aspect to determining the patient's intent and facilitate translation to the real world [156] (Fig. 3.11C). Regardless, commonly used user input methods for intention detection in assistance and rehabilitation are detailed in this subsection.

3.3.3.1 Physical user input and feedback

Extracting intent from present psychophysiological sensing modalities is extremely challenging, so many systems have resorted to simpler forms of control input such as analog switches. Analog switches like push buttons have been used to produce trigger signals in the Exo-Glove Poly [11], the system developed by Delph et al. [49], etc. In a usability evaluation study done in Ref. [11] on spinal cord injury patients with different control inputs like analog switches, EMG, and bend sensors, the patients and researchers felt the analog switches were the easiest to use when controlling the robot. Expiration switches have been used by Noritsugu et al. [79,84] and other upper- and lower-body systems to trigger the preprogrammed movements. Apart from triggered systems, sensing technologies like bend sensors, flex sensors, and strain sensors have been used in the Exo-Glove [58], the ASSIST [78,79] system, etc. The different sensing modalities employed by different systems have been mentioned in Column 9 of Table 3.1.

Joint angle [60,77] and joint velocity [74] measurements have also been used extensively in the control of wearable robots, and are sometimes combined with dynamic signals like joint torque measurements [71,89]. In systems for amplifying/augmenting the capabilities of healthy individuals and individuals with limited sensorimotor functionality, the intention is sensed by using tactile sensors [78,79], capacitive sensors [53], and similar sensing systems to generate the required movements and forces.

Physical feedback has also been provided to the user through impedance control in rehabilitation robots for the upper limb. Impedance control, typically, is used with end-effector-based robotic rehabilitation systems and is based on the idea that there should be no robotic intervention as long as the participant is adequately performing a given task. This requires consideration into the problem of path planning and trajectory generation, as each task needs to be programmatically defined. Broad and Argall [158] use a hierarchical finite state machine for path planning, while Gras [159] use learning by demonstration as a method to generate the desired movement trajectory, which could theoretically be applied to ADLs as well. If the patient fails to stay on this trajectory, the robot either creates a restoring force or amplifies the error [160]. If the rehabilitation strategy is to provide assistance to the user, a restoring force will be used; while if the patient is in the stage of challenge-based rehabilitation, error amplification may be utilized. This force could simply be proportional to the position of the end effector or joints of the robot in comparison to the desired trajectory, or could rely on velocity control in addition to position control and has been shown to be a more reliable model of motor learning [154] (Fig. 3.11A).

3.3.3.2 Muscular signals

Surface electromyography (sEMG) has commonly been used to understand biomechanical intention by sensing muscular electrical signals through the skin, making it noninvasive and compatible with wearable systems. With simple processing, sEMG signals can provide a rough idea of whether movement of a particular anatomical joint is desired, provided they are placed along the correct muscle. For this reason, a common control paradigm is to simply use sEMG as a trigger whenever the processed signal reaches a threshold. This control strategy has been shown to successfully assist the therapy of patients after stroke using the MIT-MANUS robot [157]. As a slightly more complex mechanism, proportional control has also been suggested for arm extension [161], elbow flexion [70,162], wrist [155] (Fig. 3.11B), and hand [82] assistance during rehabilitation. More in-depth processing has led to the development of biomechanical muscle models, such as the Hill-based model and myoprocessors which could provide smoother and more precise movement of the rehabilitation robot or exoskeleton [163,164].

The disadvantages of these methods correspond directly to the general disadvantages of sEMG: variability between subjects and variability over time. As the physiology of each subject is unique, the optimal node placement may not be the same for one patient as for the next, which would indicate that a bespoke solution for each individual is desirable. Because of this, personalized musculoskeletal modeling has been investigated through the collection of ultrasound data [165]. Another way of creating a somewhat individualized model has been explored through the training of neuro-fuzzy networks, which accounts for some unpredictability in personal differences, as utilized with the SUEFUL-7 exoskeleton [166]. However, even with a perfect bespoke model of a patient's muscle, the EMG signal changes significantly over time. This change may be seen because of internal factors such as muscle fatigue or small changes in bodily electrical impedance, or because of environmental factors This challenge has been handled in some research works with alternation of control strategies or switching regime models based on any information that can be gathered by the system [166,167].

In addition to controlling the robot, sEMG has also been used as a tool to evaluate the effect of the wearable robot on the wearer's function. For example, Gopura et al. [166] determined that a decrease in EMG signal magnitude while achieving similar movement indicated successful movement assistance by their exoskeleton, as the user exerted less effort. In comparison, Sasaki et al. took a longer view of evaluation using EMG and coupled it with wrist angle measurement. This group used the method of Mosso's ergograph to compare muscular fatigue between assisted and baseline completion of a repetitive lifting task, showing that their assistive wrist device was more resistant to fatigue over time [78]. Taking it even further, Pirondini et al. [168] combined EMG signals with spinal maps to study the effect of ALEx on muscle synergies, which may be key to full rehabilitation of the upper limb.

3.3.3.3 Brain—computer interface

Recently, there has been significant interest in the use of a BCI in robotic rehabilitation of neurological disease. Muscular signals may be significantly reduced after a stroke event due to the deterioration of neurons between the motor cortex and the muscles. In these cases, BCI has been proposed to act in place of the intermediary neurons. In addition, research has suggested that extraction of motor commands directly from the central nervous system promotes motor relearning more than recording through the peripheral nervous system [169]. BICs can be either invasive or noninvasive. While more information can be discerned from invasive BCI because of its direct contact with the cortex, a noninvasive solution is more desirable for safety and ease-of-use. The two main types of noninvasive BCI record either (1) electrophysiology or (2) metabolic changes [170].

The most commonly used BCI platforms that sense electrophysiology are EEG and magnetoencephalography (MEG). MEG has superior spatial and temporal resolution when compared to EEG, as well as being minimally distorted by brain lesions. It also avoids the necessity to attach electrodes onto the scalp, simplifying the procedure [170]. The main disadvantage of this system is that it is very large and immobile, which is not suitable for wearable systems. EEG systems, on the other hand, have become more wearable over time [171]. Using EEG or MEG to recognize specific movement relies on the detection of neurons reaching action potential, emitting an electrical potential and magnetic disturbance. The juxtaposition of this potential from millions of neurons results in signals such as sensorimotor rhythms, which can be seen from a decrease in amplitude of alpha and beta frequency bands with the activation of the corresponding motor areas [172]. Other methods of determining patient intent involve the detection of the P300 signal or steady-state visually evoked potentials, though both of these are typically responded to an external stimulus [173]. While these signals can provide some nuanced information, training for individual patients and long calibration times are often necessary to achieve adequate performance [174].

The main metabolic BCIs are functional magnetic resonance imaging (fMRI) and near-infrared spectroscopy (fNIRS) [170]. fMRI uses an MRI scanner to noninvasively record activity from the entire brain through a blood-oxygen level-dependent response. It consists of a magnet that surrounds the patient's body during the scan and thus it is immobile [175]. fMRI has much better spatial resolution than EEG, but it suffers from poor temporal resolution. fNIRS uses spectroscopy to detect changes in the amount of oxygenated and deoxygenated hemoglobin (HbO_2 and Hb) in different regions of the brain by gathering light information at two wavelengths in the near-infrared spectrum. As blood flows in and out of a region of the brain, the changing optical wavelength profile of the scalp indicates activation of that region [176]. Because both of these techniques only indirectly sense the neuronal signal, there is a large physiological delay (3–6 seconds) between the neurological change and the sensed signal [175]. This makes real-time response unrealistic when using these techniques.

Groups using BCI along with rehabilitation robots have typically used success in a BCI task to trigger a physical movement of the robot, such as the opening and closing of the patient's hand [170,177]. Other approaches have made use of BCI in conjunction with FES to innervate the muscles of the wearer when specific motions were imagined using motor imagery. This was accomplished for wrist and hand extension using EEG [178] and elbow flexion using NIRS [176]. In one of the most sophisticated systems to date, EEG was used in conjunction with electrooculography (EOG) and integrated into a wheelchair to facilitate and assist the grasping and manipulation of objects with the help of a hand exoskeleton in a real-world environment [156].

3.3.3.4 Context awareness and sensor fusion

When a single method of intention sensing is not sufficient, another sensing modality may provide some awareness of the situational context, which could clarify the control decisions that need to be made for the robot. This may require fusion of the sensor data between many different modalities, including physical and inertial input, muscular and brain signals, and computer vision systems or gaze detection. The method of sensor fusion could be a single fusion algorithm, unimodal or multimodal switching, or a mixture of these [173]. These concepts have been applied to prosthetics, autonomous robotic arms, and exoskeletons.

Cameras are often used for context awareness in grasping tasks, providing information about the location and size of the object of interest. For example in Ref. [179], after a myoelectric trigger indicates the basic intention of a prosthetic user, the vision system provides the final desired location of the reach, as well as a hand pose based on the size and orientation of the object. The robot then semiautonomously moves to grasp the object. In this example, data from inertial sensors are also used to track the positioning of the prosthetic hand. A vision system was similarly used in conjunction with an implantable BCI for grasping with a 7 DOF robotic arm, greatly simplifying the trajectory of the end effector and lowering the time to task completion and number of drops when compared to direct BCI control [180]. Deep learning has also been used to circumvent explicit coding of computer vision for the purpose of context aware grasping. Images of objects were instead directly translated to a predefined set of grasping techniques via a convolutional neural network [181]. The prosthetic in this case was also controlled with myoelectric signals, as was the camera that took images of the objects that were going to be grasped. This type of processing may be particularly useful for daily assistance, as humans consistently interact with objects that they have never specifically seen before, but may be similar to previously manipulated items.

Gaze detection is another feature that has been used to provide contextual information in determining user intent in assistive and rehabilitative robotics. It is often easy to understand what a person is interested in accomplishing by tracking the direction of their eyes projected onto an object in 3D space. An easy method of doing this is to track the eyes with a camera, though this has obvious constraints with regard to camera placement and user privacy and comfort. Regardless, infrared cameras have been utilized for eye tracking during robotic rehabilitation of the upper limb with the ARMin III robot in a virtual environment [182]. This has decreased the required movement constraints in rehabilitation, allowing for more natural therapy. Another method of eye tracking that may be more amenable to wearable technology is EOG, which measures changes in the corneo-retinal standing potential between two sides of the eye [183]. EOG was studied in conjunction with an assistive robot to help disabled individuals eat during mealtime, discriminating between three dishes on a table and expediting the movement of the robot to achieve assistance [184]. Such a paradigm can easily be imagined to discern the object of interest in a cluttered scene, which is typically the situation in daily life.

3.3.3.5 Shared assistance and challenge-based control

Shortly after a stroke event affecting the motor capabilities of the arm, the main concern for a patient and their physical therapist is often increasing the RoM of the affected limb.

While it has been shown that RoM can be improved simply through passive movement of the joint [185], this ignores the neurological component of movement. Voluntary drive has been shown to enhance motor learning during neurorehabilitation, and thus combining an assistive platform with knowledge of the patient's intent has the potential to simultaneously increase RoM while also facilitating the redevelopment of neural pathways [186]. Through the intention-sensing methods described in the subsections earlier and an understanding of the current ability of a patient, a rehabilitation protocol can be determined that appropriately assists or challenges the patient.

The simplest method of physical assistance during rehabilitation would be via counterbalancing, which uses weights or simple mechanisms that are able to support the arm during rehabilitation exercises [160]. However, the main objective of counterbalancing is often gravity cancelation, which may not be sufficient for some patients. Because of this, robotic devices have been developed to provide more assistance and control [160]. These robotic systems have often utilized some method of shared control, which shares the burden of movement between the robot and the patient. Shared control can be defined at the task level, in which a task is predefined by the user while a robot tracks the progression of movement, or at the signal level, where both the user and robot provide independent command signals that are compared and synthesized [158]. Another method of shared control is the partitioning of the task into different components, such as speed and direction, and giving control of one part to the user and the other to the "autonomous" robot [187]. Because a single rehabilitation patient's physical ability changes greatly over the course of their therapy, the modification of parameters for shared control has been proposed to personalize the amount of assistance each patient receives during their therapy session, through verbal feedback from the user [187]. Therefore as a patient progresses through successful rehabilitation of a nondegenerative disease, assistance can be removed over time and the body can slowly learn to rely only on itself during ADLs. In contrast, for degenerative diseases, the system could also learn to provide more assistance over time.

In the later stages of stroke rehabilitation, when basic movement is no longer the primary concern, robotic therapy may include scenarios that intensify training beyond what may be experienced in daily life. Such an approach would not only serve to strengthen muscles but may also allow easier motor adaptation to unexpected scenarios in daily life, making the therapy more robust. Different challenge-based protocols include mechanical resistance to the intended movement, constraint-induced strategies that require the mandatory use of the paretic limb, error amplification, and movement perturbation. As many hemiparetic patients are plagued by the nonuse phenomenon, encouraging use of the paretic limb and conversely discouraging the healthy one is a crucial technique to achieve full rehabilitation [33,39].

Direct mechanical resistance may be the simplest way to challenge patients during rehabilitation therapy. Increasing resistance with nonrobotic systems has been shown to improve motor function and limb strength [188,189]. For a joint-based or wearable robotic system, this could mean selectively impeding the joint of interest while allowing others to move freely or be completely locked. If using an end-effector-based robotic system, either the overall internal resistance of robotic movement could be increased or the movement of the end effector could be directionally pushed against through sensing and calculation.

Error-amplification strategies utilize a desired movement trajectory with an end-effector-based robotic system, similar to impedance-based control. However, if the user's arm strays from the preprogrammed trajectory, rather than pushing the arm toward the desired trajectory, error amplification pushes the arm farther away and forces the user to correct their own mistakes. This technique has seen significant success in persons with chronic stroke [190,191]. Physical error amplification is only one strategy, however. Visual feedback has been used to exaggerate error during rehabilitation tasks, which has also shown positive results even without any change in the physical feedback [192−194]. Another way to challenge patients during their therapy is to induce some type of perturbation as they are trying to complete a rehabilitation task. For example, a common kind of movement perturbation is the application of a force field perpendicular to a linear reaching movement, in which the force is proportional to the velocity of movement [154,195]. Many other types of perturbations could also be envisioned to challenge the rehabilitation protocol, including random force fields or timed impulses.

3.4 Challenges and Way Forward for Wearable Assistive Robotics

Advances in disciplines including material sciences, actuation, sensing, electronics, energy sources, signal processing, computer science, and machine learning combined with the enormous computational capabilities at the disposal of researchers today have enabled exoskeleton technology to come a long way from the Hardiman project of the 1960s. Although significant progress has been made, several challenges such as the exorbitant price of most systems, successful stand-alone real-world operational capabilities, limitations in the electromechanical implementation, limited operating workspace, user safety concerns, and system effectiveness, combined with limitations in intention sensing and control still plague the current crop of developed systems [5,7].

3.4.1 Challenges in Electromechanical Implementation

From the electromechanical perspective, the development of affordable, lightweight high-density power sources and wireless technology is crucial for future systems to have satisfactory functionality in real-world scenarios. Researchers also need to find creative solutions to overcome the challenge of integrating lab-based systems into a single portable stand-alone system. Today's systems are very restrictive in their capabilities and end up constraining the individual's natural ROM. An improved understanding of human biomechanics will lead to the development of bioinspired hardware which is inherently safer for HRI. Bioinspired soft-bodied structures, actuators, and sensors could potentially usher in a new age of wearable assistive robots that can be used universally across different ADL tasks and by individuals of different body types. The technology could also facilitate the development of smaller and less bulky systems that allow them to be concealed within regular clothing, and thereby help overcome the psychosocial inhibition to adopting the technology. A compact portable system would enable individuals to receive assistance/rehabilitation in the acute phase of stroke right from the hospital bed and would also facilitate at-home rehabilitation. The limited force/torque generation and power transmission capabilities have significantly restricted the application of soft technology to mostly low-force applications like finger and hand exoskeletons. Further research in this domain is essential for developing exoskeletons that would be equally accomplished at assisting high-force applications like elbow and shoulder movements as well.

3.4.2 Challenges in Intention-Sensing and Control Strategy Implementation

While bioinspired soft-bodied systems could expedite the development of the next generation of more compliant, durable, portable, affordable, and safer wearable exoskeletons, the limited intention detection and shared control capabilities of current systems is another challenging obstacle that needs to be overcome. Many exoskeletons still depend on nonintuitive and unnatural user interfaces. These inputs combined with primeval control strategies add to the cognitive load and effort required by the user. This problem is exacerbated when the user has sensorimotor impairment. According to a survey of clinicians, a majority of robotic wheelchair and prosthesis users cite poor control as a major reason for prosthesis rejection [196,197]. Users often express dissatisfaction with surrendering control to automation. Conversely, the fatigue experienced in controlling these assistive systems becomes a major detriment to continued use [196]. Therefore, there is an indisputable necessity to develop new sophisticated algorithms that incorporate information from multiple sensing modalities including vision, eye tracking, psychophysiological signals, etc. to deliver simple, intuitive intention sensing and to enable seamless shared human–robot control. These algorithms need to be able to handle the differences between individuals with different pathologies, varying severity, different ADLs, and changes experienced within a single individual over time due to phenomena such as neuroplasticity. With the development of affordable high-resolution wearable psychophysiological and kinetic-inertial sensing modalities, these algorithms could lead to the development of a class of intelligent wearable robots that become a natural extension of the user's body.

3.4.3 Commercial and Nontechnical Challenges

Many rehabilitative/assistive systems have been developed over the last two decades but less than 10% of these systems have seen translation into clinical or home use. According to a report [198], the total volume of commercially sold exoskeletons numbers around 1000, highlighting the infancy of the industry. For rehabilitation applications, the advantages of robot-aided rehabilitation is still debated as the outreach of these robotic systems has been limited [198]. The cost of the system is a major hindrance to the adoption of current commercially available rehabilitative/assistive systems. As the technology is still relatively recent, there have been additional bureaucratic and regulatory barriers and even refusal from insurance companies, making the technology even more inaccessible. Exoskeleton companies are using different strategies to bring down the cost of the systems, ranging from developing single-task systems to staying in research and not releasing expensive commercial products. The exoskeleton industry and technology is still nascent and therefore depends on government and private funding. Any activity that results in negative publicity could therefore adversely affect future research and funding opportunities and could even destroy the whole ecosystem. Recently, a few exoskeleton companies have received criticism for advertising their products in an overly optimistic manner without fully disclosing the inherent limitations of the systems, leading to user dissatisfaction, lawsuits, and bad press for the entire industry [198].

In conclusion, soft and rigid wearable robotics are solutions both crucial to managing the incredibly varied number of conditions that are encountered in the rehabilitation and assistive robotics spectrum. Rigid-bodied systems would continue to provide solutions that need high force and torque requirements and in particularly fixed rehabilitative systems. On the other hand, through the use of soft robotic technology, a new age of wearable robotics is emerging that can more adequately be applied to health care and a wider consumer base, as opposed to solely military and industry. These technologies have typically come in the form of new structural design and actuation strategies, but could also be applied to sensing mechanisms. Bioinspired designs combined with creative applications of traditional and novel actuation and intention-sensing methods will make way for more advanced control strategies for assistance and rehabilitation. Through these advancements, optimal strategies for individual-specific rehabilitation and assistance can be achieved that will improve the affected individual's daily functioning and eventually allow him/her to live more independent and normal lives.

References

[1] World Heart Organization, The atlas of heart disease and stroke: global burden of stroke, 2017. <http://www.who.int/cardiovascular_diseases/ resources/atlas/en> (accessed 15.05.17.).

[2] D.J. Reinkensmeyer, J.L. Emken, S.C. Cramer, Robotics, motor learning, and neurologic recovery, Annu. Rev. Biomed. Eng. 6 (2004) 497−525.

[3] World Health Organization, Global Health and Aging, National Institute on Aging, Washington DC, 2011.

[4] P. Heo, G.M. Gu, S. jin Lee, K. Rhee, J. Kim, Current hand exoskeleton technologies for rehabilitation and assistive engineering, Int. J. Prec. Eng. Manufact. 13 (5) (2012) 807−824.

[5] P. Maciejasz, J. Eschweiler, G.-H. Kurt, J.-T. Arne, S. Leonhardt, K. Gerlach-Hahn, et al., A survey on robotic devices for upper limb rehabilitation, J. Neuroeng. Rehabil. 11 (1) (2014) 3.

[6] P. Polygerinos, Z. Wang, K.C. Galloway, R.J. Wood, C.J. Walsh, Soft robotic glove for combined assistance and at-home rehabilitation, Rob. Auton. Syst 73 (2015) 135−143.

[7] R. Bogue, Exoskeletons and robotic prosthetics: a review of recent developments, Ind. Robot An Int. J. 36 (5) (2009) 421−427.

[8] B. Marinov, Soft Exoskeletons and Exosuits, 2015. <www.exoskeletonreport.com>.

[9] R.A.R.C. Gopura, K. Kiguchi, Mechanical designs of active upper-limb exoskeleton robots, in: Rehabilitation Robotics, 2009. ICORR, 2009, pp. 178−187.

[10] H.K. Yap, B.W.K. Ang, J.H. Lim, J.C.H. Goh, C.-H. Yeow, A fabric-regulated soft robotic glove with user intent detection using EMG and RFID for hand assistive application, in: 2016 IEEE International Conference on Robotics and Automation (ICRA), 2016, pp. 3537−3542.

[11] B.B. Kang, H. Lee, H. In, U. Jeong, J. Chung, K.J. Cho, Development of a polymer-based tendon-driven wearable robotic hand, in: Proceedings—IEEE International Conference on Robotics and Automation, 2016, pp. 3750−3755.

[12] I.A. Murray, G.R. Johnson, A study of the external forces and moments at the shoulder and elbow while performing every day tasks, Clin. Biomech. 19 (6) (2004) 586−594.

[13] A. Engin, On the biomechanics of the shoulder complex, J. Biomech. 13 (7) (1980) 575 590.

[14] F.H. Martini, M.J. Timmons, R.B. Tallitsch, P.B. Cummings, Human Anatomy, Prentice Hall, Pearson Education, San Francisco, CA, 2003.

[15] C. Pylatiuk, A. Kargov, I. Gaiser, T. Werner, S. Schulz, G. Bretthauer, Design of a flexible fluidic actuation system for a hybrid elbow orthosis, in: 2009 IEEE International Conference on Rehabilitation Robotics, ICORR 2009, 2009, pp. 167−171.

[16] C. Pylatiuk, S. Schulz, H. Vaassen, M. Reischl, Preliminary evaluation for a functional support of the elbow and shoulder joint, in: 13th International FES Society Conference, 2008, pp. 65−67.

[17] C.J. van Andel, N. Wolterbeek, C.A.M. Doorenbosch, D.(H.E.J. Veeger, J. Harlaar, Complete 3D kinematics of upper extremity functional tasks, Gait Posture 27 (1) (2008) 120−127.

[18] J.T. London, Kinematics of the elbow, J. Bone Joint Surg. Am. 63 (4) (1981) 529−535.

[19] D.J. Magermans, E.K.J. Chadwick, H.E.J. Veeger, F.C.T. Van Der Helm, Requirements for upper extremity motions during activities of daily living, Clin. Biomech. 20 (6) (2005) 591−599.

[20] C.P. Neu, J.J. Crisco, S.W. Wolfe, In vivo kinematic behavior of the radio-capitate joint during wrist flexion-extension and radio-ulnar deviation, J. Biomech. 34 (11) (2001) 1429−1438.

[21] Y. Youm, A.E. Flatt, Design of a total wrist prosthesis, Ann. Biomed. Eng. 12 (3) (1984) 247−262.

[22] L. Dovat, O. Lambercy, R. Gassert, T. Maeder, T. Milner, T.C. Leong, et al., HandCARE: a cable-actuated rehabilitation system to train hand function after stroke, IEEE Trans. Neural Syst. Rehabil. Eng. 16 (6) (2008) 582−591.

[23] V. Mathiowetz, N. Kashman, G. Volland, K. Weber, M. Dowe, S. Rogers, Grip and pinch strength: normative data for adults, Arch. Phys. Med. Rehabil. 66 (2) (1985) 69−74.

[24] K. Matheus, A.M. Dollar, Benchmarking grasping and manipulation: properties of the objects of daily living, in: IEEE/RSJ 2010 International Conference on Intelligent Robots and Systems, IROS 2010—Conference Proceedings, 2010, pp. 5020−5027.

[25] D. Caligiore, G. Pezzulo, G. Baldassarre, A.C. Bostan, P.L. Strick, K. Doya, et al., Consensus paper: towards a systems-level view of cerebellar function: the interplay between cerebellum, basal ganglia, and cortex, Cerebellum 16 (1) (2017) 203−229.

[26] S. Caproni, C. Colosimo, Movement disorders and cerebrovascular diseases: from pathophysiology to treatment, Expert Rev. Neurother 17 (5) (2016) 1−11.

[27] T. Kim, K.C. Hamade, D. Todorov, W.H. Barnett, R.A. Capps, E.M. Latash, et al., Reward based motor adaptation mediated by basal ganglia, Front. Comput. Neurosci. 11 (no) (2017).

[28] A. Kishore, T. Popa, Cerebellum in levodopa-induced dyskinesias: the unusual suspect in the motor network, Front. Neurol. 5 (2014) 157.

[29] D.J. Reinkensmeyer, E. Burdet, M. Casadio, J.W. Krakauer, G. Kwakkel, C.E. Lang, et al., Computational neurorehabilitation: modeling plasticity and learning to predict recovery, J. Neuroeng. Rehabil. 13 (1) (2016) 42.

[30] V.N.V. Goldstein M, H.J.M. Barnett, J.M. Orgogozo, N. Sartorius, L. Symon, Stroke—1989: recommendations on stroke prevention, diagnosis, and therapy, Stroke 20 (1989) 1407–1431.

[31] W.S. Harwin, J.L. Patton, V.R. Edgerton, Challenges and opportunities for robot-mediated neurorehabilitation, Proc. IEEE 94 (9) (2006) 1717–1726.

[32] T.A. Jones, Motor compensation and its effects on neural reorganization after stroke, Nat. Rev. Neurosci. 18 (5) (2017) 267–280.

[33] E. Taub, G. Uswatte, V.W. Mark, D.M.M. Morris, The learned nonuse phenomenon: implications for rehabilitation, Eura. Medicophys. 42 (3) (2006) 241–256.

[34] M.C. Cirstea, M.F. Levin, Improvement of arm movement patterns and endpoint control depends on type of feedback during practice in stroke survivors, Neurorehabil. Neural Repair 21 (5) (2007) 398–411.

[35] P. Celnik, F. Hummel, M. Harris-Love, R. Wolk, L.G. Cohen, Somatosensory stimulation enhances the effects of training functional hand tasks in patients with chronic stroke, Arch. Phys. Med. Rehabil. 88 (11) (2007) 1369–1376.

[36] R.L. Hsieh, L.Y. Wang, W.C. Lee, Additional therapeutic effects of electroacupuncture in conjunction with conventional rehabilitation for patients with first-ever ischaemic stroke, J. Rehabil. Med. 39 (3) (2007) 205–211.

[37] J.C. Chen, C.C. Liang, F.Z. Shaw, Facilitation of sensory and motor recovery by thermal intervention for the hemiplegic upper limb in acute stroke patients: a single-blind randomized clinical trial, Stroke 36 (12) (2005) 2665–2669.

[38] F. Hummel, P. Celnik, P. Giraux, A. Floel, W.H. Wu, C. Gerloff, et al., Effects of non-invasive cortical stimulation on skilled motor function in chronic stroke, Brain 128 (3) (2005) 490–499.

[39] L. Oujamaa, I. Relave, J. Froger, D. Mottet, J.Y. Pelissier, Rehabilitation of arm function after stroke. Literature review, Ann. Phys. Rehabil. Med. 52 (3) (2009) 269–293.

[40] A. Floel, U. Nagorsen, K.J. Werhahn, S. Ravindran, N. Birbaumer, S. Knecht, et al., Influence of somatosensory input on motor function in patients with chronic stroke, Ann. Neurol. 56 (2) (2004) 206–212.

[41] W. Muellbacher, C. Richards, U. Ziemann, G. Wittenberg, D. Weltz, B. Boroojerdi, et al., Improving hand function in chronic stroke, Arch Neurol 59 (2002) 1278–1282.

[42] F. Fregni, P.S. Boggio, A.C. Valle, R.R. Rocha, J. Duarte, M.J.L. Ferreira, et al., A sham-controlled trial of a 5-day course of repetitive transcranial magnetic stimulation of the unaffected hemisphere in stroke patients, Stroke 37 (8) (2006) 2115–2122.

[43] N. Takeuchi, T. Chuma, Y. Matsuo, I. Watanabe, K. Ikoma, Repetitive transcranial magnetic stimulation of contralesional primary motor cortex improves hand function after stroke, Stroke 36 (12) (2005) 2681–2686.

[44] S. Edwards, Neurological Physiotherapy, second ed, Churchill Livingstone, London, 2002.

[45] M. Xiloyannis, L. Cappello, B. Khanh Dinh, C.W. Antuvan, L. Masia, Design and Preliminary Testing of a Soft Exosuit for Assisting Elbow Movements and Hand Grasping, Springer, Cham, 2017, pp. 557–561.

[46] H.K. Yap, J.H. Lim, F. Nasrallah, J.C.H. Goh, R.C.H. Yeow, A soft exoskeleton for hand assistive and rehabilitation application using pneumatic actuators with variable stiffness, in: Proceedings—IEEE International Conference on Robotics and Automation, 2015, pp. 4967–4972.

[47] H. In, B.B. Kang, M. Sin, K.-J. Cho, Exo-Glove: a wearable robot for the hand with a soft tendon routing system, IEEE Robot. Autom. Mag 22 (1) (2015) 97–105.

[48] A. Villoslada, A. Flores, D. Copaci, D. Blanco, L. Moreno, High-displacement flexible shape memory alloy actuator for soft wearable robots, Rob. Auton. Syst 73 (2015) 91−101.

[49] M.A. Delph, S.A. Fischer, P.W. Gauthier, C.H.M. Luna, E.A. Clancy, G.S. Fischer, A soft robotic exomusculature glove with integrated sEMG sensing for hand rehabilitation, in: IEEE International Conference on Rehabilitation Robotics, 2013, pp. 1−7.

[50] A. Pedrocchi, S. Ferrante, E. Ambrosini, M. Gandolla, C. Casellato, T. Schauer, et al., MUNDUS project: multimodal neuroprosthesis for daily upper limb support, J. Neuroeng. Rehabil. 10 (1) (2013) 66.

[51] I. Galiana, F.L. Hammond, R.D. Howe, M.B. Popovic, Wearable soft robotic device for post-stroke shoulder rehabilitation: identifying misalignments, in: 2012 IEEE/RSJ International Conference on Intelligent Robots and Systems, 2012, pp. 317−322.

[52] Y.L. Park, B.R. Chen, C. Majidi, R.J. Wood, R. Nagpal, E. Goldfield, Active modular elastomer sleeve for soft wearable assistance robots, in: IEEE International Conference on Intelligent Robots and Systems, 2012, pp. 1595−1602.

[53] M. Nilsson, J. Ingvast, J. Wikander, H. Von Holst, The soft extra muscle system for improving the grasping capability in neurological rehabilitation, in: 2012 IEEE-EMBS Conference on Biomedical Engineering and Sciences, IECBES 2012, 2012, pp. 412−417.

[54] Y. Kadowaki, T. Noritsugu, M. Takaiwa, D. Sasaki, M. Kato, Development of soft power-assist glove and control based on human intent, J. Robot. Mechatronics 23 (2) (2011) 281−291.

[55] A. Borboni, M. Mor, R. Faglia, Gloreha—hand robotic rehabilitation: design, mechanical model, and experiments, J. Dyn. Syst. Meas. Control 138 (11) (2016) 111003.

[56] N.S.K. Ho, K.Y. Tong, X.L. Hu, K.L. Fung, X.J. Wei, W. Rong, et al., An EMG-driven exoskeleton hand robotic training device on chronic stroke subjects: task training system for stroke rehabilitation, in: IEEE International Conference on Rehabilitation Robotics, 2011, pp. 1−5.

[57] J. Allington, S.J. Spencer, J. Klein, M. Buell, D.J. Reinkensmeyer, J. Bobrow, Supinator extender (SUE): a pneumatically actuated robot for forearm/wrist rehabilitation after stroke, in: 2011 Annual International Conference of the IEEE Engineering in Medicine and Biology Society, 2011, pp. 1579−1582.

[58] L. Connelly, Y. Jia, M.L. Toro, M.E. Stoykov, R.V. Kenyon, D.G. Kamper, A pneumatic glove and immersive virtual reality environment for hand rehabilitative training after stroke, IEEE Trans. Neural Syst. Rehabil. Eng 18 (5) (2010) 551−559.

[59] K.Y. Tong, S.K. Ho, P.M.K. Pang, X.L. Hu, W.K. Tam, K.L. Fung, et al., An intention driven hand functions task training robotic system, in: 2010 Annual International Conference of the IEEE Engineering in Medicine and Biology, 2010, pp. 3406−3409.

[60] I. Vanderniepen, R. Van Ham, M. Van Damme, R. Versluys, D. Lefeber, Orthopaedic rehabilitation: a powered elbow orthosis using compliant actuation, in: 2009 IEEE International Conference on Rehabilitation Robotics, ICORR 2009, 2009, pp. 172−177.

[61] I. Ertas, E. Hocaoglu, D. Barkana, Finger exoskeleton for treatment of tendon injuries, in: IEEE International Conference on Rehabilitation Robotics, 2009.

[62] G. Rosati, S. Cenci, G. Boschetti, D. Zanotto, S. Masiero, Design of a single-dof active hand orthosis for neurorehabilitation, in: 2009 IEEE International Conference on Rehabilitation Robotics, 2009, pp. 161−166.

[63] A. Khanicheh, D. Mintzopoulos, B. Weinberg, A.A. Tzika, C. Mavroidis, MR_CHIROD v.2: magnetic resonance compatible smart hand rehabilitation device for brain imaging, IEEE Trans. Neural Syst. Rehabil. Eng. 16 (1) (2008) 91−98.

[64] D.E. Nathan, M.J. Johnson, J. McGuire, Feasibility of integrating FES grasp assistance with a task-oriented robot-assisted therapy environment: a case study, in: 2008 2nd IEEE RAS & EMBS International Conference on Biomedical Robotics and Biomechatronics, 2008, pp. 807−812.

[65] K. Xing, Q. Xu, J. He, Y. Wang, Z. Liu, X. Huang, A wearable device for repetitive hand therapy, in: 2008 2nd IEEE RAS & EMBS International Conference on Biomedical Robotics and Biomechatronics, 2008, pp. 919−923.

[66] A. Gupta, M.K. O'Malley, V. Patoglu, C. Burgar, Design, control and performance of RiceWrist: a force feedback wrist exoskeleton for rehabilitation and training, Int. J. Rob. Res. 27 (2) (2008) 233−251.

[67] D. Ming, J. Ueda, T. Ogasawara, Pinpointed muscle force control using a power-assisting device: system configuration and experiment, in: 2008 2nd IEEE RAS & EMBS International Conference on Biomedical Robotics and Biomechatronics, 2008, pp. 181−186.

[68] Y. Hasegawa, Y. Mikami, K. Watanabe, Y. Sankai, Five-fingered assistive hand with mechanical compliance of human finger, in: Proceedings—IEEE International Conference on Robotics and Automation, 2008, pp. 718−724.

[69] S. Balasubramanian, R. Wei, M. Perez, B. Shepard, E. Koeneman, J. Koeneman, et al., RUPERT: an exoskeleton robot for assisting rehabilitation of arm functions, in: 2008 Virtual Rehabilitation, 2008, pp. 163−167.

[70] J. Stein, K. Narendran, M. John, K. Krebs, R. Hughes, Electromyography-controlled exoskeletal upper-limb−powered orthosis for exercise training after stroke, Am. J. Phys. Med. Rehab. 86 (4) (2007) 255.

[71] Z. Fuxiang, An embedded control platform of a continuous passive motion machine for injured fingers, in: S. Kommu (Ed.), Rehabilitation Robotics, I-Tech Education Publishing, Vienna, Austria, 2007, pp. 579−606.

[72] S.H. Winter, M. Bouzit, Use of magnetorheological fluid in a force feedback glove, IEEE Trans. Neural Syst. Rehabil. Eng. 15 (1) (2007) 2−8.

[73] T.T. Worsnopp, M.A. Peshkin, J.E. Colgate, D.G. Kamper, An actuated finger exoskeleton for hand rehabilitation following stroke, in: 2007 IEEE 10th International Conference on Rehabilitation Robotics, 2007, pp. 896−901.

[74] E. Rocon, J.M. Belda-Lois, A.F. Ruiz, M. Manto, J.C. Moreno, J.L. Pons, Design and validation of a rehabilitation robotic exoskeleton for tremor assessment and suppression, IEEE Trans. Neural Syst. Rehabil. Eng. 15 (1) (2007) 367−378.

[75] T. Nef, M. Guidali, R. Riener, ARMin III—arm therapy exoskeleton with an ergonomic shoulder actuation, Appl. Bionics Biomech. 6 (2) (2009) 127−142.

[76] Q. Li, D. Wang, Z. Du, Y. Song, L. Sun, sEMG based control for 5 DOF upper limb rehabilitation robot system, in: 2006 IEEE International Conference on Robotics and Biomimetics, ROBIO 2006, 2006, pp. 1305−1310.

[77] T. Kline, D. Kamper, B. Schmit, Control system for pneumatically controlled glove to assist in grasp activities, in: 9th International Conference on Rehabilitation Robotics, 2005. ICORR, 2005, pp. 78−81.

[78] D. Sasaki, T. Noritsugu, M. Takaiwa, Development of active support splint driven by pneumatic soft actuator (ASSIST), in: Proceedings—IEEE International Conference on Robotics and Automation, 2005, pp. 520−525.

[79] T. Noritsugu, Pneumatic soft actuator for human assist technology, in: Proceedings of the 6th JFPS International Symposium on Fluid Power, 2005.

[80] C. Mavroidis, J. Nikitczuk, B. Weinberg, G. Danaher, K. Jensen, P. Pelletier, et al., Smart portable rehabilitation devices, J. Neuroeng. Rehabil. 2 (1) (2005) 18.

[81] R.C.V. Loureiro, J.M. Belda-Lois, E.R. Lima, J.L. Pons, J.J. Sanchez-Lacuesta, W.S. Harwin, Upper limb tremor suppression in ADL via an orthosis incorporating a controllable double viscous beam actuator, in: Proceedings of the 2005 IEEE 9th International Conference on Rehabilitation Robotics, 2005, pp. 119−122.

[82] M. Mulas, M. Folgheraiter, G. Gini, An EMG-controlled exoskeleton for hand rehabilitation, in: Proceedings of the 2005 IEEE 9th International Conference on Rehabilitation Robotics, 2005, pp. 371−374.

[83] A. Wege, G. Hommel, Development and control of a hand exoskeleton for rehabilitation of hand injuries, in: 2005 IEEE/RSJ International Conference on Intelligent Robots and Systems, 2005, pp. 3046−3051.

[84] T. Noritsugu, H. Yamamoto, D. Sasaki, M. Takaiwa, Wearable power assist device for hand grasping using pneumatic artificial rubber muscle, in: SICE 2004 Annual Conference, 2004, pp. 420−425.

[85] L. Lucas, M. Dicicco, Y. Matsuoka, An EMG-controlled hand exoskeleton for natural pinching, J. Robot. Mechatronics 16 (5) (2004) 1−7.

[86] H. Kobayashi, Y. Ishida, H. Suzuki, Realization of all motion for the upper limb by a muscle suit, in: RO-MAN 2004. 13th IEEE International Workshop on Robot and Human Interactive Communication (IEEE Catalog No.04TH8759), 2004, pp. 631−636.

[87] E.J. Koeneman, R.S. Schultz, S.L. Wolf, D.E. Herring, J.B. Koeneman, A pneumatic muscle hand therapy device, in: The 26th Annual International Conference of the IEEE Engineering in Medicine and Biology Society, vol. 3, 2004, pp. 2711−2713.

[88] R. Sanchez, D. Reinkensmeyer, P. Shah, J. Liu, S. Rao, R. Smith, et al., Monitoring functional arm movement for home-based therapy after stroke, in: The 26th Annual International Conference of the IEEE Engineering in Medicine and Biology Society, vol. 4, 2004, pp. 4787−4790.

[89] N.G. Tsagarakis, D.G. Caldwell, Development and control of a 'soft-actuated' exoskeleton for use in physiotherapy and training, Auton. Robots 15 (1) (2003) 21−33.

[90] M. Bouzit, G. Burdea, G. Popescu, R. Boian, The Rutgers Master II—new design force-feedback glove, IEEE/ASME Trans. Mechatron. 7 (2) (2002) 256−263.

[91] F. Ögce, H. Özyalçin, Case study: a myoelectrically controlled shoulder−elbow orthosis for unrecovered brachial plexus injury, Prosthet. Orthot. Int. 24 (3) (2000) 252−255.

[92] M.L. Turner, D.H. Gomez, M.R. Tremblay, M.R. Cutkosky, P. Alto, Preliminary tests of an arm-grounded haptic feedback device in telemanipulation, Proc. ASME IMECE Haptics Symp (1998) 1−6.

[93] PowerGrip Assisted Grasp Orthosis, Broadened Horizons, MN, USA, http://www.broadenedhorizons.com/.

[94] A.C. Lo, Clinical designs of recent robot rehabilitation trials, Am. J. Phys. Med. Rehabil. 91 (2012) S204−S216.

[95] A. Zoss, H. Kazerooni, Design of an electrically actuated lower extremity exoskeleton, Adv. Robot. 20 (2006) 967−988.

[96] General Electric Co., Hardiman I arm test, in: General Electric Report S-70−1019, Schenectady, NY, 1969.

[97] Y.L. Park, J. Santos, K.G. Galloway, E.C. Goldfield, R.J. Wood, A soft wearable robotic device for active knee motions using flat pneumatic artificial muscles, in: Proceedings—IEEE International Conference on Robotics and Automation, 2014, pp. 4805−4810.

[98] H.K. Yap, J.C.H. Goh, R.C.H. Yeow, Design and characterization of soft actuator for hand rehabilitation application, in: IFMBE Proceedings, vol. 45, 2015, pp. 367−370.

[99] A.T. Asbeck, S.M.M. Rossi, I. Galiana, Y. Ding, C.J. Walsh, Stronger, smarter, softer: next-generation wearable robots, IEEE Robot Autom. Mag. 21 (2014).

[100] Upper body mobile assistive exoskeleton catalog, <www.exoskeletonreport.com>.

[101] H. Satoh, T. Kawabata, Y. Sankai, Bathing care assistance with robot suit HAL, in: 2009 IEEE International Conference on Robotics and Biomimetics, ROBIO 2009, 2009, pp. 498−503.

[102] A. Schiele, F.C.T. van der Helm, Kinematic design to improve ergonomics in human machine interaction, IEEE Trans. Neural Syst. Rehabil. Eng. 14 (4) (2006) 456−469.

[103] C. Fleischer, K. Kondak, A. Wege, I. Kossyk, Research on exoskeletons at the TU Berlin, Advances in Robotics Research, Springer Berlin Heidelberg, Berlin, Heidelberg, 2009, pp. 335−346.

[104] A.B. Zoss, H. Kazerooni, A. Chu, Biomechanical design of the Berkeley lower extremity exoskeleton (BLEEX), IEEE/ASME Trans. Mechatron. 11 (2) (2006) 128−138.

[105] E. Guizzo, H. Goldstein, The rise of the body bots, IEEE Spectr. 42 (10) (2005) 42−48.

[106] H. Herr, R. Kornbluh, New horizons for orthotic and prosthetic technology: artificial muscle for ambulation, Smart Struct. Mater. Electroact. Polym. Actuators Devices 5385 (2004) 1−9.

[107] J.F. Veneman, R. Kruidhof, E.E.G. Hekman, R. Ekkelenkamp, E.H.F. Van Asseldonk, H. van der Kooij, Design and evaluation of the LOPES exoskeleton robot for interactive gait rehabilitation, IEEE Trans. Neural Syst. Rehabil. Eng. 15 (3) (2007) 379−386.

[108] D.P. Ferris, J.M. Czerniecki, B. Hannaford, An ankle-foot orthosis powered by artificial pneumatic muscles, J. Appl. Biomech. 21 (2) (2005) 189−197.

[109] A.T. Asbeck, S. Rossi, K. Holt, C. Walsh, A biologically inspired soft exosuit for walking assistance, Int. J. Rob. Res. 34 (2015).

[110] J. Perry, J.R. Davids, Gait analysis: normal and pathological function, J. Pediatr. Orthop. 12 (6) (1992) 815.

[111] A. Schiele, F.C.T. van der Helm, Influence of attachment pressure and kinematic configuration on pHRI with wearable robots, Appl. Bionics Biomech 6 (2) (2009) 157−173.

[112] Y. Ren, H.S. Park, and L.Q. Zhang, Developing a whole-arm exoskeleton robot with hand opening and closing mechanism for upper limb stroke rehabilitation, in 2009 IEEE International Conference on Rehabilitation Robotics, ICORR 2009, 2009, pp. 761−765.

[113] M. Cempini, M. Cortese, N. Vitiello, A powered finger-thumb wearable hand exoskeleton with self-aligning joint axes, IEEE/ASME Trans. Mechatronics 20 (2) (2015) 705−716.

[114] E.B. Brokaw, R.J. Holley, P.S. Lum, Hand spring operated movement enhancer (HandSOME) device for hand rehabilitation after stroke, in: 2010 Annual International Conference of the IEEE Engineering in Medicine and Biology, 2010, pp. 5867−5870.

[115] M.A. Ergin, V. Patoglu, ASSISTON-SE: a self-aligning shoulder-elbow exoskeleton, in: 2012 IEEE International Conference on Robotics and Automation, 2012, pp. 2479−2485.

[116] J.F. Geboers, J.H. van Tuijl, H.A.M. Seelan, M.R. Drost, Effect of immobilization on ankle dorsiflextion strength, Scand. J. Rehabil. Med. 32 (2000) 66−71.

[117] H.J. Appell, Muscular atrophy following immobilisation: a review, Sport. Med. 10 (1990) 42−58.

[118] J.F. Geboers, M.R. Drost, F. Spaans, H. Kuipers, H.A. Seelen, Immediate and long-term effects of ankle-foot orthosis on muscle activity during walking: a randomized study of patients with unilateral foot drop, Arch. Phys. Med. Rehabil. 83 (2) (2002) 240−245.

[119] B. Marinov, Reducing the cost of exoskeleton devices, 2015. <www.exoskeletonreport.com>.

[120] B. Marinov, 19 Military exoskeletons into 5 categories, 2016. <www.exoskeletonreport.com>.

[121] D. Rus, M.T. Tolley, Design, fabrication and control of soft robots, Nature 521 (7553) (2015) 467−475.

[122] S. Kim, C. Laschi, B. Trimmer, Soft robotics: a bioinspired evolution in robotics, Trends Biotechnol. 31 (5) (2013) 287−294.

[123] M. Manti, V. Cacucciolo, M. Cianchetti, Stiffening in soft robotics: a review of the state of the art, IEEE Robot. Autom. Mag. 23 (3) (2016) 93−106.

[124] R. Pfeifer, G. Gómez, Morphological Computation—Connecting Brain, Body, and Environment, Springer Berlin, Heidelberg, 2009, pp. 66−83.

[125] C. Laschi, B. Mazzolai, M. Cianchetti, Soft robotics: technologies and systems pushing the boundaries of robot abilities, Sci. Robot. 3690 (2016) 1−11.

[126] Roam Robotics, https://www.roamrobotics.com.

[127] T. Tanaka, Y. Satoli, S. Kaneko, Y. Suzuki, N. Sakamoto, S. Seki, Smart suit: soft power suit with semi-active assist mechanism—prototype for supporting waist and knee joint, in: 2008 International Conference on Control, Automation and Systems, ICCAS 2008, 2008, pp. 2002−2005.

[128] U. Jeong, H. In, H. Lee, B.B. Kang, K.-J. Cho, Investigation on the control strategy of soft wearable robotic hand with slack enabling tendon actuator, in: 2015 IEEE International Conference on Robotics and Automation (ICRA), 2015, pp. 5004−5009.

[129] Y.-L. Park, B. Chen, N.O. Pérez-Arancibia, D. Young, L. Stirling, R.J. Wood, et al., Design and control of a bio-inspired soft wearable robotic device for ankle-foot rehabilitation, Bioinspir. Biomim 9 (1) (2014) 16007.

[130] C. Gonzalez, What's the difference between pneumatic, hydraulic, and electrical actuators?, 2015. <www.machinedesign.com> (Online). Available: <http://machinedesign.com/linear-motion/what-s-difference-between-pneumatic-hydraulic-and-electrical-actuators> (accessed 23.04.17.).

[131] R. Looned, J. Webb, Z.G. Xiao, C. Menon, Assisting drinking with an affordable BCI-controlled wearable robot and electrical stimulation: a preliminary investigation, J. Neuroeng. Rehabil. 11 (1) (2014) 51.

[132] G.A. Pratt, M.M. Williamson, Series elastic actuators, IEEE/RSJ Int. Conf. Intell. Robot. Syst. Human Robot Interact. Coop. Robot. 1 (1524) (1995) 399−406.

[133] G. Rosati, J.E. Bobrow, D.J. Reinkensmeyer, Compliant control of post-stroke rehabilitation robots: using movement-specific models to improve controller performance, Proc. Asme Int. Mech. Eng. Congr. Expo. 2008 2 (2009) 167−174.

[134] C.J. Nycz, M.A. Delph, G.S. Fischer, Modeling and design of a tendon actuated soft robotic exoskeleton for hemiparetic upper limb rehabilitation, in: 2015 37th Annual International Conference of the IEEE Engineering in Medicine and Biology Society (EMBC), 2015, pp. 3889−3892.

[135] U. Jeong, H.K. In, K.J. Cho, Implementation of various control algorithms for hand rehabilitation exercise using wearable robotic hand, Intell. Serv. Robot 6 (4) (2013) 181−189.

[136] P. Polygerinos, K.C. Galloway, S. Sanan, M. Herman, C.J. Walsh, EMG controlled soft robotic glove for assistance during activities of daily living, in: IEEE International Conference on Rehabilitation Robotics, 2015, pp. 55−60.

[137] C.-P. Chou, B. Hannaford, Measurement and modeling of McKibben pneumatic artificial muscles, IEEE Trans. Robot. Autom. 12 (1) (1996) 90−102.

[138] F. Connolly, P. Polygerinos, C.J. Walsh, K. Bertoldi, Mechanical programming of soft actuators by varying fiber angle, Soft Robot. 2 (1) (2015).

[139] P. Polygerinos, K.C. Galloway, E. Savage, M. Herman, K. O'Donnell, C.J. Walsh, Soft robotic glove for hand rehabilitation and task specific training, in: Proceedings—IEEE International Conference on Robotics and Automation, 2015, pp. 2913−2919.

[140] P. Polygerinos, B. Mosadegh, A. Campo, PneuNets bending actuators. <www.softroboticstoolkit.com>.

[141] P. Polygerinos, S. Lyne, Z. Wang, L.F. Nicolini, B. Mosadegh, G.M. Whitesides, et al., Towards a soft pneumatic glove for hand rehabilitation, in: IEEE International Conference on Intelligent Robots and Systems, 2013, pp. 1512−1517.

[142] B. Mosadegh, P. Polygerinos, C. Keplinger, S. Wennstedt, R.F. Shepherd, U. Gupta, et al., Pneumatic networks for soft robotics that actuate rapidly, Adv. Funct. Mater. 24 (15) (2014) 2163−2170.

[143] I. Gaiser, S. Schulz, H. Breitwieser, G. Bretthauer, Enhanced flexible fluidic actuators for biologically inspired lightweight robots with inherent compliance, in: 2010 IEEE International Conference on Robotics and Biomimetics, 2010, pp. 1423−1428.

[144] T. Lenzi, N. Vitiello, S.M.M. De Rossi, S. Roccella, F. Vecchi, M.C. Carrozza, NEUROExos: a variable impedance powered elbow exoskeleton, in: Proceedings of the IEEE International Conference on Robotics and Automation, 2011, pp. 1419−1426.

[145] P. Polygerinos, K. Galloway, Z. Wang, F. Connolly, J.T.B. Overvelde, H. Young, Fibre-reinforced actuators. <www.softroboticstoolkit.com>.

[146] F. Connolly, C.J. Walsh, K. Bertoldi, Automatic design of fiber-reinforced soft actuators for trajectory matching, Proc. Natl. Acad. Sci. 114 (2016) 51−56.

[147] J.A. Gallego, E. Rocon, J. Ibáñez, J.L. Dideriksen, A.D. Koutsou, R. Paradiso, et al., A soft wearable robot for tremor assessment and suppression, in: Proceedings of the IEEE International Conference on Robotics and Automation, 2011, pp. 2249−2254.

[148] S. Hamid, R. Hayek, Role of electrical stimulation for rehabilitation and regeneration after spinal cord injury: an overview, Eur. Spine J. 17 (9) (2008) 1256−1269.

[149] J. Makaran, D. Dittmer, R. Buchal, D. MacArthur, The SMART wrist hand orthosis (WHO) for quadriplegic patients, J. Prosthetics Orthot 5 (3) (1993) 73.

[150] K. Andrianesis, A. Tzes, Design of an innovative prosthetic hand with compact shape memory alloy actuators, in: 21st Mediterranean Conference on Control and Automation, 2013, pp. 697−702.

[151] Y. Bar-Cohen, EAP as Artificial Muscles: Progress and Challenges, California Institute of Technology, San Diego, CA, 2004, p. 10.

[152] M. Duduta, R.J. Wood, D.R. Clarke, Multilayer dielectric elastomers for fast, programmable actuation without prestretch, Adv. Mater. 28 (36) (2016) 8058–8063.

[153] J.D.W. Madden, S. Kianzad, Twisted lines: artificial muscle and advanced instruments can be formed from nylon threads and fabric, IEEE Pulse 6 (1) (2015) 32–35.

[154] G.C. Sing, W.M. Joiner, T. Nanayakkara, J.B. Brayanov, M.A. Smith, Primitives for motor adaptation reflect correlated neural tuning to position and velocity, Neuron 64 (4) (2009) 575–589.

[155] R. Song, K.Y. Tong, X.L. Hu, X.J. Zheng, Myoelectrically controlled robotic system that provide voluntary mechanical help for persons after stroke, in: 2007 IEEE 10th International Conference on Rehabilitation and Robotics ICORR'07, 2007, pp. 246–249.

[156] S.R. Soekadar, M. Witkowski, C. Gómez, E. Opisso, J. Medina, M. Cortese, et al., Hybrid EEG/EOG-based brain/neural hand exoskeleton restores fully independent daily living activities after quadriplegia, Sci. Robot. 1 (1) (2016) 1–8.

[157] H.I. Krebs, J.J. Palazzolo, L. Dipietro, M. Ferraro, J. Krol, K. Rannekleiv, et al., Rehabilitation robotics: performance-based progressive robot-assisted therapy, Auton. Robots 15 (1) (2003) 7–20.

[158] A. Broad, B. Argall, Path planning under interface-based constraints for assistive robotics, in: Twenty-Sixth International Conference on Automated Planning and Schedule, 2016.

[159] G. Gras, V. Vitiello, G.Z. Yang, Cooperative control of a compliant manipulator for robotic-assisted physiotherapy, in: Proceedings of the IEEE International Conference on Robotics and Automation, 2014, pp. 339–346.

[160] L. Marchal-Crespo, D.J. Reinkensmeyer, Review of control strategies for robotic movement training after neurologic injury, J. Neuroeng. Rehabil. 6 (1) (2009) 20.

[161] R. Song, K. Tong, X. Hu, L. Li, Assistive control system using continuous myoelectric signal in robot-aided arm training for patients after stroke, IEEE Trans. Neural. Syst. Rehabil. Eng. 16 (4) (2008) 371–379.

[162] T. Lenzi, S.M.M. De Rossi, N. Vitiello, M.C. Carrozza, Intention-based EMG control for powered exoskeletons, IEEE Trans. Biomed. Eng. 59 (8) (2012) 2180–2190.

[163] A.V. Hill, The heat of shortening and the dynamic constants of muscle, Proc. R. Soc. B Biol. Sci. 126 (843) (1938) 136–195.

[164] J. Rosen, M.B. Fuchs, M. Arcan, Performances of hill-type and neural network muscle models-toward a myosignal-based exoskeleton, Comput. Biomed. Res. 32 (5) (1999) 415–439.

[165] L.F. de Oliveira, L.L. Menegaldo, Individual-specific muscle maximum force estimation using ultrasound for ankle joint torque prediction using an EMG-driven Hill-type model, J. Biomech. 43 (14) (2010) 2816–2821.

[166] R.A.R.C. Gopura, K. Kiguchi, Y. Yi, SUEFUL-7: a 7DOF upper-limb exoskeleton robot with muscle-model-oriented EMG-based control, in: 2009 IEEE/RSJ International Conference on Intelligent Robots and Systems. IROS 2009, 2009, pp. 1126–1131.

[167] P.K. Artemiadis, K.J. Kyriakopoulos, A switching regime model for the EMG-based control of a robot arm, IEEE Trans. Syst. Man, Cybern. Part B Cybern. 41 (1) (2011) 53–63.

[168] E. Pirondini, M. Coscia, S. Marcheschi, G. Roas, F. Salsedo, A. Frisoli, et al., Evaluation of the effects of the arm light exoskeleton on movement execution and muscle activities: a pilot study on healthy subjects, J. Neuroeng. Rehabil. 13 (2016) 9.

[169] J.M. Antelis, L. Montesano, A. Ramos, N. Birbaumer, J. Minguez, Decoding upper limb movement attempt from EEG measurements of the contralesional motor cortex in chronic stroke patients, IEEE Trans. Biomed. Eng. 64 (1) (2016) 99–111.

[170] E. Buch, C. Weber, L.G. Cohen, C. Braun, M.A. Dimyan, T. Ard, et al., Think to move: a neuromagnetic brain–computer interface (BCI) system for chronic stroke, Stroke 39 (3) (2008) 910–917.

[171] A.J. Casson, S.J.M. Smith, J.S. Duncan, E. Rodriguez-villegas, Wearable EEG: what is it, why is it needed and what does it entail? in: Conference Proceedings of the IEEE Engineering in Medicine and Biology Society, 2008, pp. 5867–5870.

[172] H. Yuan, B. He, Brain—computer interfaces using sensorimotor rhythms: current state and future perspectives, IEEE Trans Biomed Eng 61 (5) (2015) 1425—1435.

[173] D. Novak, R. Riener, A survey of sensor fusion methods in wearable robotics, Rob. Auton. Syst. 73 (2015) 155—170.

[174] F. Lotte, Signal processing approaches to minimize or suppress calibration time in oscillatory activity-based brain—computer interfaces, Proc. IEEE 103 (6) (2015) 871—890.

[175] N. Weiskopf, K. Mathiak, S.W. Bock, F. Scharnowski, R. Veit, W. Grodd, et al., Principles of a brain—computer interface (BCI) based on real-time functional magnetic resonance imaging (fMRI), IEEE Trans. Biomed. Eng. 51 (6) (2004) 966—970.

[176] K. Yanagisawa, K. Asaka, H. Sawai, H. Tsunashima, T. Nagaoka, T. Tsujii, et al., Brain—computer interface using near-infrared spectroscopy for rehabilitation, in: Control Automation and Systems (ICCAS), 2010 International Conference, 2010, pp. 2248—2253.

[177] K.K. Ang, C. Guan, K.S. Phua, C. Wang, L. Zhou, K.Y. Tang, et al., Brain—computer interface-based robotic end effector system for wrist and hand rehabilitation: results of a three-armed randomized controlled trial for chronic stroke, Front. Neuroeng. 7 (2014) 1—9.

[178] F. Meng, K. Tong, S. Chan, W. Wong, K. Lui, K. Tang, et al., BCI-FES training system design and implementation for rehabilitation of stroke patients, in: Ijcnn 2008, 2008, pp. 4103—4106.

[179] M. Markovic, S. Dosen, D. Popovic, B. Graimann, D. Farina, Sensor fusion and computer vision for context-aware control of a multi degree-of-freedom prosthesis, J. Neural Eng. 12 (6) (2015) 1—15.

[180] K. Muelling, A. Venkatraman, J. Valois, J.E. Downey, J. Weiss, S. Javdani, et al., Autonomy infused teleoperation with application to BCI manipulation, in: Proceedings of the Robotic Science and Systems, 2015.

[181] G. Ghazaei, A. Alameer, P. Degenaar, G. Morgan, K. Nazarpour, Deep learning-based artificial vision for grasp classification in myoelectric hands, J. Neural Eng. 14 (3) (2017) 1—18.

[182] D. Novak, R. Riener, Enhancing patient freedom in rehabilitation robotics using gaze-based intention detection, in: IEEE International Conference on Rehabilitation Robotics, 2013.

[183] A. Bulling, J.A. Ward, H. Gellersen, G. Troster, Eye movement analysis for activity recognition using electrooculography, IEEE Trans. Pattern Anal. Mach. Intell. 33 (4) (2011) 741—753.

[184] S. Goto, T. Sugi, M. Nakamura, Development of meal assistance orthosis for disabled persons with human intention extraction through EOG signals, in: 2006 SICE-ICASE International Joint Conference, 2006, pp. 227—232.

[185] M.H. Rahman, M.J. Rahman, O.L. Cristobal, M. Saad, J.P. Kenné, P.S. Archambault, Development of a whole arm wearable robotic exoskeleton for rehabilitation and to assist upper limb movements, Robotica 33 (2014) 1—21.

[186] M. Lotze, C. Braun, N. Birbaumer, S. Anders, L.G. Cohen, Motor learning elicited by voluntary drive, Brain 126 (4) (2003) 866—872.

[187] D. Gopinath, S. Jain, B.D. Argall, Human-in-the-loop optimization of shared autonomy in assistive robotics, IEEE Robot. Autom. Lett. 2 (1) (2017) 247—254.

[188] M.M. Ouellette, N.K. LeBrasseur, J.F. Bean, E. Phillips, J. Stein, W.R. Frontera, et al., High-intensity resistance training improves muscle strength, self-reported function, and disability in long-term stroke survivors, Stroke 35 (6) (2004) 1404—1409.

[189] S.L. Morris, K.J. Dodd, M.E. Morris, Outcomes of progressive resistance strength training following stroke: a systematic review, Clin. Rehabil. 18 (1) (2004) 27—39.

[190] J.L. Patton, M.E. Stoykov, M. Kovic, F.A. Mussa-Ivaldi, Evaluation of robotic training forces that either enhance or reduce error in chronic hemiparetic stroke survivors, Exp. Brain Res. 168 (3) (2006) 368—383.

[191] J.L. Patton, M. Kovic, Fa Mussa-Ivaldi, Custom-designed haptic training for restoring reaching ability to individuals with poststroke hemiparesis, J. Rehabil. Res. Dev. 43 (5) (2006) 643—656.

[192] Y. Wei, P. Bajaj, R. Scheldt, J. Patton, Visual error augmentation for enhancing motor learning and rehabilitative relearning, in: Proceedings of the 2005 IEEE 9th International Conference on Rehabilitation Robotics, 2005, pp. 505−510.

[193] B.R. Brewer, R. Klatzky, and Y. Matsuoka, Initial therapeutic results of visual feedback manipulation in robotic rehabilitation, in: International. Workshop on Virtual Rehabilitation, 2006, pp. 160−166.

[194] B.R. Brewer, R.L. Klatzky, Y. Matsuoka, Visual-feedback distortion in a robotic rehabilitation environment, Proc. IEEE 94 (9) (2006) 1739−1750.

[195] Ra Scheidt, D.J. Reinkensmeyer, Ma Conditt, W.Z. Rymer, Fa Mussa-Ivaldi, Persistence of motor adaptation during constrained, multi-joint, arm movements, J. Neurophysiol. 84 (2) (2000) 853−862.

[196] B.D. Argall, Machine learning for shared control with assistive machines, in: ICRA Workshop on Autonomous Learning: From Machine Learning to Learning in Real World Autonomous Systems, 2013, pp. 1−5.

[197] E.A. Biddiss, T.T. Chau, Upper limb prosthesis use and abandonment: a survey of the last 25 years, Prosthet. Orthot. Int. 31 (3) (2007) 236−257.

[198] B. Marinov, Lack of communication is the greatest challenge to the exoskeleton industry, 2016. <www.exoskeletonreport.com>.

CHAPTER 4

Upper Limb Wearable Exoskeleton Systems for Rehabilitation: State of the Art Review and a Case Study of the EXO-UL8—Dual-Arm Exoskeleton System

Yang Shen, Peter Walker Ferguson, Ji Ma and Jacob Rosen

Department of Mechanical and Aerospace Engineering, University of California Los Angeles (UCLA), Los Angeles, CA, United States

4.1 Background Information on Upper Limb Stroke Rehabilitation

As one of the leading causes of severe long-term disability [1], stroke (ischemic and hemorrhagic) results in 795 K new patients every year (16.9 M worldwide) in addition to the existing 6.6 M stroke patients (33 M worldwide). Hemiparesis (one-sided weakness) or hemiplegia (one-sided paralysis) frequently occurs with spasticity (stiff or tight muscles) and joint/muscle coupling affects 80% of stroke victims. The specific functionality and severity depend on the brain trauma position and size. The patients' participation in activities of daily living (ADLs) is affected, creating a burden on themselves, their families, and society [2]. Of particular interest is that movement capabilities of stroke patients are normally more severely affected on one side, depending on which brain hemisphere has trauma. The bilateral training mode discussed later is based on this observation.

Many poststroke patients are able to regain some capabilities after rehabilitation training. However, due to the limitation of time/skills of human physical therapists, stroke survivors often do not receive sufficient training and do not recover the capabilities they should. Rehabilitation robots, which always have contact with the human's body and are thus "wearable," have the potential to automate the training process and increase the exercise dose while reducing the service cost. As average life expectancy is lengthened by improved medical treatment, the absolute amount of stroke survivors is increasing and rehabilitation wearables are expected to have a promising market.

Wearable Technology in Medicine and Health Care.
DOI: https://doi.org/10.1016/B978-0-12-811810-8.00004-X

In general, rehabilitation wearables can be categorized based on the body parts to be trained: hand and fingers, upper limbs, and lower limbs. Wearables in each category must meet different challenges: devices for hand and fingers require fine movement with low torque output; robots for upper limbs address different symptoms including weakness, spasticity, and/or limited range of motion; and lower limb training devices could focus on balance or energy utilization efficiency. While the wearables for lower limb rehabilitation have gained significant attention in recovering locomotion abilities [3,4], upper limb rehabilitation training devices should receive at least equal attention as participation in ADLs is greatly dependent on upper limb capabilities.

4.2 State of the Art in Upper Limb Rehabilitation Wearables

Our review, in this part, focuses on upper limb rehabilitation wearables excluding standalone hand exoskeletons. These devices are summarized in Tables 4.1 and 4.2.

Early upper limb rehabilitation wearables were based on end-effector robots that transmit forces and torques to patients at the contact point. Since the 1990s, numerous clinical trials have been conducted on end-effector upper limb rehabilitation wearables indicating improved treatment outcomes when compared to traditional therapy [5–9]. Following these successes, upper limb robotic rehabilitation systems have gained acceptance and naturally evolved from end-effector robots, to single-arm exoskeletons, to dual-arm and full-body exoskeletons.

The MIT-MANUS [10–13], commercialized as the InMotionArm (Interactive Motion Technologies, Inc., Cambridge, MA, United States), is a direct-drive five bar–linkage SCARA robot. The robot is attached to the patient's forearm and produces horizontal planar translations. Additional attachments have been developed to enable active control of forearm pronation/supination, wrist flexion/extension, and wrist abduction/adduction. The system is used with robotic therapy games to motivate and coordinate therapeutic tasks, a strategy adopted by the majority of upper limb robotic rehabilitation systems.

The upper limb motion assist system developed by AIST [14] and NeReBot [15] maneuvers the patient's arm by changing the lengths of three cables suspending orthoses/splints worn by the patient. The upper limb mobile assist system by AIST consists of two such orthoses placed on the forearm near the elbow and the wrist. By changing the positions of both orthoses, two rotations and three translations of the forearm can be controlled. The NeReBot is a cable-driven robot featuring a single splint attached to the entire forearm actuated by three motors.

The GENTLE/s [16] and ACT^{3D} [17] both feature a HapticMaster robot [18] connected with a forearm orthosis. The HapticMaster enables each device three active translational degrees of freedom (DOFs) of the forearm. The GENTLE/s system also features a passive elbow orthosis suspended from above by cables for gravity compensation. The ACT^{3D} provides adjustable active gravity compensation.

Table 4.1: Existing upper limb wearables.

Device	Use	Mechanical Design	Control Method	Active (Passive) DOF					
				T	S	E	F	W	H
Dynamometers									
Biodex System 4 Pro [23]	R, TR	Actuated by an electric motor	Isokinetic resistance mode, eccentric mode, passive motion mode, isometric mode, isotonic mode, or position control	1	—	—	—	—	—
HUMAC NORM [24]	R, TR	Actuated by a brushless motor	Isometric testing, isokinetic testing, passive mode, isometric mode, isokinetic mode, or isotonic mode.	1	—	—	—	—	—
End-Effector Robots									
MIT-MANUS [10–13]	R	Actuated by brushless DC motors	Impedance control, EMG trigger	2 + 3	0	0	2 + 1	0 + 2	0
Upper limb motion assist system by AIST [14]	A	Actuated by DC servo motors with pulleys	Position control or velocity control	5	0	0	5	0	0
GENTLE/s [16,18]	R, HF	Actuated by brushed DC motors with antibacklash lead screw spindles	Position, impedance, or admittance control	3 (4)	0	(1)	3 (3)	0	0
ACT^{3D} [17,18]	R, HF	Actuated by brushed DC motors with antibacklash lead screw spindles	Force control	3 (3)	0	0	3 (3)	0	0
NeReBot [15]	R	Actuated by brushless motors with cable transmissions	Position control	3	0	0	3	0	0
iPAM System [19]	R	Pneumatically actuated	Admittance control	6 (6)	3 (3)	0	3 (1)	(2)	0
Bi-Manu-Track [20]	R	Actuated by electric motors	Impedance control	1 (1) ×2	0	0	1 or (1) ×2	(1) or 1 ×2	0
MIME [21]	R	Actuated by servo motors	Position control	6 × 2	0	0	6 × 2	0	0
KINARM [22]	R			2 × 2	1	1	0	0	0

(Continued)

Table 4.1: (Continued)

Device	Use	Mechanical Design	Control Method	Active (Passive) DOF					
				T	S	E	F	W	H
Single-Arm Exoskeletons									
MULOS [29]	A	Actuated by electric motors with cable transmissions, bevel gearboxes, and timing belts	Velocity or force control	5	3	1	1	0	0
L-Exos[31–33]	R, HF	Actuated by permanent magnet DC motors with tendon transmissions	Impedance or direct force control	4 (1)	3	1	(1) − (1) +1	0	0 +2
SARCOS Master Arm [25,26]	TO	Hydraulically actuated	Impedance control. Local control of each joint.	10	3	1	1	2	3
SRE [46]	R	Actuated by braided pneumatic muscle actuators	Joint position control, joint torque control, or impedance control	7	3	1	1	2	0
MEDARM [43]	R	Actuated by electric motors with cable-driven transmissions and timing belts		6	5	1	0	0	0
IntelliArm [41,42]	R	Actuated by servomotors	"Intelligent stretch," "back-drivable," "assistive," or "resistive"	8 (2)	4 (2)	1	1	1 (2)	1
RUPERT IV [47,48]	R	Actuated by pneumatic muscle actuators	PID-based feedback, iterative learning controller-based feedforward control	5	2	1	1	1	0
BONES [34–36]	R	Actuated by pneumatic cylinder actuators	Model-based adaptive control	4 + 2	3	1	0 + 1	0 + 1	0
ABLE [37,38]	R, A, TO, HF	Actuated by DC motors with screw-and-cable transmissions	Hybrid force-position control	7	3	1	1	2	0
ARMin III [39]	R	Actuated by brushed DC motors with harmonic drive transmissions	Impedance control	4 + 2	3	1	0 + 1	0 + 1	0
MGA [40]	R	Actuated by brushless DC motors with harmonic drive transmissions	"Composite" control groups joints. Each group can use impedance, admittance, or position control.	5 (1)	4	1	(1)	0	0

	Use	Actuation	Control	T	S	E	F	W	H
SUEFUL-7 [30]	A	Actuated by motors with pulleys, cable drives, spur gears, and bevel gears	Muscle-model-oriented EMG-based impedance control	7	3	1	1	2	0
RehaBot [49]	R	Actuated by serial elastic actuators and direct-drive motors	Force control	7	3	1	1	2	0
Exorn [44,45]	R, A	Actuated by DC geared and brushless DC servo motors	Position control or EMG-based control	10	6	1	1	2	0
ETS-MARSE [50]	R	Actuated by brushless DC motors with harmonic drive transmissions	Computed torque control or EMG-based control	7	3	1	1	2	0
Dual-Arm Exoskeletons									
EMY [52]	BMI	Actuated by brushed DC motors with screw-cable-systems and gearboxes	Position, velocity, or torque control	8	2 × 2	1 × 2	1 × 2	0	0
CAPIO [53]	HF, TO	Actuated by serial elastic actuators	Zero force, inverse dynamic, force feedback, or determinate force control	20 (4)	3 (1) × 2	2 × 2	1 × 2	2 × 2	0
Recupera-Reha [54]	R	Actuated by brushless DC motors, servo motors, and serial elastic actuators	Position, torque, or EMG/EEG-based control	12 (2)	3 × 2	1 × 2	1 × 2	(1) × 2	1 × 2

Use: R, rehabilitation; A, motion assistance; HF, haptic feedback; P, power augmentation; TO, teleoperation; BMI, evaluation of BMI; TR, training (athletic/strength). DOF: T, total; S, shoulder; E, elbow; F, forearm; W, wrist; H, hand; + / −, DOFs of optional attachments; × 2 for dual arm device.

Table 4.2: Graphical summary of existing upper limb rehabilitation robots.

The iPAM system [19] features two rigid 3D robot arms connected to the patient at the upper arm and wrist. The system can therefore actively control the positions of upper arm and forearm, but both connection points passively permit all orientation DOFs.

Bi-Manu-Track [20], MIME [21], and KINARM [22] are dual-arm robotic systems and are thus capable of bimanual therapy, a desirable feature that is not achievable with a single-arm system. Bi-Manu-Track is a portable reconfigurable device limited to one active and one passive DOF between forearm pronation/supination and wrist flexion/extension. MIME consists of 6-DOF Puma-560 robots and position digitizers attached at each forearm. KINARM is a planar device that mechanically supports the weight of the arm while actuating two-DOF horizontal motions.

An additional notable class of rehabilitation robot that can be used for the upper limbs is the dynamometer. Dynamometers such as the Biodex System 4 Pro [23] and the HUMAC NORM [24] feature a single motor that can be repositioned and connected to various attachments to target specific motions.

End-effector robots have been shown to be effective in rehabilitation, and several have even found commercial success. However, these robots suffer from several critical limitations.

End-effector robots typically have significantly reduced ranges of motion when compared to the human arm. For the workspace of an end-effector robot to encompass the workspace of the human arm, the robot must be very large because the base of the robot must be outside of the reach of the arm to prevent collisions. In addition, the robot would need to reach each part of the workspace of the human arm without physically overlapping with the user.

End effectors move individual points of the human arm. The human arm is a redundant manipulator with seven DOFs, so controlling position and/or orientation of a point on the arm does not control the configuration of the entire arm. Consequently, it is challenging for an end-effector rehabilitation robot to target a specific joint motion for therapy. To the best of the authors' knowledge, there is no end-effector rehabilitation robot that can determine and control all of the DOFs of the human arm.

To circumvent these and other limitations, a large number of upper limb exoskeleton robots have been developed. Upper limb exoskeletons are structured in an anthropometric fashion that supports the partial/full range of motion of the human arm. They are designed to be worn by the user, and are attached at multiple locations. Although this can significantly complicate the design of the robot, it enables much larger ranges of motion and the ability to target specific joint motions for therapy. Exoskeletons can broadly be categorized by application, number of DOFs, and whether the exoskeleton is worn on one or both arms.

The SARCOS Master Arm [25,26] and SAM [27,28] are single-arm exoskeletons designed for teleoperation. The Sarcos Master Arm and SAM have the seven main DOFs of the human arm: shoulder flexion/extension, shoulder abduction/adduction, shoulder internal/external rotation, elbow flexion/extension, forearm pronation/supination, wrist flexion/extension, and wrist abduction/adduction. SAM is a wearable portable system, weighing just 7 kg.

MULOS [29] uses cable transmissions at the shoulder joints, a bevel gearbox at the elbow, and a timing belt at the forearm. SUEFUL-7 [30] features offset centers of rotation at the wrist to match the slightly offset joint axes of the wrist and a moving center of rotation at the shoulder joint to more accurately match movements of the shoulder. These systems are designed to provide assistance with ADLs.

L-Exos [31−33] has a passive forearm DOF, but an attachment makes it active and adds two hand DOFs (thumb and forefinger). L-Exos can apply a 100 N force on the palm in any direction enabling its use as a haptic feedback device for virtual reality (VR). BONES [34−36] uses a parallel mechanism for a spherical joint at the shoulder and a serially placed actuator for the elbow DOF. An attachment can add the forearm DOF and wrist flexion/

extension. ABLE [37,38] features screw-and-cable transmission systems that enable the motor to be placed along the limb parallel to the cable. This permits ABLE to have a highly compact design compared to systems with transversal motors or beveled gearboxes.

In order to account for the human shoulder not being a perfect spherical joint, several exoskeletons have been designed with additional or offset shoulder DOFs. ARMin III [39] couples the shoulder elevation angle with a vertical translation of the shoulder, and has an attachable active forearm pronation/supination and wrist flexion/extension module. MGA [40] has an extra vertical translation shoulder DOF. IntelliArm [41,42] has not only the added active vertical translation and but also two passive horizontal translation shoulder DOFs. MEDARM [43] replaces the standard three-DOF shoulder mechanism with two rotational DOFs at the sternoclavicular joint and three rotational DOFs at the glenohumeral joint. Exorn [44,45] is a portable exoskeleton designed to have all the DOFs of the human arm including two at the shoulder girdle and four at the glenohumeral joint.

SRE [46] is a seven-DOF rehabilitation exoskeleton that has a singularity when the arm is parallel to the ground due to the shoulder joint design. RUPERT IV [47,48] is a five-DOF portable exoskeleton. RehaBot [49] is a commercially developed upper limb exoskeleton that is part of a larger rehabilitation system. ETS-MARSE [50] is a rehabilitation exoskeleton designed for use with electromyography (EMG)-based control.

The earlier single-arm exoskeletons feature a wide range of designs with varying complexities targeting various joints. However, single-arm exoskeletons are inherently incapable of performing tasks requiring coordination between both arms. More importantly, bilateral movement training has been shown to be more effective in specific aspects of stroke rehabilitation than unilateral movement training [51]. To perform bilateral actions, it is therefore necessary to use a dual-arm exoskeleton. Due in part to the complexity of dual-arm systems, they tend to be more recently developed, and there are far fewer, compared to single-arm exoskeletons.

EMY [52] is a dual-arm exoskeleton with active DOFs of shoulder internal/external rotation, shoulder flexion/extension, elbow flexion/extension, and forearm pronation/supination. It features the same screw-cable-system for actuation that ABLE uses. The forearm DOF is achieved by a parallel structure of three rods on ball-joints connecting a rotating arch to a fixed arch. EMY is designed specifically for the evaluation of Brain Machine Interface.

CAPIO [53] is a dual-arm exoskeleton with 20 active DOFs, including four on the back and an extra translational DOF at each elbow. CAPIO uses serial elastic actuators and is designed for use as a haptic feedback device and teleoperation.

The modular upper limb portion of the full-body Recupera-Reha [54] system is a recent dual-arm exoskeleton designed for stroke rehabilitation. It has six active DOFs, including one for hand grasp, and one passive DOF for wrist flexion/extension for each arm. The shoulder mechanism uses brushless DC motors, while the elbow and forearm DOFs are actuated by two different custom serial elastic actuators.

4.3 Dual Arm Exoskeleton System EXO-UL8 Case Study

Clinical trials in stroke rehabilitation training bring inspiration to features of new rehabilitation robots. One promising training protocol is the so-called "mirror-image" bilateral training [55], during which the patient moves his/her healthy arm and unhealthy arm simultaneously. This training method may accelerate the recovery of poststroke motor capability as bilateral mirror movements are thought to stimulate the crosstalk between two brain hemispheres. While it is difficult for traditional physical therapists to simultaneously control both arms of a patient in the same movement pattern, multi-DOF powered exoskeletons are intrinsically capable of doing the task. Among the many exoskeletons with a multi-DOF feature on one arm, a few have expanded the design to a symmetric, dual-arm system (Table 4.1). The "EXO-UL8," a dual arm exoskeleton system developed by Bionics Lab, University of California Los Angeles (UCLA), is the fourth generation in this upper limb rehabilitation exoskeleton robot series (Fig. 4.1). It is also the second generation that comes with two arms. In EXO-UL8, contact forces are measured by force sensors placed between the braces (upper arm, forearm, palm, and fingers) and the exoskeleton structure. Joint angles are measured by encoders located on the shafts of the joints. Two PCs connected via UDP are dedicated to low-level real-time control, VR rendering, and data collection.

4.3.1 Mechanical Design

According to the possible efficiency of bilateral rehabilitation training, since the third generation, the system has two arms facilitating unilateral and bilateral training modes. In this section, we focus on the features of either side of the system. The coupling between two arms will be discussed in the Section 4.3.2.

4.3.1.1 Range of motion

The EXO-UL8, like its predecessor, was kinetically designed to overlap with 99% of a healthy human arm workspace (Fig. 4.2). The shoulder joint was designed to eliminate singular configurations within the workspace and was repositioned at the edge of the arm workspace [56]. Single-DOF hand grippers were added to increase the total number of DOFs to 8 for each arm and to enable reach-and-grasp motions that are critical to the recovery of the motor control system following stroke [57–60]. Furthermore, each link is adjustable in length in a telescopic fashion to accommodate a wide range of anthropometric arm dimensions (5%–95%). Each joint includes mechanical limits to prevent motion beyond anatomical limits.

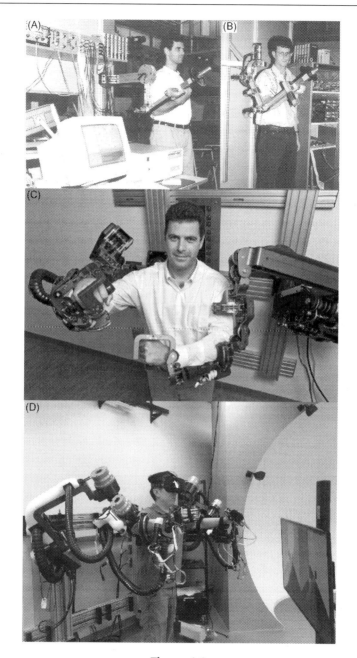

Figure 4.1
Generations of the upper limb exoskeleton system. (A) EXO-UL1, a one-DOF (elbow joint) powered exoskeleton developed as a proof of concept for using myosignals as the primary command signals; (B) EXO-UL3, a three-DOF (two at shoulder, one at elbow) powered exoskeleton; (C) EXO-UL7, a dual-arm powered exoskeleton (cable-driven) with seven DOFs on each arm; (D) EXO-UL8, a dual-arm powered exoskeleton with seven DOFs and a one-DOF haptic gripper on each side, actuated by harmonic motors.

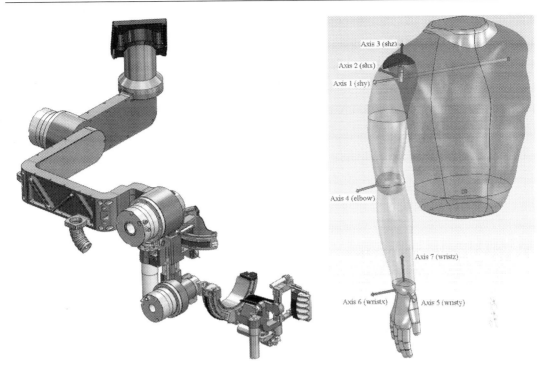

Figure 4.2
The EXO-UL8 and the corresponding DOFs marked on the human arm.

4.3.1.2 Actuation mechanisms

Instead of continuing the cable-driven actuation mechanism in the previous generation (EXO-UL7), the new exoskeleton uses electric drives. There are several reasons for using electric motors, (1) increased torque outputs enable abnormal movement correction as well as gravity compensation; (2) more accurate bottom-layer control can be achieved without unwanted compliance/delay; (3) acceptable torque—volume ratio. For each arm, three harmonic drive (Harmonic Drive Systems, Inc., Japan) servo systems are equipped with encoders and brakes to facilitate movement for the first three DOFs at the shoulder joint and to enable freezing functionality at emergency configurations. The servo systems were sized to support the joint torques developed as a result of the weights of the mechanical arm, the operator's arm as well as a 5 kg payload held by the hand. In addition, the servo systems can produce joint torques that are equivalent in magnitude to the ones produced by a healthy human and they allow the EXO-UL8 to move with the linear and angular velocities recorded in ADLs [2]. Five DC motors (Maxon Motor, Switzerland) are used to realize the five remaining DOFs (one at the elbow, three at the wrist, one at the hand gripper). A set of four force/torque (F/T) sensors are located at the physical interaction points between the human operator and the exoskeleton system: three multiaxis F/T sensors (ATI mini 40) are located on the upper arm, lower arm, and palm, between a brace and the corresponding exoskeleton link; one single-axis force sensor is incorporated into the exoskeleton hand for sensing grasping forces applied by the fingers. Anodized aluminum links are custom made and all cables are covered with 3D-printed carbon fiber–coated shells.

4.3.2 Control Strategies

To realize the seamless integration of human arms and the exoskeleton, a comprehensive controller (Fig. 4.3) was developed including: (1) admittance control, as the foundation of the control approach translating forces applied by the operator arm and hand on the various F/T sensors into joint angle changes [61]; (2) gravity and friction compensation, as a component of the control algorithm that compensates gravity and friction through feedforward-model-based prediction that is fed into the joint torques; (3) swivel prediction (human arm redundancy resolution) that aims to position the elbow joint at an appropriate swivel angle; and (4) other force fields used to provide patients additional assistance during training. To keep the system simple and easy-to-use, no contextual information like EEG or EMG is used and thus the controller is more complex.

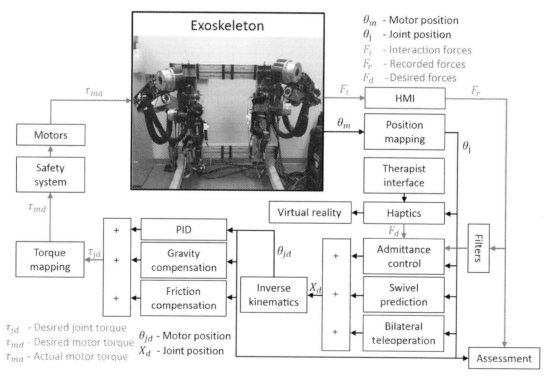

Figure 4.3
Block diagram of the EXO-UL8 controller.

1. Applying Pattern from Human Motion: Admittance Control

   Admittance control in task space is the fundamental servo control scheme of the system. It maps forces and torques applied on the multiaxis F/T sensors located at the contact interface from the operator's arm/hand to the exoskeleton system, and provides linear and angular velocity according to these input commands to the system in a way that aims to set the interaction contact forces and torques to zero [62,63].

2. Assistive Mode: Gravity Compensation

   Among all the elements of the equation of motion (inertia, velocity, gravity), experimental results indicate that during ADL of the human arm and hand gravitational loads are the largest. As such a gravity compensation algorithm is used to compensate the gravitational loads generated at the joint space due to the gravity field. Gravitational loads generated by the weight of the exoskeleton itself as well as of the human operator's arm are fed forward into the control system of the exoskeleton [64].

3. Redundancy Resolution and Joint Synergies

   The human arm with its seven DOFs is considered a redundant system. Therefore it is necessary for the control system to estimate the swivel angle (elbow joint position with respect to an axis connecting the shoulder and wrist). The manipulability ellipsoid-based redundancy resolution criterion provides an estimation of the elbow's swivel angle within an error range of ± 5 degrees (Fig. 4.4) [65]. The approach is further generalized by synthesizing multiple criteria [66]. This feature helps stroke patients in (1) freely using their functional arm to control the exoskeleton with less resistance during bilateral movement training; (2) getting posture correction (unwanted compensation removal) during unilateral movement training. In addition, illustrated in Fig. 4.5, the modeling of joint synergies observed from stroke patients provides a reference in training task trajectory design and interaction force limit calculation [67,68].

4. Force Fields

   Shown in Fig. 4.6, different force fields should be implemented when using different training modes.

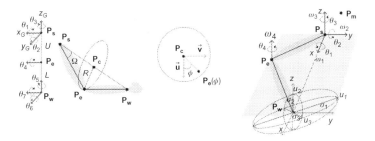

Figure 4.4
Redundancy resolution calculation based on the end-effector's manipulability [65].

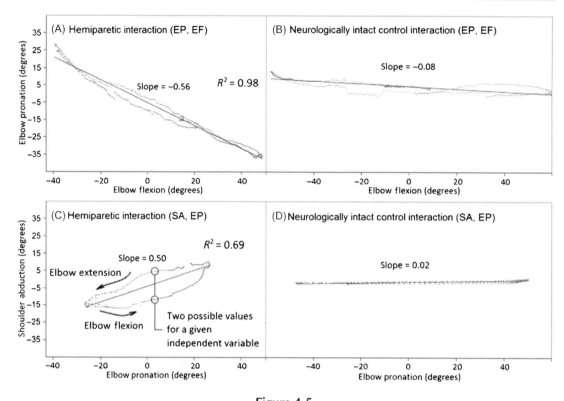

Figure 4.5
Joint synergies observed on a stroke patient (A and C) compared with those on a healthy one (B and D). SA, shoulder abduction; EF, elbow flexion; EP, elbow pronation.

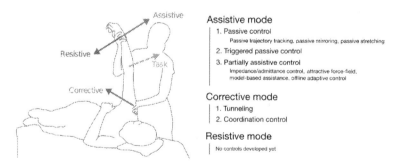

Figure 4.6
Global strategies for robotic-mediated rehabilitation and current implementations on exoskeletons [69].

Force fields are task-specific and implemented as either repulsion or attraction fields within a VR task, or a combination of both. For example, in a reach-to-grasp task while moving along a specific trajectory, an attraction force field is placed on the target, and the trajectory is wrapped with a radial repulsion force field (tunneling) that opposes deviation of the hand from the trajectory.

Our recent pilot study on asymmetric dual-arm training has validated the assistive force functionality on EXO-UL8 [70]. Another theoretical study of ours has introduced the Arm Postural Stability Index (APSI) and shown that resistive forces change the human arm redundancy resolution, thus the controlling strategy could be altered based on magnitude/joint position [71].

5. System Safety

A straightforward approach to ensure system safety is to set limiting thresholds of joint position and its first and second order derivatives with respect to time. In EXO-UL8, a multimodal safety management has been embedded into the system at the hardware, software, and operational levels. As mentioned earlier in mechanical design, the exoskeleton with brakes on all the actuators covers 99% of the joint range of a healthy person, and a subset of that movement range is constrained by rubber stops. Based on the encoder readings, the joint velocity and acceleration are also limited by the controller. In addition to these settings, an enabling pad controlled by the physical therapist and an emergency stop are embedded into the loop to freeze the system at any point in time. A 3 Hz low-pass Butterworth filter is applied to the force processing level in order to eliminate the system instability resulting from a human's unintentional vibration.

4.3.3 Virtual Reality Tasks

Based on ADL, as well as the previously measured daily activity data including the range of motion in each joint of a healthy human [2], a set of 18 VR tasks were developed. The exoskeleton system or a Kinect camera can be used to interact with these tasks in VR. The motivation for these two approaches is that, while patients will only have access to the exoskeleton system during therapeutic sessions in a clinical setting, they could easily utilize a Kinect to do at-home training between appointments. In this way, the treatment is part of a continuum bridging the clinic and home settings. The VR tasks are categorized based on whether the user interacts with static or dynamic VR objects. Furthermore, each category includes diagnostic and treatment tasks utilizing either a single DOF (moving in a single plane) or multiple DOFs (moving in space). A graphical user interface allows the therapist to control key parameters of each task and to set individual levels of difficulty for each patient. The typical parameters that can be changed are locations and size of the interactive objects as well the speed they move in space.

Haptic feedback can be calculated and transferred back to each joint of the exoskeleton based on collision detection. Most of the tasks are designed to be operated in both unilateral and bilateral modes. In the bilateral mode, a patient uses the healthy arm to move the exoskeleton and the same movement is copied to the other side, enabling the disabled arm to realize mirror-image movement; uncoupled bilateral mode is used for advanced asymmetric dual-arm manipulation, such as opening a bottle or folding clothes.

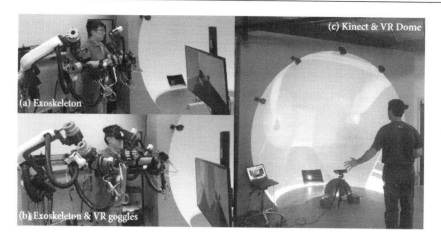

Figure 4.7
Modes of operation in virtual reality: (A) bilateral mode of the exoskeleton; (B) using exoskeleton with VR goggles; (C) using the immersive VR 3D dome with a Kinect v2 camera.

Using the previously mentioned controller, our robotic exoskeleton can train severely impaired patients in different modes including unilateral and bilateral. The coupling between the two arms is further modified to stiff/spring/damping models. Different viewports are used to improve the VR interaction experience, including a head mounted device with a 100-degree field of view and an immersive 3D dome (Fig. 4.7).

References

[1] D. Mozaffarian et al., Heart Disease and Stroke Statistics—2016 Update: A Report From the American Heart Association, 2016.

[2] J. Rosen, J.C. Perry, N. Manning, S. Burns, B. Hannaford, The human arm kinematics and dynamics during daily activities—toward a 7 DOF upper limb powered exoskeleton,"in International Conference on Advanced Robotics (2005) 532–539.

[3] A.M. Dollar, H. Herr, Lower extremity exoskeletons and active orthoses:challenges and state-of-the-art, IEEE Trans. Robot. 24 (1) (2008) 144–158.

[4] T. Yan, M. Cempini, C.M. Oddo, N. Vitiello, Review of assistive strategies in powered lower-limb orthoses and exoskeletons, Rob. Auton. Syst. 64 (2015) 120–136.

[5] H.I. Krebs, N. Hogan, B.T. Volpe, M.L. Aisen, C. Diels, Overview of clinical trials with MIT-MANUS: a robot-aided neuro-rehabilitation facility, Technol. Heal. Care 7 (6) (1999) 419–423.

[6] P.S. Lum, C.G. Burgar, P.C. Shor, M. Majmundar, M. Van der Loos, Robot-assisted movement training compared with conventional therapy techniques for the rehabilitation of upper-limb motor function after stroke, Arch. Phys. Med. Rehabil. 83 (7) (2002) 952–959.

[7] S. Coote, B. Murphy, W. Harwin, E. Stokes, The effect of the GENTLE/s robot-mediated therapy system on arm function after stroke, Clin. Rehabil. 22 (5) (2008) 395–405.

[8] S. Hesse, C. Werner, M. Pohl, S. Rueckriem, J. Mehrholz, M.L. Lingnau, Computerized arm training improves the motor control of the severely affected arm after stroke: a single-blinded randomized trial in two centers, Stroke 36 (9) (2005) 1960–1966.

[9] C.G. Burgar, et al., Robot-assisted upper-limb therapy in acute rehabilitation setting following stroke: Department of Veterans Affairs multisite clinical trial, J. Rehabil. Res. Dev. 48 (4) (2011) 445.

[10] S.K. Charles, H.I. Krebs, B.T. Volpe, D. Lynch, N. Hogan, Wrist rehabilitation following stroke: initial clinical results, in: Proceedings of the 9th International Conference on Rehabilitation Robotics, 2005. ICORR 2005, 2007, pp. 13−16.

[11] N. Hogan, H.I. Krebs, J. Charnnarong, P. Srikrishna, A. Sharon, MIT—MANUS: a workstation for manual therapy and training, in: Proceedings of the IEEE International Workshop on Robot and Human Communication, 1992, pp. 161−165.

[12] H.I. Krebs, et al., Rehabilitation robotics: pilot trial of a spatial extension for MIT-Manus, J. Neuroeng. Rehabil. 1 (2004) 1−15.

[13] L. Dipietro, M. Ferraro, J.J. Palazzolo, H.I. Krebs, B.T. Volpe, N. Hogan, Customized Interactive Robotic Treatment for Stroke: EMG-Triggered Therapy 13 (3) (2005) 325−334.

[14] K. Homma, S. Hashino, T. Arai, An upper limb motion assist system: experiments with arm models, in: Proceedings 1998 IEEE/RSJ Int. Conf. Intell. Robot. Syst. Innov. Theory, Pract. Appl. (Cat. No.98CH36190), vol. 2, no. October, 1998, pp. 758−763.

[15] G. Rosati, P. Gallina, S. Masiero, A. Rossi, Design of a new 5 d.o.f. wire-based robot for rehabilitation, in: Proceedings of the 2005 IEEE 9th International Conference on Rehabilitation Robotics, 2005, pp. 430−433.

[16] R.U.I. Loureiro, F. Amirabdollahian, M. Topping, B. Driessen, Upper limb robot mediated stroke therapy—GENTLE/s approach, Auton. Robots 15 (1) (2003) 35−51.

[17] M.D. Ellis, T.M. Sukal-moulton, S. Member, J.P.A. Dewald, Impairment-based 3-D robotic intervention improves upper extremity work area in chronic stroke: targeting abnormal joint torque coupling with progressive shoulder abduction loading, IEEE Trans. Robot. 25 (3) (2009) 549−555.

[18] R.Q. Van Der Linde, P. Lammertse, HapticMaster—a generic force controlled robot for human interaction, Ind. Robot An Int. J. 30 (6) (2003) 515−524.

[19] P.R. Culmer, et al., A control strategy for upper limb robotic rehabilitation with a dual robot system, IEEE/ASME Trans. Mechatr. 15 (4) (2010) 575−585.

[20] S. Hesse, G. Schulte-Tigges, M. Konrad, A. Bardeleben, C. Werner, Robot-assisted arm trainer for the passive and active practice of bilateral forearm and wrist movements in hemiparetic subjects, Arch. Phys. Med. Rehabil. 84 (6) (2003) 915−920.

[21] C.G. Burgar, P.S. Lum, P.C. Shor, H.F. Machiel Van der Loos, Development of robots for rehabilitation therapy: the Palo Alto VA/Stanford experience, J. Rehabil. Res. Dev. 37 (6) (2000) 663−673.

[22] A.M. Coderre, et al., Assessment of upper-limb sensorimotor function of subacute stroke patients using visually guided reaching, Neurorehabil. Neural Repair 24 (6) (2010) 528−541.

[23] Biodex. [Online]. Available from: <http://www.biodex.com/physical-medicine/products/dynamometers/system-4-pro> (accessed 08.04.07).

[24] HUMAC NORM. [Online]. Available from: <http://www.csmisolutions.com/products/isokinetic-extremity-systems/humac-norm> (accessed 08.04.07).

[25] M. Mistry, P. Mohajerian, S. Schaal, An exoskeleton robot for human arm movement study, in: Proceedings of the 2005 IEEE/RSJ International Conference on Intelligent Robots and Systems, 2005, pp. 4071−4076.

[26] D. McMonagle, Robotic Hands and Arms Developed by Raytheon SARCOS, 2010. [Online]. Available from: <https://sspd.gsfc.nasa.gov/workshop_2010/day2/Don_McMonagle/100325_SatServ_Workshop_Raytheon_w-markings-B.pdf>

[27] P. Letier *et al.*, SAM: a 7-DOF portable arm exoskeleton with local joint control, in: Proceedings of the 2008 IEEE/RSJ International Conference on Intelligent Robots and Systems, IROS, 2008, pp. 3501−3506.

[28] J. Rebelo, T. Sednaoui, E.B. Den Exter, T. Krueger, A. Schiele, Bilateral robot teleoperation: a wearable arm exoskeleton featuring an intuitive user interface, IEEE Robot. Autom. Mag. 21 (4) (2014) 62−69.

[29] G.R. Johnson, D.A. Carus, G. Parrini, S.S. Marchese, R. Valeggi, The design of a five-degree-of-freedom powered orthosis for the upper limb, Proc. Inst. Mech. Eng. Part H J. Eng. Med. 215 (3) (2001) 275–284.

[30] R.A.R.C. Gopura, K. Kiguchi, Y. Li, SUEFUL-7: a 7DOF upper-limb exoskeleton robot with muscle-model-oriented emg-based control, in: 2009 IEEE/RSJ International Conference on Intelligent Robots and Systems, IROS 2009, 2009, pp. 1126–1131.

[31] A. Frisoli, F. Rocchi, S. Marcheschi, A. Dettori, F. Salsedo, M. Bergamasco, A new force-feedback arm exoskeleton for haptic interaction in virtual environments, in: Proceedings of the First Joint Eurohaptics Conference and Symposium on Haptic Interfaces for Virtual Environment and Teleoperator Systems, 2005, pp. 195–201.

[32] A. Frisoli et al., Robotic assisted rehabilitation in virtual reality with the L-EXOS, in: Proceedings of the 7th International Conference on Disability, Virtual Reality and Associated Technologies with Artabilitation (ICDVRAT 2008), 2008, pp. 253–260.

[33] S. Marcheschi, A. Frisoli, C.A. Avizzano, M. Bergamasco, A method for modeling and control complex tendon transmissions in haptic interfaces, in: Proceedings of the 2005 IEEE International Conference on Robotics and Automation, 2005, no. April, pp. 1773–1778.

[34] J. Klein et al., Biomimetic orthosis for the neurorehabilitation of the elbow and shoulder (BONES), in: Proceedings of the 2nd Biennial IEEE/RAS-EMBS International Conference on Biomedical Robotics and Biomechatronics, BioRob 2008, 2008, pp. 535–541.

[35] M.-H. Milot, et al., A crossover pilot study evaluating the functional outcomes of two different types of robotic movement training in chronic stroke survivors using the arm exoskeleton BONES, J. Neuroeng. Rehabil. 10 (2013) 112.

[36] E.T. Wolbrecht, V. Chan, D.J. Reinkensmeyer, J.E. Bobrow, Optimizing compliant, model-based robotic assistance to promote neurorehabilitation, IEEE Trans. Neural Syst. Rehabil. Eng. 16 (3) (2008) 286–297.

[37] P. Garrec, J.P. Friconneau, Y. Méasson, Y. Perrot, ABLE, an innovative transparent exoskeleton for the upper-limb, in: 2008 IEEE/RSJ International Conference on Intelligent Robots and Systems, IROS, 2008, pp. 1483–1488.

[38] Haption, ABLE Exoskeleton, vol. 33, no. 0, p. 53210. Available from: < https://www.haption.com/en/products-en/able-en.html > .

[39] T. Nef, M. Guidali, R. Riener, ARMin III—arm therapy exoskeleton with an ergonomic shoulder actuation, Appl. Bionics Biomech. 6 (2) (2009) 127–142.

[40] C. Carignan, J. Tang, S. Roderick, Development of an exoskeleton haptic interface for virtual task training, in: Proceedings of the 2009 IEEE/RSJ International Conference on Intelligent Robots and Systems, 2009, pp. 3697–3702.

[41] Y. Ren, H.S. Park, L.Q. Zhang, Developing a whole-arm exoskeleton robot with hand opening and closing mechanism for upper limb stroke rehabilitation, in: Proceedings of the 2009 IEEE International Conference on Rehabilitation Robotics, 2009, pp. 761–765.

[42] L.Q. Zhang, H.S. Park, Y. Ren, Developing an intelligent robotic arm for stroke rehabilitation, in: Proceedings of the 2007 IEEE 10th International Conference on Rehabilitation Robotics, 2007, pp. 984–993.

[43] S.J. Ball, I.E. Brown, S.H. Scott, MEDARM: a rehabilitation robot with 5DOF at the shoulder complex, in: Proceedings of the 2007 IEEE/ASME International Conference on Advanced Intelligent Mechatronics, 2007, pp. 1–6.

[44] S.K. Manna, S. Bhaumik, A bioinspired 10 DOF wearable powered arm exoskeleton for rehabilitation, J. Robot. 2013 (2013) 1–15.

[45] S.K. Manna, D. Kumar, S. Bhaumik, Design & analysis of a portable exoskeleton structure for rehabilitation—a mechatronic approach, in: Proceedings of the International Conference on Research and Development Prospectus on Engineering and Technology (ICRDPET), 2013, no. March, pp. 214–220.

[46] S. Kousidou, N. Tsagarakis, D.G. Caldwell, C. Smith, Assistive exoskeleton for task based physiotherapy in 3-dimensional space, in: Proceedings of the First IEEE/RAS-EMBS International Conference on Biomedical Robotics and Biomechatronics, 2006, BioRob 2006, 2006, vol. 2006, pp. 266–271.

[47] R. Wei, S. Balasubramanian, L. Xu, and J. He, Adaptive iterative learning control design for RUPERT IV, in: Proceedings of the 2nd Biennial IEEE/RAS-EMBS International Conference on Biomedical Robotics and Biomechatronics, BioRob 2008, 2008, pp. 647–652.

[48] S. Balasubramanian *et al.*, Rupert: an exoskeleton robot for assisting rehabilitation of arm functions, in: Proceedings of 2008 Virtual Rehabilitation, 2008, pp. 163–167.

[49] J. Hu *et al.*, An advanced rehabilitation robotic system for augmenting healthcare, in: International Conference of the IEEE Engineering in Medicine and Biology Society, EMBS, 2011, pp. 2073–2076.

[50] M.H. Rahman, C. Ochoa-Luna, M. Saad, EMG based control of a robotic exoskeleton for shoulder and elbow motion assist, J. Autom. Control Eng. 3 (4) (2015) 270–276.

[51] H. Kim, et al., Kinematic data analysis for post-stroke patients following bilateral versus unilateral rehabilitation with an upper limb wearable robotic system, IEEE Trans. Neural Syst. Rehabil. Eng. 21 (2) (2013) 153–164.

[52] B. Morinière, A. Verney, N. Abroug, P. Garrec, Y. Perrot, EMY: a dual arm exoskeleton dedicated to the evaluation of Brain Machine Interface in clinical trials, in: Proceedings of the 2015 IEEE/RSJ International Conference on Intelligent Robots and Systems, IROS, 2015, pp. 5333–5338.

[53] M. Mallwitz, N. Will, J. Teiwes, E.A. Kirchner, The capio active upper body exoskeleton and its application for teleoperation, in: Proceedings of the 13th Symposium on Advanced Space Technologies in Robotics and Automation, 2015, pp. 1–8.

[54] E.A. Kirchner *et al.*, Recupera-Reha: exoskeleton technology with integrated biosignal analysis for sensorimotor rehabilitation, in: 2. Transdisziplinäre Konf. "Technische Unterstützungssysteme, die die Menschen wirklich wollen". Transdisziplinäre Konf. "Technische Unterstützungssysteme, die die Menschen wirklich wollen", December 12–13, Hamburg, Germany, 2016, pp. 504–517.

[55] J.H. Cauraugh, J.J. Summers, Neural plasticity and bilateral movements: a rehabilitation approach for chronic stroke, Prog. Neurobiol. 75 (5) (2005) 309–320.

[56] C. Carignan, M. Liszka, S. Roderick, Design of an arm exoskeleton with scapula motion for shoulder rehabilitation, in: Proceedings of 2005 International Conference on Advanced Robotics, ICAR '05, 2005, pp. 524–531.

[57] H. Kim, Z. Li, D. Milutinović, J. Rosen, Resolving the redundancy of a seven DOF wearable robotic system based on kinematic and dynamic constraint, in: Proceedings of International Conference on Robotics and Automation, 2012, pp. 305–310.

[58] Z. Li, K. Gray, J.R. Roldan, D. Milutinovic, J. Rosen, The joint coordination in reach-to-grasp movements, in: IEEE/RSJ International Conference on Intelligent Robots and Systems, 2014, pp. 906–911.

[59] N. Friedman, et al., Retraining and assessing hand movement after stroke using the MusicGlove: comparison with conventional hand therapy and isometric grip training, J. Neuroeng. Rehabil. 11 (1) (2014) 76.

[60] G. Alon, Loss of upper extremity motor control and function affect women more than men, J. Nov. Physiother. Phys. Rehabil 1 (2014) 19–24.

[61] J.C. Perry, J. Rosen, Design of a 7 degree-of-freedom upper-limb powered exoskeleton, in: Proceedings of IEEE/RAS-EMBS International Conference on Biomedical Robotics and Biomechatronics, 2006, pp. 805–810.

[62] L.M. Miller, Comprehensive Control Strategies for a Seven Degree of Freedom Upper Limb Exoskeleton Targeting Stroke Rehabilitation, University of Washington, Seattle, WA, 2012.

[63] H. Kim, L.M. Miller, Z. Li, J.R. Roldan, J. Rosen, Admittance control of an upper limb exoskeleton— reduction of energy exchange, in: Engineering in Medicine and Biology Society (EMBC), 2012 Annual International Conference of the IEEE, 2012, pp. 6467–6470.

[64] L.M. Miller, Gravity Compensation for a 7 Degree of Freedom Powered Upper Limb Exoskeleton, University of Washington, Seattle, WA, 2006.

[65] H. Kim, L.M. Miller, A. Al-Refai, M. Brand, J. Rosen, Redundancy resolution of a human arm for controlling a seven DOF wearable robotic system, in: Proceedings of the Annual International Conference of the IEEE Engineering in Medicine and Biology Society, 2011, pp. 3471–3474.

[66] Z. Li, H. Kim, D. Milutinović, J. Rosen, Synthesizing redundancy resolution criteria of the human arm posture in reaching movements, in: Redundancy in Robot Manipulators and Multi-Robot Systems, 2013, pp. 201–240.

[67] M. Simkins, A.H. Al-Refai, J. Rosen, Upper limb joint space modeling of stroke induced synergies using isolated and voluntary arm perturbations, IEEE Trans. Neural Syst. Rehabil. Eng. 22 (3) (2014) 491–500.

[68] A. Feldman, Y. Shen, J. Rosen, Modeling of joint synergy and spasticity in stroke patients to solve arm reach tasks, in: 2017 IEEE Signal Processing in Medicine and Biology Symposium (SPMB), Philadelphia, PA, 2017, pp. 1–3.

[69] T. Proietti, V. Crocher, A. Roby-Brami, N. Jarrasse, Upper-limb robotic exoskeletons for neurorehabilitation: a review on control strategies, IEEE Rev Biomed Eng. 9 (2016) 4–14.

[70] Y. Shen, J. Ma, B. Dobkin, J. Rosen, Asymmetric dual arm approach for post stroke recovery of motor functions utilizing the EXO-UL8 exoskeleton system: A pilot study, in: 2018 40th Annual International Conference of the IEEE Engineering in Medicine and Biology Society (EMBC), Honolulu, HI, 2018, pp. 1–7.

[71] Y. Shen, B.P.Y. Hsiao, J. Ma, J. Rosen, Upper limb redundancy resolution under gravitational loading conditions: Arm postural stability index based on dynamic manipulability analysis, in: 2017 IEEE-RAS 17th International Conference on Humanoid Robotics (Humanoids), Birmingham, 2017, pp. 332–338.

CHAPTER 5

Lower Limb Exoskeleton Robot to Facilitate the Gait of Stroke Patients

Ling-Fung Yeung and Raymond Kai-Yu Tong

Department of Biomedical Engineering, Faculty of Engineering, The Chinese University of Hong Kong, Hong Kong

5.1 Gait Abnormality in Stroke

Stroke is caused by intracranial hemorrhage or thrombosis, which cuts off arterial supply to brain tissue and usually damages the motor pathway of the central nervous system (CNS) affecting one side of the body. Reduced descending neural drive to the paretic limb leads to hemiplegic gait in stroke patients [1]. About half of the stroke survivors lost their walking capacity immediately after stroke and only around 40% of these patients could regain independent walking after rehabilitation, with certain degree of muscle weakness and spasticity [2].

Typical hemiplegic gait patterns are characterized by lengthened lower limb due to the weakness in flexors and spasticity in extensors [3], which is illustrated in Fig. 5.1. In stance phase, the spasticity in ankle plantarflexor muscles can produce excessive plantarflexion moment that inhibits knee flexion and prevents the leg from moving forward over the foot, which effectively reduce forward momentum and the leaf spring effects of ankle joint in push-off. In swing phase [4], the inability to activate knee flexor muscles and ankle dorsiflexor muscles could fail foot clearance, which is potentially dangerous as the foot pointing downward (drop foot) might trip on the ground and would cause fall incident [5]. Weak hip flexors during swing would limit the placement of the swinging foot ahead of the hip, contributing to shorter step length. Forefoot placement instead of heel strike at initial contact would be unable to prepare the ankle dorsiflexors for the eccentric shock absorption, causing abrupted loading response with foot slapping sound.

Common compensatory mechanisms work on the proximal joints around the hip, including (1) steppage gait: the elevation of pelvis on the swinging leg with hip abduction of the contralateral standing leg, which lead to inclination of trunk toward unaffected side, and/or (2) circumductory gait: hip abduction and circumduction of the swinging leg, moving the

Wearable Technology in Medicine and Health Care.
DOI: https://doi.org/10.1016/B978-0-12-811810-8.00005-1

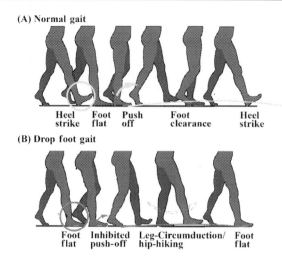

Figure 5.1

(A) Normal gait pattern, (B) Typical drop foot gait begins with forefoot contact, which would lead to abrupted loading response producing foot slapping sound. In mid-stance, spastic plantarflexors could inhibit forward ankle rocker and reduced push-off. During swing phase, the lengthened leg caused by spastic ankle plantarflexion would lead to poor foot clearance, which could be compensated by affected leg circumduction and unaffected side hip hiking.

foot through an arc away from the body. Compensatory gait patterns utilize unaffected side or alternative muscle synergies to accomplish the otherwise difficult walking tasks, but often result in gait asymmetry with higher risk of falling [1,6].

5.2 Gait Rehabilitation in Stroke: Current Practices

Stroke rehabilitation is based on the theory of neuroplasticity, in which the brain is continuously remodeling the neural circuitry in response to the external environment. Through animal modeling and human studies, neuroscientists found that skilled training could enhance cortical excitability and stimulate functional recovery in case of stroke [7]. The nature of brain plasticity is task-specific and dynamic, meaning high-intensity and repetitive skilled training that drives the use of a specific motor control can improve the targeted function, while the lack of use can lead to further degradation [8,9]. In addition, the induction of neuroplasticity requires sufficient motivation and attention with salient feedback to promote voluntary engagement in tasks, simply forcing the passive use of impaired function will not lead to sustainable effects [7]. Key components of gait recovery are summarized in Fig. 5.2.

Conventional gait training encourages stroke patients to practice over-ground walking with manual support and posture correction from therapists. Early treatment within

Figure 5.2

Major drive forces of gait rehabilitation toward neuroplasticity post-stroke. *Adapted from J.A. Kleim, T.A. Jones, Principles of experience-dependent neural plasticity: implications for rehabilitation after brain damage, J. Speech. Lang. Hear. Res. 51 (1) (2008) S225-S239.*

3-month after stroke onset (sub-acute phase) is the best recovery period when the effects of plasticity fully unfold [8]. Parallel bar and body-weight-supported treadmill training (BWSTT) are often used to ensure safety in gait training, but it would be physically exhausting for both therapists and patients to repeat hundreds of complex gait cycles in a single training session. Two types of automatic gait trainers have been developed to provide robotic assistance: exoskeleton robot and end-effector robot.

Lokomat (Hocoma AG, Switzerland) [10] is a treadmill-based exoskeleton robot that controls hip and knee joints through predefined trajectories; end-effector robot like G-EO System (Rehab Technology AG, Switzerland) [11] actuate only the feet placed on a movable support symmetrically simulating foot trajectory in gait. These existing gait trainers passively move the patients on a fixed treadmill with body weight support. Level of assistance can be adjusted and customized on each side for training on asymmetric gait pattern. Nevertheless, through about 20-year of applications, only limited published evidence proved efficacy of these automatic gait trainers over conventional gait training; except studies showed robot-assisted gait training (RAGT) enhanced training intensity and dosage, which could clearly relieve the physical burden of therapists [11,12].

5.3 Gait Rehabilitation in Stroke: Powered Exoskeleton

Recent advancement of wearable and portable lower limb powered exoskeleton opened up new possibilities in gait rehabilitation. Apart from the augmented training intensity [12], wearable powered exoskeletons offer mobility in over-ground gait training and task variations such as stair ambulation [13,14]. Enabling patients to walk freely could enhance active participation, voluntary gait engagement, and salient feedback [7]. Other advantages of exoskeleton robots include the ability to collect quantitative gait data for evaluation of patient performances and to generate sensory stimulation synchronized to gait pattern for inducing motor relearning.

5.3.1 Full Lower Limb Exoskeleton

Most existing lower limb powered exoskeletons were initially designed for augmenting healthy human locomotion in military or weight-lifting purposes, or for gait assistance in non-ambulatory paraplegics such as patients with spinal cord injury (SCI) [15]. A common feature of the existing designs is a bilateral suit that worn over whole legs and pelvis with rigid torso, partially or completely supporting the body weight, or requiring crutches to ensure balance. Most of the existing designs focus on actuating large proximal joints like hip and knee, with passive assistance at ankle joint. Researches of SCI patients walking in powered exoskeletons demonstrated that the devices could improve their activity level and provide passive assistance; hence, regular gait training using the powered exoskeleton could have positive impact on quality of life, ability to walk, cardiovascular endurance, and motor neurological status [16,17].

Major differences between devices lay in their control strategies. A brief summary of reported devices is presented in Table 5.1. ReWalk (ReWalk Robotics, USA) [16,17], Vanderbilt exoskeleton (Indego, Parker Hannifin, USA) [13,18], and Ekso-GT (Ekso Bionics, USA) [20,21] detected forward trunk tilting and shifting in center of gravity to initiate walking steps; Indego also used electric motors in combination with functional electrical stimulation (FES) to enervate paralyzed muscles [18]. Rex (Rex Bionics, New Zealand) [22] used joystick to control movement directions. Hybrid Assisted Limb (HAL-5, Cyberdyne, Japan) [23−25] used electromyogram (EMG) to interpret user intended lower limb movements, which was then translated into actions. The H2 exoskeleton (Technaid S.L., Spain) [26], used lower limb joint angles as feedback to passively control the leg following reference trajectories based on assist-as-needed algorithm.

Among the abovementioned full lower limb exoskeleton, only three devices have been evaluated on post-stroke population [27]. Randomized controlled trial (RCT) study of unilateral version of HAL-5 showed RAGT led to improved gait independency in sub-acute stroke patients ($n = 22$) [23], and faster gait speed in chronic stroke patients ($n = 18$) [25].

Table 5.1: Summary of reported existing full lower limb powered exoskeletons.

Device	Joint	Weight (kg)	Portability	Clinical Availability	Robotic Output	Control Method	Performance Evaluation
ReWalk, from ReWalk Robotics, USA [16,17]	H, K	23.3	Y	Y	Assistive; Brushless DC motor at hip, knee; passive support at ankle	Remote control to determine walking mode, tilt sensor to detect weight shift and to initiate body movement; position control with angular feedback from motor	Pre–post study through 6-month training, SCI patient (n = 1) improved gait independency and quality of life/SCI patients (n = 8) increased physical activities.
Indego, from Parker Hannifin, USA [18,19]	H, K	12	Y	Y	Assistive; Brushless DC motor at hip, knee; with passive support at ankle; FES stimulation at thigh	Sensor to detect weight shift and to initiate body movement; torque control using torque feedback from motor	Pre–post study through 5-session training. Improved walking speed and distance in SCI patient (n = 16). Showed feasibility to enable level walk and stair ambulation.
Ekso-GT, from Ekso Bionics, USA [20,21]	H, K	23	Y	Y	Assistive; hydraulic powered motor at hip and knee; passive support at ankle	Remote control to determine walking mode, tilt sensor to detect weight shift and to initiate body movement; Position control with angular feedback from motor	Pre–post study through 12-session training for 4 weeks. Improved walking capacity and lower limb functionality in chronic stroke (n = 11); improved trunk control in sub-acute stroke (n = 12).
Rex, from Rex Bionic, New Zealand [22]	H, K	?	Y	Y	Assistive; Brushless DC motor at hip, knee, and ankle	Joystick to control movement directions	Tested in SCI patients (n = 20). Study showed feasibility and safety in using REX to perform functional exercises standing.
HAL-5, from Cyberdyne, Japan [23–25]	H, K	10	Y	Y	Assistive; Brushless DC motor at hip and knee	Proportional myoelectric control used EMG electrode at wearer's thigh, control together with feedback from angular sensor at hip and knee joints, and pressure sensor under feet	RCT in chronic stroke patients through 12-session training for 4 weeks. HAL group (n = 11) had more improvement in gait independency than conventional group (n = 11).
H2, from Technaid S. L, Spain [26]	H, K, A	11	Y	Y	Assistive; Brushless DC motor at hip, knee, and ankle	Position control with assist-as-needed following reference trajectories, used potentiometer to sense angular position, strain gauge to measure joint torque, foot switch to detect foot contact	Pre–post study through 12-session training for 4 weeks. Chronic stroke patients (n = 3) showed adaptive improvement in gait trajectories across sessions.

SCI, spinal cord injury; RCT, randomized controlled trial; H, hip; K, knee; A, ankle; Y, yes; N, no.

Pre—post study of Ekso-GT showed wearing the bionic suit for 12-session gait training improved walking capacity and lower limb functionality in chronic stroke patients ($n = 11$), and improved trunk control in sub-acute stroke patients ($n = 12$) [21]. Pre—post study of H2 exoskeleton showed chronic stroke patients ($n = 3$) could adapt to the reference gait trajectories across 12 training sessions over 5 weeks [26].

A remark in the control of powered exoskeleton using user intention obtained directly from EMG signals (as in HAL-5 using EMG signal of the thigh muscles) could be feasible and promising in gait training for motor relearning in a bottom-up approach toward brain plasticity. However, researchers should be aware of the post-stroke impaired motor control would render abnormal EMG pattern resulting from muscle weakness, spasticity, and co-contraction [28], especially in smaller plantarflexor muscles at ankle joint which is susceptible to spasticity [29]. That might be one of the reasons why the actuation at ankle joint was often left out in the development of powered exoskeleton for full lower limb design.

In contrast to SCI patients, post-stroke patients usually require unilateral assistance on paretic side only, with wide range of gait independency (six-category in Functional Ambulatory Category, FAC) and different levels of disability and compensation strategies at hip, knee, and ankle joints, respectively [1]. Thus bilateral exoskeletons that can power the legs on both sides might not be suitable for post-stroke patients, and perturbation on the unaffected side may even induce undesirable compensatory gait patterns [30]. Preferable design of lower limb powered exoskeletons for post-stroke population would be unilateral on the paretic side with customizable modular robots for hip, knee, and ankle joints that can be combined or separated depending on the level of disability.

5.3.2 Modular Lower Limb Exoskeleton

Drop foot is a typical gait abnormality among post-stroke patients and is a major contributor of the compensatory gait pattern, as described earlier Section 5.1 in this chapter. Ankle—foot-orthosis (AFO) is commonly applied at affected ankle of post-stroke patients to aid foot clearance and heel strike. While supporting weak ankle dorsiflexors and mitigating spasticity at plantarflexors could potentially correct the gait instability and asymmetry, power assistance at the weak affected hip and knee joints were mostly intended to increase walking speed and to reduce metabolic cost of walking [31]. A brief summary of reported modular devices is presented in Table 5.2.

Examples of hip and knee exoskeleton are the Stride Management Assist (SMA, Honda, Japan) [32] which used hip joint angle as feedback to aid symmetry in bilateral hip movement for optimal step length; and the elastic knee exoskeleton developed in MIT [33] with the addition of elastic element at knee joint to enhance long-distance running.

Semi-active AFO equipped with sensors and controllable braking mechanism that can modulate the impedance of ankle joint dynamically, for adapting to different gait phases

Table 5.2: Summary of reported existing modular lower limb powered exoskeletons.

Device	Joint	Weight (kg)	Portability	Clinical Availability	Robotic Output	Control Method	Performance Evaluation
Modular Lower Limb Exoskeleton for hip and knee							
Stride Management Assist SMA, from Honda [32]	H	2.8	Y	Y	Assistive; Brushless DC motor at hip	Used feedback from hip joint angle to generate assistive torque in hip joint to regulate asymmetry.	RCT in chronic stroke patients, through 18-session training for 6–8 weeks. SMA group ($n = 25$) had more increase in affected side step length and reduced spatial asymmetry than control ($n = 25$).
Clutch-spring knee exoskeleton, from MIT [33]	K	0.71	Y	N	Resistive; leaf spring knee brace to augment running, and clutch knee lock to provide stiffness at stance phase	FSM, used encoder at knee joint and IMU to identify gait phases	Tested in healthy ($n = 5$), the elastic knee might reduce knee joint stiffness and metabolic demand in trained long-distance runner.
Semi-active AFO							
i-AFO, from Osaka University of Japan [34–36]	A	0.99	Y	N	Resistive; MR damper provided varying ankle joint damping during gait cycle	FSM, used rotatory potentiometer to measure ankle joint angle, and used accelerometer to detect foot contact	Tested in hemiplegic ($n = 1$), achieved good foot clearance in gait analysis
MR damper AFO, from Halmstad University [37]	A	?	Y	N	Resistive; MR damper provided varying ankle joint damping during gait cycle	FSM, used goniometer to measure ankle joint angle, to detect level walking and stair ambulation	Tested in Healthy ($n = 3$), successfully switch between foot control states of level walk and stair ambulation
Active AFO							
Active AFO, from MIT [38]	A	2.6	N	N	Assistive; SEA consisted of brushless DC motor in series with spring with variable stiffness	FSM, used rotatory potentiometer to measure ankle joint angle, force transducer under foot sole to measure ground reaction force, and foot switch to detect heel strike	Tested in drop foot patients ($n = 2$) and Healthy ($n = 3$). Variable joint impedance to minimize the occurrence of foot slap could reduce gait asymmetry.

(*Continued*)

Table 5.2: (Continued)

Device	Joint	Weight (kg)	Portability	Clinical Availability	Robotic Output	Control Method	Performance Evaluation
Active AFO, from University of Michigan [39]	A	?	N	N	Assistive; APM consisted of inflatable bladder that shortened and produced tension when filled with compressed air	Used foot switch to trigger plantarflexion assistance from APM during push-off; or EMG-controlled, used soleus EMG level to activate APM, to amplify soleus muscle output	Tested in healthy ($n = 3$), foot-switch control showed feasibility to assist push-off/ tested in healthy ($n = 6$): EMG control showed the orthosis could reshape wearer's gait pattern
Active AFO, from Yonsei University [40]	A	?	N	N	Assistive; SEA consisted of brushless DC motor in series with constant spring	FSM, used rotary potentiometer to measure ankle joint angle, FSR under foot sole to detect ground contact	Tested in healthy ($n = 5$). Active AFO generated assistive torque reduced gait abnormality compared with rigid AFO.
Robotic Tendon AFO, from Arizona State University [41]	A	1.75	N	N	Assistive; SEA consisted of brushless DC motor in series with constant spring	Used position feedback of the SEA spring to control ankle moment following reference pattern trajectories	Pre–post study through 9-session training for 3 weeks. Chronic stroke patients ($n = 3$). showed improvement in gait performance with longer walking distance, faster speed.
PPAFO from University of Illinois [42]	A	1.9 on leg, 1.2 on waist	Y	N	Assistive; Pneumatic powered bidirectional rotatory actuator	FSM, used force sensors under foot sole to detect foot loading events	Tested in healthy ($n = 3$) and patient with impaired plantarflexor ($n = 1$), correctly timed ankle assistance facilitated walking.
WALL-X, from Ghent University [43]	A	0.76	N	N	Assistive; APM consisted of inflatable bladder that shortened and produced tension when filled with compressed air	FSM, used heel switch to detect foot contact, and algorithm to trigger assistance at predicted % stride time	Tested in healthy ($n = 10$) showed reduced metabolic cost of walking at optimized assistance timing.

Device	Joint	Mass (kg)			Assistance	Sensing/control	Testing & results
Autonomous leg exoskeleton, from MIT [44]	A	1.2 on leg, 1.0 on waist	Y	N	Assistive; fiber glass strut with unidirectional actuator	FSM, used instrumented insole to detect ground contact, used optical encoder to measure angular position	Tested in healthy ($n = 7$), walking with the autonomous leg exoskeleton reduced metabolic cost
Soft Exosuit, from Harvest University [45]	A	0.9 on leg, 3.2 on waist	Y	N	Assistive; Bowden cable-based mechanical power transmission to ankle joint	Used load cells and potentiometer to monitor motor output, gyroscopes to detect gait events, with pre-set activation timing	Tested in healthy ($n = 9$), walking with soft exosuit facilitate ankle motion, increased propulsion, and reduced metabolic cost.
Anklebot, from MIT [46]	A	1.6	N	N	Assistive; two linear actuator at ankle	Seated performance-based training, required repetitive ankle movements to control a video game completing tasks.	Pre−post study through 18-session training for 6 weeks. Chronic stroke patients ($n = 8$). showed improved ankle control and walking speed.

SEA, series elastic actuator; APM, artificial pneumatic muscle; MR damper, magneto-rheological damper; RCT, randomized controlled trial; H, hip; K, knee; A, ankle; Y, yes; N, no; FSM, finite state machine.

controlled by Finite State Machine (FSM), typically using magneto-rheological (MR) damper to resist drop foot during swing phase for foot clearance, and then release the ankle for forward rocker motion during stance phase. Examples are the MR damper AFOs developed in Osaka University [34–36], and in Halmstad University of Sweden [37]. They acted as AFO with controllable impedance, but no active power assistance was added into the system to assist walking.

Active AFO equipped with sensors and actuator that can generate power for both motion control and propulsion assistance in walking. A non-exhaustive list of actuation mechanisms including: serial elastic actuator (SEA) such as the Active AFO developed in MIT [38], in Yonsei University of South Korea [40], and Robotic Tendon AFO developed in Arizona State University [41], which consists of linear actuator in series with spring for better force control and shock absorption; artificial pneumatic muscle (APM) developed in University of Michigan [39], and the Wall-X developed in Ghent University [43] used inflatable bladders that can be shortened or expanded to produce tension as it was attached across joint; pneumatic rotatory actuator developed in University of Illinois [42] used compressed air to power a bidirectional motor; active autonomous ankle developed by MIT [44], and Soft Exosuit developed in Harvard University [45] used cable transmission attached to the orthosis to pull the foot in plantarflexion or dorsiflexion directions.

Large variety of robotic AFO designs have been investigated, but they rarely or not yet succeeded to push forward into practical clinical application [47,48]. Many researches demonstrated the immediate effects of using the robotic AFO to assist ankle joint movement, reduced gait abnormality, increased walking speed, or reduced metabolic cost of walking (see Table 5.2), but most of these feasibility testing were still confined in laboratory setting with small sample size, or on unimpaired population only. Major challenges in the development process are the heavy weight, bulkiness, and the power requirement of an untethered portable device at the ankle joint [49]. Added mass at the ankle joint distal to lower limb would be susceptible to greater moment of inertia, which would lead to mechanical joint impedance with greater energy expenditure during walking [50–52]. Hence, recent attempt to use lightweight materials in orthosis, to distribute weight of device across the leg and waist, or to transfer torque from remote power source to ankle joint using cables would be promising designs of portable exoskeleton robots.

5.4 Design of the Wearable Exoskeleton Lower Limb for Stroke Rehabilitation

Review of the lower limb powered exoskeleton showed few of the existing devices were designed specifically for post-stroke population, or only preliminary feasibility tests on the immediate effects of walking in the exoskeletons. Among the reviewed devices, only HAL-5 [23] and SMA [32] were evaluated in RCT settings for clinical application in post-stroke population; however, they only provided powered assistance in hip or knee joints for enhancing body upright support and leg swinging movement, or with passive assistance at ankle. On the other hand, devices that generated active assistance at ankle joint (like active AFO) were considered to be heavy weight and tethered [47,49,53], which significantly affect their portability and feasibility in practical application, for patients would have to wear it for long period of training session.

Stroke patients could have large between-subject variation in abnormal gait pattern depending on level of disability, severity, and recovery. Complications in lower limb joints such as muscle spasticity and weakness would lead to different type of compensatory gait patterns seem in individual stroke patients [1]. Therefore, the level of power assistance provided to the wearer could be calibrated based on assist-as-needed on the three lower limb joints. The modular design enabled clinicians to customize the gait training tailor-made for stroke patients, for instance, stroke patients with drop foot gait abnormality could be prescribed with the ankle robot module, for patients with weak knee muscles and difficulty to support on the knee, the knee robot module with knee lock could be added on.

Chronic stroke patients (stroke onset > 6 months) often attained certain degree of gait independency (FAC ≥ 4, able to walk on level surface without manual contact, but require stand-by supervision) after hospital discharge [2]. Instead of bounding on treadmill-based body-weight-supported gait trainer, powered exoskeleton that allowed over-ground free walking (with supervision) would be more preferable for in-patient gait rehabilitation of chronic stroke patients, considering the patients could experience more terrain variations for generalization to daily walking tasks.

In an attempt to translate the existing designs of lower limb powered exoskeleton from laboratory to practical clinical application, a Wearable Exoskeleton Lower Limb (WELL) has been specially designed for post-stroke patients with hemiplegic gait [54] (Fig. 5.3). The following functional requirements were considered in the design:

- *Modular design*: Customizable to most stroke patients
- *Lightweight and portable*: Energetically autonomous
- *Intention-driven gait initiation*: Capable of detecting walking intention based on gait pattern signals
- *High-intensity and repetitive gait training*: Powered assistance during swing phase
- *Body weight support*: Capable of supporting upright posture by locking the knee during stance phase

5.4.1 Mechanical Design

5.4.1.1 Modular robots

The WELL was designed using a modular approach, which was a linked system of power-assisted rotary mechanism to be worn on three joints on the affected side (Fig. 5.4):

1. *Hip joint*: To assist the hip swinging forward (hip flexion) with longer step length;

Figure 5.3
Stroke patients walking in the Wearable Exoskeleton Lower Limb (WELL).

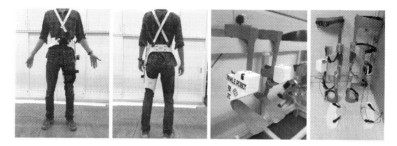

Figure 5.4
The Wearable Exoskeleton Lower Limb (WELL) can be separated into three modules wearable on the hip, knee, and ankle joints, respectively, with independent sensor and control systems that can be synchronized.

2. *Knee joint*: To provide foot clearance during swing phase by assisting knee flexion, and to support stance phase by locking the extended knee joint; and
3. *Ankle joint*: To provide foot clearance during swing phase by correcting dropped foot, and to generate propulsion during push-off.

Each of the robot modules was a one degree-of-freedom (1 DOF) machine that could acquire gait pattern as a control signal from its embedded sensors, and output assistive torque through its rotary axis to aid the movement of wearers' joint. The hip, knee, and ankle robot modules could either be separated and work with the controller individually, or mechanically and electronically interconnected to work in synchronized fashion.

5.4.1.2 Customized to wearer

Each of the robot module consisted of two orthotic braces and an articulated joint. The shank brace could be shared by the knee and ankle robot to form a knee ankle foot orthosis (KAFO). The braces were made of thick Carbon Fiber Reinforced Polymer (CFRP) shell which was strong yet lightweight and thin. Geometry of the braces was conformed to the

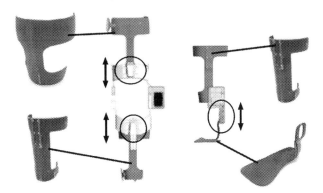

Figure 5.5
Carbon fiber reinforced braces were made conformed to leg curvature. Telescopic slide rail adjustment with locking mechanism [red circles (black in print version)] to accommodate limb length variation on knee robot and ankle robot.

leg curvature. Thick paddings were inserted between leg and brace to ensure comfort while firmly secured around the leg. The foot piece can be secured by inserting into wearer's shoe. The braces had a number of adjustment features that enabled the robot adapting to a wide range of body size, such as adjustable ratchet-ladder buckles and straps to accommodate limb circumferences variation and the telescopic slide rail with locking mechanism at different length to accommodate limb length variation (Fig. 5.5).

5.4.1.3 Power-saving mechanical design

To simplify the design aiming for mass production in clinical application, most of the components were off-the-shelf products that were readily available on the market, or materials and supplies commonly accessible in orthotic workshop. A list of required materials for fabrication included carbon fiber lamination fabrics and acrylic resin, metal bars, servomotor with gearbox, and electromechanical knee locking joint.

Brushless DC motors (Dynamixel MX106, Robotis, South Korea) with built-in PID control, were installed at the articulated joints on the lateral side of hip, knee, and ankle. The servomotor could output maximum torque at 10 Nm, maximum angular speed 55 rpm with gear reduction ratio 1:225, operating at 12 V 5.2 A. Mechanical angle limiters were positioned to prevent excessive range of motion that would cause injury to wearer. 3D printed plastic box shielded the wearer's leg from movable parts of servomotor and gears.

For ankle joint with severe plantarflexion spasticity, torque output as high as 13.5 Nm was required to move the spastic ankle to 20 degree dorsiflexion from dropped foot position [29]. A gear transmission system was used to amplify the motor output to 16.6 Nm with gear ratio 1:1.67, so a small motor could be used at ankle joint (Fig. 5.6). The gearbox also

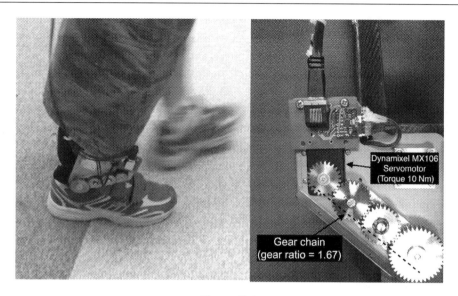

Figure 5.6
Patient walked with the robotic ankle (left), and the gearbox at ankle to transfer the amplified motor output (right).

allowed the motor to be positioned at anterior tibia to minimize interference with ankle movement.

The knee joint would have to withstand considerable weight loading during stance phase, but free to swing after terminal stance. A method to reduce power requirement while providing sufficient upright support to standing was to incorporate a controllable locking mechanism to the knee joint. The electromechanical knee lock (Neuro Tronic W.3 system knee joint, Fior & Gentz, Germany) worked by a spring pressing a pawl lock against a toothed ring, the lock could be released by pulling down the plunger using magnetized solenoid (Fig. 5.7). The electromechanical lock was locked in default under safe power-off mode. Thereby the motor could be idle during stance phase, and the torque output of 10 Nm was enough to power the knee during swinging motion.

5.4.2 Control Method

5.4.2.1 Electronic design

The computation was performed using Arduino Pro Mini with ATmega328-5V-16MHz microprocessor unit (MPU) (Atmel, USA). Gait pattern signals were sampled at 30 Hz, low-pass filtered at 4 Hz cut-off frequency, in which 30 Hz sampling rate can preserve 99% of the signal power in gait [55]. The MPU acquired sensory feedback from three servomotors, three Inertial Measurement Units (IMUs), and two Force Sensitive Resistors

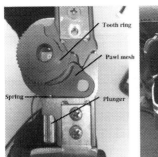

Figure 5.7
Interior of the electromechanical knee lock (left), Brushless DC motor at lateral side of robotic knee (middle), and electromechanical knee lock at medial side of robotic knee (right).

(FSRs); then controlled the three servomotor outputs and the electromechanical knee lock. The three robot modules were connected through Serial Peripheral Interface Bus (SPI) using a master—slave architecture. Hence, the individual module could operate independently. A block diagram of signal flow is presented in Fig. 5.8.

The circuit board and battery were located inside a control box mounted in a waist bag carried by the wearer (Fig. 5.9A). A 12 V 1800 mAh Lithium Polymer (LiPo) battery supplied power to the circuit board, the servomotor, and the electromechanical lock. The battery had capacity enough for 5-hour operation without recharging based on power drawn from the electronics. A 8-core Category-6 cable was used to connect the control box from waist down to servomotors and sensors, whereas the RJ45 connector with lock prevent the cable from accidentally disconnected during vigorous leg movement of walking. This robotic system could function fully independently, with option to communicate in wireless to computer or smart phone using Bluetooth. The wireless communication allowed therapists to conveniently record sensor signals and perform system configuration.

The combined weight of the whole lower limb exoskeleton was measured 3.8 kg (control box: 0.5 kg, hip module = 1.6 kg, knee module = 1.2 kg, ankle module = 0.5 kg).

Two FSR (FSR-402, Interlink Electronics, USA) placed underneath the foot piece at forefoot and heel, respectively, as ON/OFF foot switches (Fig. 5.9B). Resistance across the FSR dropped if it was loaded, which could be measured using voltage divider and be identified by passing a certain threshold. Hysteresis was added to the threshold crossing detection algorithm to avoid unstable polarity switching in foot loading. Since drop foot is common in post-stroke patients, characterized by forefoot strike at initial contact instead of heel strike, the sensitivity of the threshold detection could be configured to shift the weighting toward forefoot sensors in case of patients with severe drop foot gait abnormality.

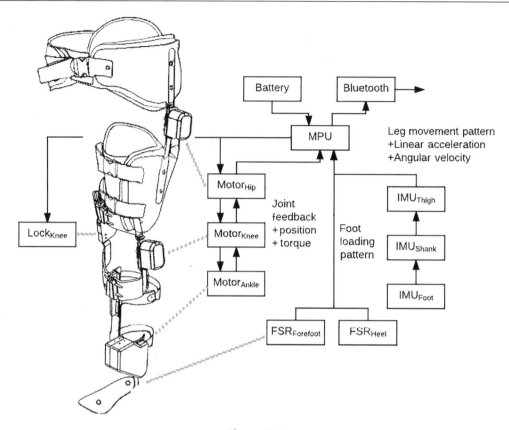

Figure 5.8
Block diagram showing signal flow of the Wearable Exoskeleton Lower Limb (WELL).

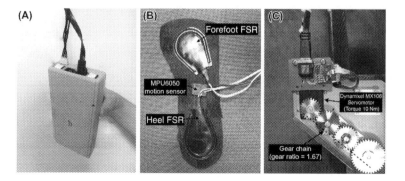

Figure 5.9
Control box with RJ45 cable connecting to motors and sensors (A), FSR under forefoot and heel position (B), and IMU on shank brace with orientation indicated in X and Y directions (C).

Three IMU (MPU6050 6-axis MotionTracking, InvenSense, USA) with integrated accelerometer and gyroscope, mounted on the thigh, shank, and foot brace, respectively (Fig. 5.9C). The motion sensor was oriented to measure linear acceleration X (backward) and Y (upward), and angular velocity Z (forward rotation) of the affected shank in the sagittal plane. The gait pattern of shank movement could be used to classify wearer's walking intention in the next step, at the instant just before push-off at pre-swing phase. The gait pattern classifier was trained using labeled data set, currently including three walking conditions: over-ground walking, stair ascending, and stair descending.

5.4.2.2 Intention-driven control algorithm

The WELL was controlled by a FSM (Fig. 5.10A). The robot operation started with stance phase, when the stroke patient was standing with loaded affected side. Complete unloading of both Heel AND Forefoot sensors initiated the pre-swing phase. The foot unloading could be triggered by affected leg toe-off, or by body weight shifting as a result of the unaffected side stepping forward.

In the short instance of the pre-swing phase, the gait pattern classifier analyzed the leg movement pattern to classify the current step into: level walk, stair ascend, or stair descend. The classifier was generated based on data set from previous training sessions. An example of the classifier using a simple Decision Tree algorithm was presented in Fig. 5.10B. Using two gait features acquired at pre-swing phase (shank tilting angle and shank angular velocity), the classification accuracy could reach 98.9% with only one stair ascend instance was misclassified as level walk. Previous studies demonstrated the feasibility of using accelerometer and gyroscope to classify over-ground walking and stair ambulation in drop foot patients [56,57]. Then in the swing phase, the motor output generated predefined torque assistance (see Table 5.3) according to the classified walking condition.

In the pre-swing phase, if the force sensor detected the foot loading in Heel sensor was relieved prior to the Forefoot sensor, i.e., the foot was preparing to push-off, powered plantarflexion assistance would be generated for assisting push-off propulsion.

In the swing phase, the FSM would enter into loading response if the robot detected any loading in either Heel sensor (heel strike) OR Forefoot sensors (forefoot contact as in dropped foot gait). The powered joints would immediately return to initial standing position and released torque output. The electromechanical knee lock would be activated (if Knee Robot module was connected) before the FSM entered stance phase once again (Fig. 5.10C).

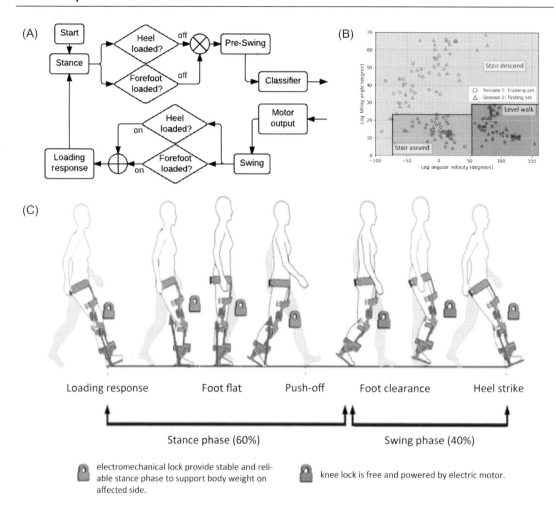

Figure 5.10

Finite state machine (FSM) of the robotic leg. During pre-swing phase, the Decision Tree classified three walking conditions (level walk, stair ascend, and stair descend) with accuracy of 98.9% (top). Electromechanical lock supported the knee in stance phase and released the knee in swing phase (bottom).

Table 5.3: Torque assistance generated by the motor in different gait phase.

Gait Phase	Hip Joint	Knee Joint	Ankle Joint	
Initial contact	Neutral	Neutral	Neutral	Either Heel ON or Forefoot ON
Loading response	Free	Lock	Free	
Mid-stance	Free	Lock	Free	
Push-off	Flexion	Flexion	Plantarflexion	Heel OFF prior to Forefoot OFF
Pre-swing	Flexion	Flexion	Dorsiflexion	Both Heel OFF and Forefoot OFF
Mid-swing	Extension	Extension	Dorsiflexion	
Terminal swing	Extension	Extension	Dorsiflexion	

References

[1] C. Beyaert, R. Vasa, G.E. Frykberg, Gait post-stroke: pathophysiology and rehabilitation strategies, Neurophysiol. Clin 45 (4−5) (2015) 335−355.

[2] S.H. Jang, The recovery of walking in stroke patients: a review, Int. J. Rehabil. Res. 33 (4) (2010) 285−289.

[3] D.A. Winter, Biomechanics and Motor Control of Human Movement, Wiley, New Jersey, 2009.

[4] S. Moore, K. Schurr, A. Wales, A. Moseley, R. Herbert, Observation and analysis of hemiplegic gait: swing phase, Aust. J. Physiother. 39 (4) (1993) 271−278.

[5] J.L. Burpee, M.D. Lewek, Biomechanical gait characteristics of naturally occurring unsuccessful foot clearance during swing in individuals with chronic stroke, Clin. Biomech. (Bristol, Avon). 30 (10) (2015) 1102−1107.

[6] V. Weerdesteyn, M. de Niet, H.J.R. van Duijnhoven, A.C.H. Geurts, Falls in individuals with stroke, J. Rehabil. Res. Dev. 45 (8) (2008) 1195.

[7] E.J. Plautz, G.W. Milliken, R.J. Nudo, Effects of repetitive motor training on movement representations in adult squirrel monkeys: role of use versus learning, Neurobiol. Learn. Mem. 74 (1) (2000) 27−55.

[8] J.A. Kleim, T.A. Jones, Principles of experience-dependent neural plasticity: implications for rehabilitation after brain damage, J. Speech. Lang. Hear. Res. 51 (1) (2008) S225−S239.

[9] S.H. Kreisel, M.G. Hennerici, H. Bäzner, Pathophysiology of stroke rehabilitation: the natural course of clinical recovery, use-dependent plasticity and rehabilitative outcome, Cerebrovasc. Dis. 23 (4) (2007) 243−255.

[10] M.F. Ng, R.K. Tong, L.S. Li, A pilot study of randomized clinical controlled trial of gait training in subacute stroke patients with partial body-weight support electromechanical gait trainer and functional electrical stimulation six-month follow-up, Stroke 39 (1) (2008) 154−160.

[11] S. Mazzoleni, A. Focacci, M. Franceschini, A. Waldner, C. Spagnuolo, E. Battini, et al., Robot-assisted end-effector-based gait training in chronic stroke patients: a multicentric uncontrolled observational retrospective clinical study, Neurorehabilitation 40 (4) (2017) 483−492.

[12] L.H. Thomas, B. French, J. Coupe, N. McMahon, L. Connell, J. Harrison, et al., Repetitive task training for improving functional ability after stroke: a major update of a cochrane review, Stroke. 48 (4) (2017) e102−e103.

[13] R.J. Farris, H.A. Quintero, M. Goldfarb, Performance evaluation of a lower limb exoskeleton for stair ascent and descent with paraplegia, Conf. Proc. IEEE. Eng. Med. Biol. Soc. (2012) (2012) 1908−1911.

[14] S. Federici, F. Meloni, M. Bracalenti, M.L. De Filippis, The effectiveness of powered, active lower limb exoskeletons in neurorehabilitation: a systematic review, NeuroRehabilitation 37 (3) (2015) 321−340.

[15] A.M. Dollar, H. Herr, Lower extremity exoskeletons and active orthoses: challenges and state-of-the-art, IEEE Trans. Robot. 24 (1) (2008) 144−158.

[16] K. Raab, K. Krakow, F. Tripp, M. Jung, Effects of training with the ReWalk exoskeleton on quality of life in incomplete spinal cord injury: a single case study, Spinal. Cord. Ser. Cases 2 (2016) 15025.

[17] P. Asselin, S. Knezevic, S. Kornfeld, C. Cirnigliaro, I. Agranova-Breyter, W.A. Bauman, et al., Heart rate and oxygen demand of powered exoskeleton-assisted walking in persons with paraplegia, J. Rehabil. Res. Dev. 52 (2) (2015) 147−158.

[18] K.H. Ha, H.A. Quintero, R.J. Farris, M. Goldfarb, Enhancing stance phase propulsion during level walking by combining FES with a powered exoskeleton for persons with paraplegia, Conf. Proc. IEEE. Eng. Med. Biol. Soc. 2012 (2012) 344−347.

[19] C. Hartigan, C. Kandilakis, S. Dalley, M. Clausen, E. Wilson, S. Morrison, et al., Mobility outcomes following five training sessions with a powered exoskeleton, Top. Spinal. Cord. Inj. Rehabil. 21 (2) (2015) 93−99.

[20] S.A. Kolakowsky-Hayner, J. Crew, S. Moran, A. Shah, Safety and feasibility of using the EksoGT™ bionic exoskeleton to aid ambulation after spinal cord injury, J. Spine 4 (3) (2013).

[21] F. Molteni, G. Gasperini, M. Gaffuri, M. Colombo, C. Giovanzana, C. Lorenzon, et al., Wearable robotic exoskeleton for over-ground gait training in sub-acute and chronic hemiparetic stroke patients: preliminary results, Eur. J. Phys. Rehabil. Med. 53 (5) (2017) 676–684.

[22] N. Birch, J. Graham, T. Priestley, C. Heywood, M. Sakel, A. Gall, et al., Results of the first interim analysis of the RAPPER II trial in patients with spinal cord injury: ambulation and functional exercise programs in the REX powered walking aid, J. Neuroeng. Rehabil. 14 (1) (2017) 60.

[23] H. Watanabe, N. Tanaka, T. Inuta, H. Saitou, H. Yanagi, Locomotion improvement using a hybrid assistive limb in recovery phase stroke patients: a randomized controlled pilot study, Arch. Phys. Med. Rehabil. 95 (11) (2014) 2006–2012.

[24] S. Kubota, Y. Nakata, K. Eguchi, H. Kawamoto, K. Kamibayashi, M. Sakane, et al., Feasibility of rehabilitation training with a newly developed wearable robot for patients with limited mobility, Arch. Phys. Med. Rehabil. 94 (6) (2013) 1080–1087.

[25] T. Yoshimoto, I. Shimizu, Y. Hiroi, M. Kawaki, D. Sato, M. Nagasawa, Feasibility and efficacy of high-speed gait training with a voluntary driven exoskeleton robot for gait and balance dysfunction in patients with chronic stroke: nonrandomized pilot study with concurrent control, Int. J. Rehabil. Res. 38 (4) (2015) 338–343.

[26] M. Bortole, A. Venkatakrishnan, F. Zhu, J.C. Moreno, G.E. Francisco, J.L. Pons, et al., The H2 robotic exoskeleton for gait rehabilitation after stroke: early findings from a clinical study, J. Neuroeng. Rehabil. 12 (2015) 54.

[27] D.R. Louie, J.J. Eng, Powered robotic exoskeletons in post-stroke rehabilitation of gait: a scoping review, J. Neuroeng. Rehabil. 13 (1) (2016) 53.

[28] N. Neckel, M. Pelliccio, D. Nichols, J. Hidler, Quantification of functional weakness and abnormal synergy patterns in the lower limb of individuals with chronic stroke, J. Neuroeng. Rehabil. 3 (2006) 17.

[29] Z. Zhou, Y. Sun, N. Wang, F. Gao, K. Wei, Q. Wang, Robot-assisted rehabilitation of ankle plantar flexors spasticity: a 3-month study with proprioceptive neuromuscular facilitation, Front. Neurorobot. 10 (2016) 16.

[30] J.L. Emken, R. Benitez, D.J. Reinkensmeyer, Human–robot cooperative movement training: learning a novel sensory motor transformation during walking with robotic assistance-as-needed, J. Neuroeng. Rehabil. 4 (2007) 8.

[31] A.L. Hsu, P.F. Tang, M.H. Jan, Analysis of impairments influencing gait velocity and asymmetry of hemiplegic patients after mild to moderate stroke, Arch. Phys. Med. Rehabil. 84 (8) (2003) 1185–1193.

[32] C. Buesing, G. Fisch, M. O'Donnell, I. Shahidi, L. Thomas, C.K. Mummidisetty, et al., Effects of a wearable exoskeleton stride management assist system (SMA®) on spatiotemporal gait characteristics in individuals after stroke: a randomized controlled trial, J. Neuroeng. Rehabil. 12 (2015) 69.

[33] G. Elliott, G.S. Sawicki, A. Marecki, H. Herr, The biomechanics and energetics of human running using an elastic knee exoskeleton, IEEE. Int. Conf. Rehabil. Robot (2013) 6650418.

[34] J. Furusho, T. Kikuchi, M. Tokuda, T. Kakehashi, K. Ikeda, S. Morimoto, et al., Development of shear type compact MR brake for the intelligent ankle–foot orthosis and its control; research and development in NEDO for practical application of human support robot, IEEE. Int. Conf. Rehabil. Robot (2007) 89–94.

[35] H. Naito, Y. Akazawa, K. Tagaya, T. Matsumoto, M. Tanaka, An ankle–foot orthosis with a variable-resistance ankle joint using a magnetorheological-fluid rotary damper, J. Biomech. Sci. Eng. 4 (2) (2009) 182–191.

[36] T. Kikuchi, S. Tanida, K. Otsuki, T. Yasuda, J. Furusho, Development of third-generation intelligently controllable ankle–foot orthosis with compact MR fluid brake, IEEE International Conference on Robotics and Automation (ICRA) (2010) 2209–2214.

[37] W. Svensson, U. Holmberg, Ankle–foot-orthosis control in inclinations and stairs, 2008, IEEE Conference on Robotics Automation and Mechatronics (2008) 301–306.

[38] J.A. Blaya, H. Herr, Adaptive control of a variable-impedance ankle–foot orthosis to assist drop-foot gait, IEEE Trans. Neural Syst. Rehabil. Eng. 12 (1) (2004) 24–31.

[39] G.S. Sawicki, K.E. Gordon, D.P. Ferris, Powered lower limb orthoses: applications in motor adaptation and rehabilitation, 9th International Conference on Rehabilitation Robotics (2005) 206−211.

[40] S. Hwang, J. Kim, J. Yi, K. Tae, J. Ryu, Y. Kim, Development of an active ankle foot orthosis for the prevention of foot drop and toe drag, International Conference on Biomedical and Pharmaceutical Engineering (2006) 418−423.

[41] J. Ward, T. Sugar, J. Standeven, J.R. Engsberg, Stroke survivor gait adaptation and performance after training on a powered ankle foot orthosis, IEEE International Conference on Robotics and Automation, ICRA (2010) 211−216.

[42] K.A. Shorter, Y. Li, E.A. Morris, G.F. Kogler, E.T. Hsiao-Wecksler, Experimental evaluation of a portable powered ankle−foot orthosis, Conf. Proc. IEEE. Eng. Med. Biol. Soc. (2011) (2011) 624−627.

[43] P. Malcolm, P. Fiers, V. Segers, I. Van Caekenberghe, M. Lenoir, D. De Clercq, Experimental study on the role of the ankle push off in the walk-to-run transition by means of a powered ankle−foot-exoskeleton, Gait Posture 30 (3) (2009) 322−327.

[44] L.M. Mooney, H.M. Herr, Biomechanical walking mechanisms underlying the metabolic reduction caused by an autonomous exoskeleton, J. Neuroeng. Rehabil. 13 (2016) 4.

[45] L.N. Awad, J. Bae, K. O'Donnell, S.M.M. De Rossi, K. Hendron, L.H. Sloot, et al., A soft robotic exosuit improves walking in patients after stroke, Sci. Transl. Med. 9 (400) (2017).

[46] L.W. Forrester, A. Roy, R.N. Goodman, J. Rietschel, J.E. Barton, H.I. Krebs, et al., Clinical application of a modular ankle robot for stroke rehabilitation, NeuroRehabilitation 33 (1) (2013) 85−97.

[47] M. Alam, I.A. Choudhury, A. Bin Mamat, Mechanism and design analysis of articulated ankle−foot orthoses for drop-foot, Sci. World J. (2014) 867869.

[48] A.C. Lo, Clinical designs of recent robot rehabilitation trials, Am. J. Phys. Med. Rehabil 91 (11 Suppl. 3) (2012) S204−S216.

[49] K.A. Shorter, J. Xia, E.T. Hsiao-Wecksler, W.K. Durfee, G.F. Kogler, Technologies for powered ankle−foot orthotic systems: possibilities and challenges, IEEE/ASME Trans. Mechatron. 18 (1) (2013) 337−347.

[50] S. Rossi, A. Colazza, M. Petrarca, E. Castelli, P. Cappa, H.I. Krebs, Feasibility study of a wearable exoskeleton for children: is the gait altered by adding masses on lower limbs? PLoS One. 8 (9) (2013) e73139.

[51] I. Khanna, A. Roy, M.M. Rodgers, H.I. Krebs, R.M. Macko, L.W. Forrester, Effects of unilateral robotic limb loading on gait characteristics in subjects with chronic stroke, J. Neuroeng. Rehabil. 7 (1) (2010) 1.

[52] R.C. Browning, J.R. Modica, R. Kram, A. Goswami, The effects of adding mass to the legs on the energetics and biomechanics of walking, Med. Sci. Sports. Exerc. 39 (3) (2007) 515−525.

[53] A. Pennycott, D. Wyss, H. Vallery, V. Klamroth-Marganska, R. Riener, Towards more effective robotic gait training for stroke rehabilitation: a review, J. Neuroeng. Rehabil. 9 (2012) 65.

[54] L.F. Yeung, C. Ockenfeld, M.K. Pang, H.W. Wai, O.Y. Soo, S.W. Li, et al., Design of an exoskeleton ankle robot for robot-assisted gait training of stroke patients, IEEE. Int. Conf. Rehabil. Robot (2017) (2017) 211−215.

[55] E.K. Antonsson, R.W. Mann, The frequency content of gait, J. Biomech. 18 (1) (1985) 39−47.

[56] H.Y. Lau, K.Y. Tong, The reliability of using accelerometer and gyroscope for gait event identification on persons with dropped foot, Gait Posture 27 (2) (2008) 248−257.

[57] H.Y. Lau, K.Y. Tong, H. Zhu, Support vector machine for classification of walking conditions using miniature kinematic sensors, Med. Bio. Eng. Comput. 46 (6) (2008) 563−573.

CHAPTER 6

Wearable Sensors for Upper Limb Monitoring

Joo Chuan Yeo[1] and Chwee Teck Lim[1,2,3]

[1]Department of Biomedical Engineering, National University of Singapore, Singapore
[2]Mechanobiology Institute, National University of Singapore, Singapore [3]Biomedical Institute for Global Health Research and Technology, National University of Singapore, Singapore

6.1 Introduction

The human hand is one of the most sophisticated biological motor systems, with the largest number of mechanical degrees of freedom compared to other body parts and primates [1]. The intrinsic complexity is required to allow variability in finger movements for different functional tasks. The ability of the fingers to work together to perform daily activities, such as buttoning a shirt, turning a door knob, or using the utensils, is therefore an intrinsic measure of quality of life and functional status. However, in order for individual fingers to coordinate smoothly for functional tasks, simultaneous control is essential [2,3]. This is highly dependent on the sensorimotor pathway initiated from the central nervous system to generate motor commands all the way to the peripheral afferents [4]. Furthermore, the entire pathway is highly cascaded to allow fine muscular control. In fact, there are more than 34 extrinsic and intrinsic muscles spanning across the palm and in the forearm, reflecting its high degree of complexity and intricacy. Furthermore, apart from finger synchronization, a combination of hand—eye coordination, arm stability, and grip control is required [4—6].

Consequently, musculoskeletal disorders affecting even the slightest portion of the sensorimotor pathway may produce poor force control and reduce hand dexterity. For example, tendonitis, trigger finger, and carpal tunnel syndrome affect different parts of the muscles and nerve system, but display similar detrimental effects to the hand function. Likewise, autoimmune diseases, such as rheumatoid arthritis, psoriasis, and scleroderma, may cause pain and stiffness of joint owing to inflammation, therefore limiting its range of motion [7]. Similarly, damage to the central nervous system results in ineffective motor commands, which may be manifested in the extremities. As a result, patients suffering from neurological diseases, such as stroke, cerebral palsy, and other degenerative neurological disorders, possess certain characteristic involuntary gait and hand movements. For example, elderly people suffering from Parkinson's disease display motor symptoms of tremor, rigidity, bradykinesia, and gait variability [8−10]. A characteristic symptom of Parkinson's disease is the involuntary hand tremor between 4 Hz and 6 Hz [11]. Therefore researchers have also explored the possibilities of monitoring hand movement as means of prediction of neurological diseases. Hence, the epitome of upper limb monitoring relies on accurate, reliable, and real-time finger dynamics as an objective assessment of rehabilitation progress and even disease diagnosis.

However, accurate measurements of finger forces and movements are extremely challenging. Firstly, our individual fingers can translate as far as 24 mm [12], and may be flexed over 60 degree [13]. At the other end of the spectrum, finger movements related to tremor could be as small as 0.1 mm [14]. With regards to force perception, our fingers can distinguish weights as little as 5 mg, but could easily withstand forces up to 6 kg during object grasping [15,16]. The broad range of forces and movement could hardly be achieved with conventional systems without compromising its accuracy, resolution, and reliability. In fact, current gold standards have been limited to simple devices, such as hand dynamometers and goniometers, which provide only static measurements with atypical and unnatural movements. Studies have shown that ergonomic design of the hand dynamometer, such as the handle grip span [17], handle shape [18], could also affect users' comfort and its corresponding grip power. Finally, as our daily activities involve many different postures and grip controls, the ideal assessment is one that is continuous and unobtrusive to obtain a realistic assessment of the hand function and its dexterity.

Wearable sensors are hence attractive, as it enables imperceptibility during assessment. In this chapter, we provide a review of some of the sensors and the key enabling technologies in the existing literature. Next, we examine the form factors of the sensors to confer its wearability. Subsequently, we looked into the roles of these sensors, especially in terms of upper limb monitoring and disease sensing. Finally, we examine the strengths and gaps of the existing sensors, and propose some opportunities for the next-generation wearable sensors to fulfill its requirements in healthcare applications.

6.2 Sensing Technologies

The key to wearable technology lies in the sensors quantifying biological parameters arising from the body. In this section, we focus on the main sensing principles, which can be divided into optical, biopotential, gravitational, or pressure sensing solutions. Fig. 6.1 provides a schematic overview of these enabling technologies and a brief description of its characteristics. Importantly, depending on the sensing parameters, these devices may be attached to the upper limb to capture various biological parameters. Furthermore, unobtrusive measurements could be achieved either by embedding them in fabric sleeves, integrating them into wristwatches, or directly adhering them to the skin.

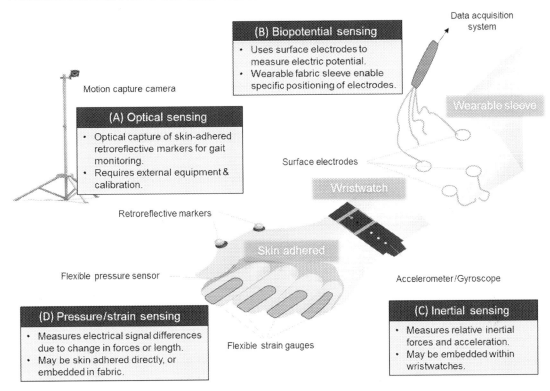

Figure 6.1: Key sensing methods governing wearable technology.
(A) Optical sensing, comprising motion capture camera, and retroreflective markers.
(B) Biopotential sensing, comprising data acquisition system, and surface electrodes. (C) Inertial sensing, including accelerometers, gyroscopes, magnetometers, and barometers. (D) Pressure or strain sensing, including force sensor, pressure sensor, and strain gauges.

Motion capture is probably the first instance in wearable technology, existing as early as 1980s [19,20]. Started off previously in the research laboratories with the purpose of performing locomotion and gait analysis, it has since been adopted by the entertainment industry to produce realistic animations. Typically, as many as six optical cameras are installed within a defined space in the laboratory to capture retroreflective markers on the skin surface. These markers reflect light back to its source with minimum scattering, and provide high precision and accuracy within sub-millimeter range [21]. Due to its high accuracy, motion capture systems have been used extensively in research to investigate the effect of numerous parameters (e.g., age, gender, footwear, load, and floor conditions) on gait control [22,23]. Wearable retroreflective suits are especially useful in recognition of joints and landmarks, and measuring its corresponding joint angle. More recently, researchers have demonstrated the possibility of using image processing techniques to derive facial recognition and corresponding distance using a combination of standard optical RGB and infrared cameras. Through image processing, optical data is rendered by several algorithms using background subtraction, edge detection, and depth estimation. The rendered images may then be stacked to provide the relevant spatiotemporal information [24]. The recent achievements led to the ability to remove the need of retroreflective markers, enabling a markerless camera-based system for motion capturing. Its simplicity was quickly recognized by the commercial companies and has gained much popularity in the augmented and virtual reality.

Biopotential parameters have also been extensively utilized to determine muscle functional status. Specifically, motor neurons generate muscle activation through the relay of an electric potential, creating an electromyography (EMG) signal. The asynchronous pattern of activation providing a smooth execution could be picked up by conductive electrodes. By observing this electrical signal and its properties (i.e., speed, amplitude, and time delay), abnormalities in muscle contraction may be obtained. Here, needle electrodes probing within the muscle fiber provide the most ideal signal recording, but it is rarely implemented due to its tedious and invasive procedures. Dry surface electrodes are therefore preferred with its possibility to integrate into fabric or textile for easy donning. However, the strength of the EMG signal depends on many factors—the muscle potential, subcutaneous skin layer, surface electrodes, electrolyte–skin interface, motion artifacts, among others [25]. Signal amplification and conditioning are also important considerations in decoding the relevant information.

Currently, wearables comprising inertial sensors have seen the most commercial success, especially in the fitness and wellness space. These include accelerometers, gyroscopes, magnetometers, and barometers. Generally, the operating principle depends on a seismic mass deflecting a mechanical beam due to inertial force, angular rotation, or a change in relative pressure [26]. The mechanical deflection is then measured electrically to produce an output signal proportional to the change in force or pressure. The use of inertial measurement units have long been demonstrated to be useful for detection of human movements (e.g., sitting, lying, standing, and walking) [27–29]. Furthermore, through data analytics, parameters such as arm swing and postural orientation may be used as physical activity estimation [26,30], body motion measurements [31–33], or fall detection [34]. For example, Madrigal et al. demonstrated the suitability of an inertial measurement unit to evaluate the flexo-extension movement of shoulder joint [35]. With improved sensitivity and even lower cost expected over the near future, the inertial sensing technology enables a greater diversity of applications and set the pace towards Internet of Things phenomenon.

Pressure or strain sensing has been receiving increasing attention as it provides direct measurements related to the forces or pressures. This is particularly useful in healthcare applications, where mechanotransduced signals arising from the body, or as a result of interactions from the external environment may be elucidated through these sensors. Typically, these microelectromechanical systems comprise conductive elements configured to mechanically deflect in response to applied forces. Following these mechanical deformations, the changes in electrical signals may be measured and its correlation obtained. Depending on the sensing elements, transduced electrical parameters include piezoelectricity [36−38], triboelectricity [39,40], capacitance [41−44], and resistance [45−53]. For example, quartz crystals contain electric dipole moments that induce changes in polarization under mechanical stress, leading to accumulation of electric charges, thus creating the piezoelectric effect. Similarly, capacitive sensors utilize the deflection of a diaphragm to alter the dielectric properties between two plates of a capacitor. Due to the minute deformations of the sensing elements, signal conditioning electronics are often required on its interface to increase its detection sensitivity.

6.3 Wearable Form Factors

Wearable technology has gained immense popularity recently in lifestyle and healthcare products, such as fitness tracking [54,55], rehabilitation [56−58], fall detection [59,60], and even wound healing [61]. The popularity of wearable technology is set to continue to rise exponentially over the next decade. IDTechEx forecasted that the wearable technology market will rise to a market of $74 billion with over 3 billion wearable devices by 2025. Importantly, these applications revolutionized the way we monitor and access patient status through remote continuous monitoring. Wearable technology generally possesses few characteristics. Firstly, they have to be conformable to the skin curvatures. Alternatively, the sensors ought to be mechanically flexible, stretchable, lightweight, and thin, such that it would not impose on the user's movements [62]. Here, the key objective lies in the embodiment of sensing elements within wearable form factors.

Several strategies have been developed to achieve this objective. The miniaturization of microelectromechanical systems has enabled it to be sufficiently small to be positioned on relatively flat topographies on the body. In fact, the relentless driving forces of microelectronics fabrication technologies pushing the boundaries of Moore's Law have significantly reduced the form size of these electronics [63], enabling them to assume product forms that were previously unattainable. Pressure sensors, inertial sensors, and optical sensors occupying tiny footprints in millimeter scales are commercially available. As a result, this allows the possibility of embedding these sensors in wristwatches or existing fashion accessories. Its small feature size has therefore increased its wearability dramatically.

The convergence of textile and electronics has also led to a plethora of wearable textile-based devices [64]. Many techniques have been used to realize e-textiles. For example, by using conductive threads that can be woven into textile, electrical interconnects between sensing elements may be achieved. These conductive threads may be fabricated using direct wire drawing [64], or by functionalizing fibers using combinations of PEDOT:PSS [65,66], silver nanowires [67,68], carbon nanotubes [69−71], graphene [72,73], and liquid metal

[74,75]. Similarly, printed electronics involving direct printing of conductive inks unto woven fabrics has been demonstrated in many research laboratories [76−79]. By enabling direct metallization in conventional inkjet or screen printing processes, functional elements including RFID tags, flexible displays, thin film transistors, and sensors may be incorporated within fabrics.

Apart from embedding electronics in accessories, direct adhesion unto skin is highly advantageous as it minimizes the necessity of additional accessories to be worn. However, when considering wearability on the upper limb, the requirements are very demanding. Our fingers are highly curved with few nonflat topographies, and the finger joints may be stretched as much as 100%. Therefore electronics need new form factors to integrate seamlessly unto our bodies. Flexible and stretchable electronic materials are key to this transformation. As such, increasingly more wearables incorporating flexible strain gauges and pressure sensors have been realized [62,80−82]. Researchers have explored several methods to confer such flexibility and stretchability to electronics. For instance, stiff sensor materials are made in ultrathin formats that it overcomes the mechanical stiffness to create a flexible element [83]. Intrinsic stretchability while maintaining the electrical conductivity has been demonstrated possible using various nanomaterials. Some examples include silver (Ag) and gold (Au) nanocomposites [44,51,53,84−86], carbon nanotubes (CNTs) [42,43,45,47,52,87], and graphene [48,88−92].

Alternatively, deterministic geometry of structural elements embedded in elastomers could be optimized to improve its stretchability. Designs include wavy geometries [93−95], mesh networks [96,97], and serpentine geometries [98−100]. By altering the geometry, localized stress effects were reduced, allowing high strain even of intrinsically stiff materials without breakage. In addition, a few researchers have also looked into shape reconfigurable conductive elements that possess the softness yet high electrical properties suitable for the transduction of mechanical forces into electrical signals. For example, several groups have developed various ways to create a soft, stretchable, yet conductive elastomer through micropatterning and the deposition of a liquid metallic alloy to achieve sensing capabilities [75,101−104].

Overall, wearable technology is propelled by the advances of sensors exhibiting superior performance or novel form factors. With these key enabling technologies introduced, we shall now explore the utility of sensors encompassing these technologies for healthcare applications in the upper limb.

6.4 Wearable Sensors for Upper Limb Monitoring

As introduced previously, upper limb impairment is the primary obstacle to reacquire functional tasks in activities of daily living. To evaluate sensorimotor control, both kinematic analysis and force sensing are necessary. Typically, this involves the measurement of various parameters, such as the range of motion, the speed, the time delay, the reaction forces, and the muscle strength. More importantly, the analysis of these parameters aims to validate the upper limb's dexterity, agility, and force control. In this section, we focus on several components of the upper extremities that are essential in our daily activities, namely, the hand, the finger digits, and the forearm.

6.5 Finger Movement

Among the finger digits, the thumb is the most interesting as it possesses higher degree of freedom of movements, owing to its morphological anatomy. In contrast to the other digits, the thumb follows unique movements and therefore plays a crucial role in prehensile tasks, involving grasping, seizing or holding [105−107]. Specifically, the thumb engages in circumduction and opposition movements via the carpometacarpal joint to enable a clamping force on the objects [108]. In recent years, with the advancement of mobile technology, the thumb is also frequently used in mobile gadgets in typing, tapping, and swiping associated activities [109]. Inevitably, excessive use of mobile gadgets results in repetitive strain injuries to the thumb leading to higher incidence of musculoskeletal disorders, such as thumb tendonitis or De Quervain's tenosynovitis [110].

Research efforts to measure finger motion have therefore increased in recent years. In particular, the demand for virtual reality and augmented reality has led to the demand for camera-based motion analysis system as a tool for finger dynamics measurements. Typically, surface markers are used to represent the underlying bone segment, where researchers demonstrated a high correlation between camera-based motion analysis and using fluoroscopy methods [111]. As a result, range of motion between fingers may be determined. However, despite its high accuracy, the entire setup typically requires extensive camera equipment setup (Fig. 6.2A). Furthermore, in the case of upper limb monitoring, a higher resolution of the finger movement is required. As such, it involves a labor-intensive process of marker attachment to the bony landmarks on the hand (Fig. 6.2B) [112]. Given the proximity of retroreflective markers on the fingers, marker occlusion and misalignment are also apparent. Therefore the use of marker-based motion capture is often confined to research environment, owing to its bulky setup and limited ease of use. Markerless motion-sensing systems have previously been reported, with commercially available systems boasting of static positional accuracy below 0.5 mm [113]. For example, Microsoft Kinect technology utilizes this concept to visualize finger movements and has demonstrated to be viable and comparable to conventional systems by several research groups [24,114−117]. Even more recently, a Creative Senz3D camera optimized for short-range gesture interaction has been used and provided a better alternative for hand movements [118]. Fig. 6.2C shows a schematic of the setup requiring only a single camera system setup. Inset further shows the actual camera, comprising a HD 720p image sensor and a 3D depth

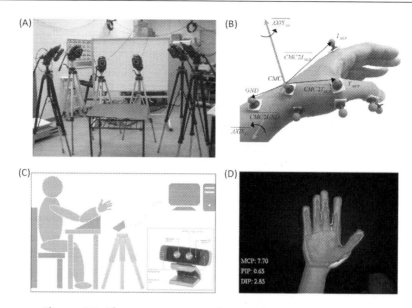

Figure 6.2: Finger movement using motion capture systems.
(A) Typical marker-based camera system setup. (B) Retroflective markers attached to bony landmarks on the hand. (C) Schematic showing single camera markerless system. Inset shows the actual camera with image sensor and depth sensor. (D) Image representation of the hand based on the camera sensors. *(B) Adapted from W. Park, et al., A soft sensor-based three-dimensional (3-D) finger motion measurement system, Sensors 17 (2) (2017), with permission. (D) Adapted from T. Pham, et al., A non-contact measurement system for the range of motion of the hand, Sensors 15 (8) (2015), with permission.*

sensor. Fig. 6.2D further shows an example of the hand image segmentation obtained from the camera, providing clear representation of the actual hand. Finger range of motion was further determined to be well correlated with the universal goniometer. However, despite its simplicity, motion artifacts and finger occlusion are still apparent. This is especially difficult to overcome in upper limb monitoring, for example in object grasping, where the fingers are hidden from view of the camera. Furthermore, due to the limitations of the single camera system, the proposed model assumed a hinge joint movement for the fingers, rendering a poor estimate for the thumb rotation and circumduction.

Apart from motion-based sensing systems, wearable glove systems embedded with soft strain sensors have been proposed. In particular, applied force or strain due to flexion/extension could be easily measured through the stretchable strain sensors. To this end, several researchers have detected finger flexion/extension using stretchable elastomeric composites comprising carbon nanotubes [45,119], metallic nanoparticles [53,85,120], graphene [88,121], and conductive liquids [122–124]. Yet, some of these strain sensors suffer from reduced accuracy with pronation/supination and could suffer from cross-talk effect [125]. Rather than measuring individual finger movements, the dynamic movement of the thumb could prove to be a better assessment. In our group, we created a combination of flexible pressure sensor and stretchable strain gauge embedded within a fabric glove to measure forces and motion respectively (Fig. 6.3A) [126]. Importantly, by screen-printing metallic ink on a silicone rubber substrate with a specific pattern, unique electronic signatures may be obtained from different thumb movements. For example, in the case of thumb rotation, it follows a sinusoidal waveform where larger range of motion can be detected with a larger change in the normalized electrical impedance (Fig. 6.3B). However, for thumb bending, the peak waveforms are distinctly different. Fig. 6.3C shows the dynamic electrical profile of the sensor during thumb flexion and extension. Thumb flexion and extension about the metacarpal joint result in sharp kinks on the stretchable strain gauge, leading to sharper peaks. Next, by creating microchannels with liquid metallic alloy to create the pressure sensor, applied forces can be limited to regions of interest. Fig. 6.3D

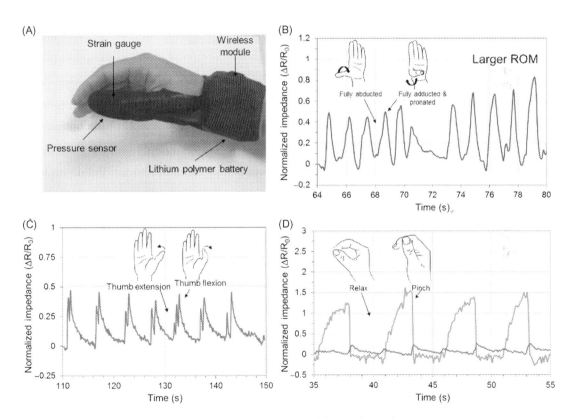

Figure 6.3: Finger movement using wearable strain and pressure sensors.
(A) Glove with embedded with a wireless system comprising of pressure sensor and stretchable strain gauge. (B) Dynamic electrical profile of the strain sensor during thumb rotation.
(C) Dynamic electrical profile of the strain sensor during thumb bending. (D) Dynamic electrical profile of the pressure sensor during lateral pinch exercises. *Adapted from J.C. Yeo, et al., Tactile sensorized glove for force and motion sensing, in: 2016 IEEE Sensors, 2016, with permission.*

shows the dynamic electrical profile of both the pressure sensor and strain gauge when the user performs lateral pinch exercises. Of interest, the electrical resistance of the pressure sensor increases at every pinch, but returns to the baseline when the thumb relaxes. Similarly, the strain gauge corresponded well with the pinching movements and forces. The demonstration of this data glove shows the potential to provide quantitative kinematics and forces related to the thumb, which can be used to monitor the rehabilitative progress of the thumb.

6.6 Fingertip Reaction Forces

The detection of reaction forces on the fingers is especially useful to determine intrinsic muscles control after rehabilitation. Object grasping is essential for daily activities and requires adequate force control to prevent slipping but not excessively large to cause damage to the object [127]. It is estimated that object manipulation requires contact forces ranging between 2 N and 10 N [128]. The object size may also change the dynamics of object manipulation, but primarily, it involves the palm, the index finger, and the thumb. The relationship between finger contact forces and object grasping is therefore of interest to many researchers.

However, few technologies can fulfill this requirement. Motion capture-based systems and inertial measurement units are ineffective in measuring actual fingertip forces due to the miniscule movements. During object grasping, the smooth muscles are activated to provide control to the fingers. Therefore EMG could be used to extract the action potential signal of different muscles during specific movements. The intensity of the signal peaks is also a good approximation of the forces applied. However, our fingers comprise several muscle groups innervated within the individual digit. For example, the index finger alone has seven intrinsic and extrinsic muscles inserted, indicating high complexity in the muscle groups activation [1]. This further implied that a high number of electrodes are necessary to provide accurate representations, constraining wearability and ease of use. Furthermore, the same movement may be initiated by different muscle groups in different tasks, indicating nonrepeatability. Taking account of these requirements, wearable force sensors are the most ideal to perform the quantification of fingertip forces in this aspect.

Many researchers have previously developed instruments comprising force sensors of different configurations to measure the contact forces [15,16,129,130]. However, due to the limitation of the sensors in form and function, these sensors are often embedded within the object of interest and are positioned in such a way to reduce variability. Few have demonstrated the use of wearable sensors to measure object grasping of natural objects in its native environment. To achieve this, our group has demonstrated this possibility by embedding thin-film microfluidic pressure sensors on a glove, as shown in Fig. 6.4A. In this study, three locations were identified in object grasping, specifically, distal phalange of

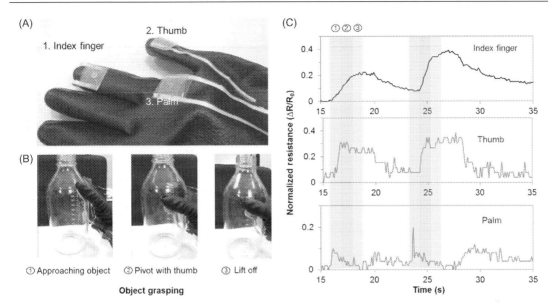

Figure 6.4: Fingertip reaction forces during object grasping.
(A) Glove with embedded pressure sensors. (B) Object grasping with the data glove showing the relative motions of approaching object, pivoting with thumb, and lifting off object. (C) Dynamic electrical profile of the pressure sensors showing the respective motions during object grasping. *Adapted from J.C. Yeo, et al., Wearable tactile sensor based on flexible microfluidics, Lab Chip 16 (17) (2016) 3244–3250, with permission.*

index finger, right anterior trapezoid of palm, and the distal phalange of thumb. In our experiments, the interaction of the finger during object grasping could be elucidated based on the electrical responses of the flexible pressure sensors. Fig. 6.4B presents the motions that corresponded to the object grasping motions, namely "approaching the object," "pivot with thumb," and "lift off." Based on the electrical profile shown in Fig. 6.4C, the hand approached the object until the palm came into contact with the device. In particular, a simple object grasping task may be analyzed based on the changes in the electrical resistance arising from the reaction forces. More importantly, this could potentially be used as a rehabilitative or diagnostic medical screening tool to determine key contact forces in object grasping movements.

6.7 Forearm Grip Strength

Forearm grip strength is a good approximation of overall muscle status and is highly preferred owing to its simplicity in execution [131]. In fact, it is an essential component in physical examination as it can serve as both clinical and prognostic measures, especially in older adults [131,132]. Moreover, in patients, grip strength can be used to predict hospitalization length [133,134], and may be used to measure rehabilitation progression

[135]. However, despite its high diagnostic utility, considerable variation in current equipment provides inconsistent results, making comparisons between studies difficult [136]. One of the reasons could be owing to different measurement methodologies, such as hydraulic, pneumatic, or strains [136]. Furthermore, the conventional hand dynamometer for measuring grip strength is bulky and limited to static measurements. Again, the motion sensing and inertial sensing technologies offer limited capabilities in measuring grip strength.

Myoelectric signals can be detected using surface EMG electrodes attached to the forearm. Similar to finger movements, contractions of arm muscles could be picked up by these electrodes to quantify the corresponding forces [137−139]. Makino and Shinoda demonstrated this possibility with a wearable EMG interface armband comprising an array of densely packed electrodes [140]. More recently, wearable EMG platform comprising 24 pad electrode arrays, signal amplifier, and wireless communication platform has been demonstrated for use to analyze functional grasp activities [141]. Correspondingly, when the subject performs a palmar grasp, the contraction of the muscle fibers producing an electrical impulse which would picked up by the wearable sensor system. An EMG colormap indicating the muscle activity during the functional activity is shown in Fig. 6.5B. Potentially, this demonstrates an interesting wearable suitable for assessing muscle strength.

Apart from EMG measurements, stretchable conductive sensors may also serve the purpose of measuring grip strength by determining the strain observed in various clenching gestures [142]. One possibility was demonstrated by our group, using a unique liquid-based graphene-oxide nanosuspension occupying the microchannels formed by two different silicone elastomers [92].

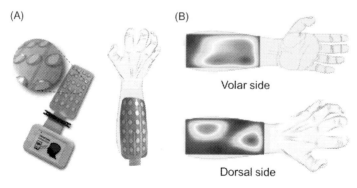

(A) (B)

Volar side

Dorsal side

Figure 6.5: Wearable surface electromyography electrodes for grasping measurements.
(A) EMG data acquisition produced by Tecnalia Serbia, comprising a 24 pad array electrode, signal amplifier platform, and wireless communication module. (B) Representative EMG maps of grasping activity of a typical subject, denoting high EMG signal and low or no EMG output from corresponding muscle zones. *Adapted from L. Popović Maneski, et al., Stimulation map for control of functional grasp based on multi-channel EMG recordings, Med. Eng. Phys. 38 (11) (2016) 1251−1259, with permission.*

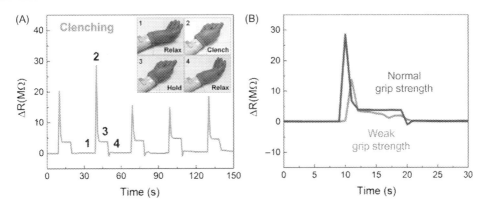

Figure 6.6: Wearable device for grip strength sensing and differentiation.
(A) Relative change in the resistance response of the device against the dynamic fist clenching motion. Inset shows the actual pressure sensor attached to the wrist for the detection of the clenching movement. (B) Characteristic responses of normal handgrip and weak handgrip. *Adapted from Kenry, et al., Highly flexible graphene oxide nanosuspension liquid-based microfluidic tactile sensor, Small 12 (12) (2016) 1593−1604, with permission.*

By attaching the sensor to the forearm, clenching gestures could be detected based on the change in electrical resistance following clench−hold−relax movements (Fig. 6.6A). The results showed that different actions may be recognized in these movements, and is repeatable. Accordingly, both intra and inter comparisons of the dynamic electrical profile following grip studies of patients could be used to determine the functional status of the patient's arm. In particular, a person with weaker grip strength will have lower peak values and would not be able to sustain the grip hold for long periods, as shown in Fig. 6.6B. This also demonstrates the utility of dynamic measurements providing time-based measurements for better comparisons.

In all, we provided an introduction of how wearable sensors may be used on different parts of the upper limb to measure movements, reaction forces, and even strength studies. The thumb, fingers, hand, and forearm were especially highlighted as they represent higher degrees of freedom in movement and are actively involved in object manipulation. Apart from these locations, other body parts such as the elbow, shoulder, neck, and head are important for consideration during rehabilitation monitoring. A combination of sensors covering different parts of the body will be ideal in assessing functional dexterity. In the next section, we looked into some of the challenges in this area, and propose future trends of the wearable sensors.

6.8 Design Challenges and Future Trends

Wearable sensors have seen tremendous growth in recent years, aided by the possibilities of low-cost, high performance electronics within a small footprint. In other aspects, existing sensors are adopting novel flexible and stretchable form factors, opening up many new possibilities in diverse applications such as consumer electronics, robotics, manufacturing, automotive, and aerospace [143]. In the healthcare sector, there exists a paradigm shift for remote continuous patient screening through wearable sensor systems. In spite of the optimistic outlook, there remains challenges in existing sensors when incorporating within a truly wearable system. Here, we provide a brief outlook on the abovementioned sensor technologies, highlighting its challenges and opportunities, and propose some of the applications that extend beyond joint angle monitoring.

Inertial measurement units are now fast approaching maturity with the exponential volume of production of MEMS inertial sensors [144]. With its high demand, an entire ecosystem comprising material suppliers, manufacturing equipment, metrology services, and software integration has been in place to support the facilities. Chip scale integration further expanded the possibilities with the integration of heterogeneous technologies in reduced form factor, lower power consumption, and lower manufacturing cost. Already, MEMS inertial sensors have seen much commercial adoption and implementation in various clinical trials. Furthermore, inertial sensors were previously limited to triaxial sensing, i.e., measuring acceleration about the three orthogonal axes. With the current technology allowing the inclusion of triaxial angular rate sensor, and triaxial magnetometer within the same compact unit, acceleration and rotation about the orthogonal axes can be independently measured, achieving 6- and 9-axis sensing. Importantly, various motions such as orientation, tilt, vibration could be measured with high precision and sensitivity [145]. With the inclusion of these aspects, applying inertial sensors for assessing individual fingers movement and analyzing complex gestures may soon be a reality. However, even with reduced footprint, silicon-based devices are a material mismatch to our skin and continue to impose discomfort to the user. Therefore imperceptibility continues to be key motivation in materials research for these sensors.

Interestingly, optical-based sensors have also fast transformed from the macro to minuscule levels. Instead of using artificial markers as positional trackers, researchers have also adopted optical sensors to measure changes in internal structures [146]. For example, researchers have demonstrated optical measurements through photoplethysmogram, a blood volume measurement based on changes in light absorption based on the vessel deformation. Through detection of blood flow, corresponding heart rate, and even oxygen saturation can be estimated. Wearable sensors comprising of microLEDs have demonstrated this possibility in recent years and fast gaining popularity [147]. Similarly, biopotential sensors have gradually experienced a transition shift towards more complicated forms of sensing. Electrodermal sensing was introduced as early as 1960s [148], but have recently gained traction, with its possibility to measure emotional levels, skin temperature, and skin hydration among others [149,150].

Further advances in flexible and stretchable electronics continue to enable new breakthroughs in form factor, thereby providing better conformability and imperceptibility.

With the progression of electronics that are almost skin-like in both appearance and mechanical properties, it is expected that this new form factor will open up even newer functions and applications, such as artificial intelligence or biochemical sensing. To do this, the gap between function and form factor has to be rapidly bridged. Demand for sensors that are sensitive enough to measure physiological signs over long periods of time will continue to be on the rise. Furthermore, existing accessories such as signal conditioning electronics and batteries are still lagging in achieving similar flexibility as that of the sensor. Therefore self-powered or low-powered sensors are increasingly of interest as it seeks to overcome the limitations of existing battery technology [149].

Beyond upper limb joint angle tracking and movement, wearable sensors on the upper limb have tremendous potential for disease sensing and monitoring. In particular, movement disorder, such as hand tremor, is especially prevalent in certain neurological diseases. Typically, this involves involuntary oscillatory movements, and could be difficult to diagnose visually. For example, essential tremor is a pathologic form of tremor often characterized by asymmetric distal extremities flexion—extension oscillation between 5 Hz and 10 Hz and most noticeable in hand raised postures [151,152]. In contrast, parkinsonian tremor, affecting over 10 million people worldwide, occurs between 4 Hz and 6 Hz, with a characteristic pronation—supination or "pill-rolling" movement that is present both at rest and at task specific movements [11,152,153]. On the other hand, physiologic tremor has significantly higher tremor frequency (8—12 Hz) with lower amplitude and irregular frequency [153,154]. Evidently, the frequency of oscillation, the tremor characteristics, and the activities that activate tremor may be used to differentiate between different neurological diseases. A combination of accelerometers, motion capture, surface EMG and strain gauges may be used together to identify trajectory speed, type of motion, muscular contractions, posture, tremor amplitude, and frequency. Potentially, this highlights the possibility of extending tele-rehabilitation to remote or tele-diagnosis.

Similarly, the monitoring of patient's vital parameters, such as heart rate, breathing rate, pulse rate, electrocardiogram, blood pressure, glucose level, oxygen saturation, lung tidal volume, and skin temperature, is of interest especially with rising ageing population. Again, this often requires the inclusion of more than one sensor to perform multiple measurements and analysis. Therefore researchers are also working to develop sensors capable of multiple functionalities within the same footprint. Next-generation wearables will almost certainly encompass more than one sensor. Even more suitably, wearable chemical and biological sensors that enable continuous monitoring through sweat secretion have also been reported. Indeed, the future trend is geared towards an artificial electronic patch that behaves similar to skin mechanically, chemically, and even biologically. This is especially useful for monitoring of elderly and chronically diseased patients and detection of adverse health events. Apart from sensors research, wireless platforms are also essential to facilitate remote sensing while ensuring seamless data transmission. Taken together, the integration of sensor technologies, electronics miniaturization, thin-film battery, and wireless communication platform form the key enablers to wearable sensors for biomedical and healthcare applications. Finally, the incorporation of sensors alongside research towards energy harvesting, data analytics, machine learning, and low power computing will further propel the possibilities of wearables beyond current imagination.

6.9 Conclusion

The human upper limb is designed for use in many occupational tasks, and its functional capacity is a key determinant in quality of life. Especially crucial is the hand function which is extremely complex owing to the many articulations present. Yet it is primarily involved in prehensile and object manipulation, which accounts for more than 50% of daily activities. In fact, individual finger digits, together with the wrist, forearm, elbow, and shoulder, play unique roles in object manipulation. Therefore rather than measuring individual joint movements, a system capable of quantifying functional upper limb analysis serves better purpose in evaluating hand dexterity in its entirety. The role of wearable sensors has been especially crucial in providing objective assessment to these functional activities. These have been demonstrated through the deployment of wearable sensors using optical, inertial, biopotential, pressure, and strain sensing mechanisms. Several wearable technologies have served well in enabling the measurement of physical parameters such as individual fingers' position, acceleration, range of motion, frequency, and reaction forces. The analysis of upper limb dynamics has proven to be useful in terms of rehabilitation monitoring and even neurological disease detection. With the advent of wearable chemical and biological sensors, this information could extend to biomonitoring for applications towards personalized diagnosis and medicine.

References

[1] M.H. Schieber, M. Santello, Hand function: peripheral and central constraints on performance, J. Appl. Physiol. 96 (6) (2004) 2293.

[2] J.A. Martin, et al., Age and grip strength predict hand dexterity in adults, PLoS One 10 (2) (2015) e0117598.

[3] A.M. Dollar, Classifying Human Hand Use and the Activities of Daily Living, in: R. Balasubramanian, V.J. Santos (Eds.), The Human Hand as an Inspiration for Robot Hand Development, Springer International Publishing, Cham, 2014, pp. 201–216.

[4] R.S. Johansson, et al., Eye–hand coordination in object manipulation, J. Neurosci. 21 (17) (2001) 6917–6932.

[5] J. Hermsdörfer, et al., Grip force control during object manipulation in cerebral stroke, Clin. Neurophysiol. 114 (5) (2003) 915–929.

[6] M. Santello, J.F. Soechting, Force synergies for multifingered grasping, Exp. Brain Res. 133 (4) (2000) 457–467.

[7] M. Packer, et al., Hand impairment and functional ability: a matched case comparison study between people with rheumatoid arthritis and healthy controls, Hand Therapy 21 (4) (2016) 115–122.

[8] J.D. Schaafsma, et al., Gait dynamics in Parkinson's disease: relationship to Parkinsonian features, falls and response to levodopa, J. Neurol. Sci. 212 (1) (2003) 47–53.

[9] A. Winogrodzka, et al., Rigidity and bradykinesia reduce interlimb coordination in Parkinsonian gait, Arch. Phys. Med. Rehabil. 86 (2) (2005) 183–189.

[10] J.M. Hausdorff, et al., Impaired regulation of stride variability in Parkinson's disease subjects with freezing of gait, Exp. Brain Res. 149 (2) (2003) 187–194.

[11] J. Jankovic, Parkinson's disease: clinical features and diagnosis, J. Neurol. Neurosurg. Psychiatry 79 (4) (2008) 368.

[12] B. Ellis, A. Bruton, A study to compare the reliability of composite finger flexion with goniometry for measurement of range of motion in the hand, Clin. Rehabil. 16 (5) (2002) 562–570.

[13] M.C. Hume, et al., Functional range of motion of the joints of the hand, J. Hand Surg. 15 (2) (1990) 240–243.

[14] S. Calzetti, et al., Frequency/amplitude characteristics of postural tremor of the hands in a population of patients with bilateral essential tremor: implications for the classification and mechanism of essential tremor, J. Neurol. Neurosurg. Psychiatry 50 (5) (1987) 561.

[15] R.G. Radwin, et al., External finger forces in submaximal five-finger static pinch prehension, Ergonomics 35 (3) (1992) 275–288.

[16] A.A. Amis, Variation of finger forces in maximal isometric grasp tests on a range of cylinder diameters, J. Biomed. Eng. 9 (4) (1987) 313–320.

[17] S.-J. Lee, et al., Handle grip span for optimising finger-specific force capability as a function of hand size, Ergonomics 52 (5) (2009) 601–608.

[18] Y.-K. Kong, D.-M. Kim, The relationship between hand anthropometrics, total grip strength and individual finger force for various handle shapes, Int. J. Occup. Safe. Ergon. 21 (2) (2015) 187–192.

[19] R.B. Davis, Clinical gait analysis, IEEE Eng. Med. Biol. Magaz. 7 (3) (1988) 35–40.

[20] K.D. Taylor, et al., An automated motion measurement system for clinical gait analysis, J. Biomech. 15 (7) (1982) 505–516.

[21] M. Windolf, N. Götzen, M. Morlock, Systematic accuracy and precision analysis of video motion capturing systems—exemplified on the Vicon-460 system, J. Biomech. 41 (12) (2008) 2776–2780.

[22] W. Tao, et al., Gait analysis using wearable sensors, Sensors 12 (2) (2012).

[23] X. Qu, J.C. Yeo, Effects of load carriage and fatigue on gait characteristics, J. Biomech. 44 (7) (2011) 1259–1263.

[24] A. Schmitz, et al., Accuracy and repeatability of joint angles measured using a single camera markerless motion capture system, J. Biomech. 47 (2) (2014) 587–591.

[25] R. Merletti, et al., Technology and instrumentation for detection and conditioning of the surface electromyographic signal: state of the art, Clin. Biomech. 24 (2) (2009) 122–134.

[26] C.-C. Yang, Y.-L. Hsu, A review of accelerometry-based wearable motion detectors for physical activity monitoring, Sensors 10 (2010) 8.

[27] V.T. Inman, H.D. Eberhart, The major determinants in normal and pathological gait, J Bone Joint Surg Am 35 (3) (1953) 543–558.

[28] M. Ermes, et al., Detection of daily activities and sports with wearable sensors in controlled and uncontrolled conditions, IEEE Trans. Inform. Technol. Biomed. 12 (1) (2008) 20–26.

[29] K. Aminian, et al., Physical activity monitoring based on accelerometry: validation and comparison with video observation, Med. Biol. Eng. Comput. 37 (3) (1999) 304–308.

[30] N.F. Butte, U.L.F. Ekelund, K.R. Westerterp, Assessing physical activity using wearable monitors: measures of physical activity, Med. Sci. Sports Exerc. 44 (1S) (2012).

[31] M.F. Gago, et al., The effect of levodopa on postural stability evaluated by wearable inertial measurement units for idiopathic and vascular Parkinson's disease, Gait Posture 41 (2) (2015) 459–464.

[32] D. Novak, et al., Toward real-time automated detection of turns during gait using wearable inertial measurement units, Sensors 14 (10) (2014).

[33] A. Leardini, et al., Validation of the angular measurements of a new inertial-measurement-unit based rehabilitation system: comparison with state-of-the-art gait analysis, J NeuroEng Rehabil 11 (1) (2014) 136.

[34] M.N. Nyan, F.E.H. Tay, E. Murugasu, A wearable system for pre-impact fall detection, J. Biomech. 41 (16) (2008) 3475−3481.

[35] J.A. Barraza Madrigal, et al., Evaluation of suitability of a micro-processing unit of motion analysis for upper limb tracking, Med. Eng. Phys. 38 (8) (2016) 793−800.

[36] J. Zhou, et al., Flexible piezotronic strain sensor, Nano Lett. 8 (9) (2008) 3035−3040.

[37] D. Mandal, S. Yoon, K.J. Kim, Origin of piezoelectricity in an electrospun poly(vinylidene fluoride-trifluoroethylene) nanofiber web-based nanogenerator and nano-pressure sensor, Macromol. Rapid Commun. 32 (11) (2011) 831−837.

[38] K.Y. Lee, M.K. Gupta, S.-W. Kim, Transparent flexible stretchable piezoelectric and triboelectric nanogenerators for powering portable electronics, Nano Energy 14 (2015) 139−160.

[39] J.-W. Jeong, et al., Materials and optimized designs for human-machine interfaces via epidermal electronics, Adv. Mater. 25 (47) (2013) 6839−6846.

[40] Y. Yang, et al., Human skin based triboelectric nanogenerators for harvesting biomechanical energy and as self-powered active tactile sensor system, ACS Nano 7 (10) (2013) 9213−9222.

[41] S.C.B. Mannsfeld, et al., Highly sensitive flexible pressure sensors with microstructured rubber dielectric layers, Nat. Mater. 9 (10) (2010) 859−864.

[42] D.J. Cohen, et al., A highly elastic, capacitive strain gauge based on percolating nanotube networks, Nano Lett. 12 (4) (2012) 1821−1825.

[43] S.Y. Kim, et al., Highly sensitive and multimodal all-carbon skin sensors capable of simultaneously detecting tactile and biological stimuli, Adv. Mater. 27 (28) (2015) 4178−4185.

[44] J. Wang, et al., A highly sensitive and flexible pressure sensor with electrodes and elastomeric interlayer containing silver nanowires, Nanoscale 7 (7) (2015) 2926−2932.

[45] T. Yamada, et al., A stretchable carbon nanotube strain sensor for human-motion detection, Nat. Nano 6 (5) (2011) 296−301.

[46] C. Pang, et al., A flexible and highly sensitive strain-gauge sensor using reversible interlocking of nanofibres, Nat. Mater. 11 (9) (2012) 795−801.

[47] D.J. Lipomi, et al., Skin-like pressure and strain sensors based on transparent elastic films of carbon nanotubes, Nat. Nano 6 (12) (2011) 788−792.

[48] B. Zhu, et al., Microstructured graphene arrays for highly sensitive flexible tactile sensors, Small 10 (18) (2014) 3625−3631.

[49] N.T. Tien, et al., A flexible bimodal sensor array for simultaneous sensing of pressure and temperature, Adv. Mater. 26 (5) (2014) 796−804.

[50] L. Pan, et al., An ultra-sensitive resistive pressure sensor based on hollow-sphere microstructure induced elasticity in conducting polymer film, Nat. Commun. (2014) 5.

[51] S. Gong, et al., A wearable and highly sensitive pressure sensor with ultrathin gold nanowires, Nat. Commun. (2014) 5.

[52] E. Roh, et al., Stretchable, transparent, ultrasensitive, and patchable strain sensor for human−machine interfaces comprising a nanohybrid of carbon nanotubes and conductive elastomers, ACS Nano 9 (6) (2015) 6252−6261.

[53] M. Amjadi, et al., Highly stretchable and sensitive strain sensor based on silver nanowire−elastomer nanocomposite, ACS Nano 8 (5) (2014) 5154−5163.

[54] B. Ka, Remote monitoring foot inserts used to enhance sports performance through increased range of motion, Int. J. Phys. Med. Rehabil. 03 (2015) 05.

[55] Y. Menguc, et al., Wearable soft sensing suit for human gait measurement, Int. J. Robot. Res. 33 (14) (2014) 1748−1764.

[56] Y. Joo, et al., Silver nanowire-embedded PDMS with a multiscale structure for a highly sensitive and robust flexible pressure sensor, Nanoscale 7 (14) (2015) 6208−6215.

[57] E. Sardini, M. Serpelloni, V. Pasqui, Wireless wearable T-shirt for posture monitoring during rehabilitation exercises, IEEE Trans. Instrum. Meas. 64 (2) (2015) 439−448.

[58] Y.L. Park, et al., Design and control of a bio-inspired soft wearable robotic device for ankle−foot rehabilitation, Bioinspir. Biomim. 9 (1) (2014) 016007.

[59] T. Shany, et al., Sensors-based wearable systems for monitoring of human movement and falls, IEEE Sens. J. 12 (3) (2012) 658−670.

[60] O. Aziz, S.N. Robinovitch, An analysis of the accuracy of wearable sensors for classifying the causes of falls in humans, IEEE Trans. Neural Syst. Rehabil. Eng. 19 (6) (2011) 670−676.

[61] Y. Hattori, et al., Multifunctional skin-like electronics for quantitative, clinical monitoring of cutaneous wound healing, Adv. Healthcare Mater. 3 (10) (2014) 1597−1607.

[62] J.C. Kenry, Yeo, C.T. Lim, Emerging flexible and wearable physical sensing platforms for healthcare and biomedical applications, Microsyst. Nanoeng. 2 (2016) 16043.

[63] H.G. Craighead, Nanoelectromechanical systems, Science 290 (5496) (2000) 1532.

[64] M. Stoppa, A. Chiolerio, Wearable electronics and smart textiles: a critical review, Sensors 14 (2014) 7.

[65] M. Hamedi, R. Forchheimer, O. Inganas, Towards woven logic from organic electronic fibres, Nat. Mater. 6 (5) (2007) 357−362.

[66] M.Z. Seyedin, et al., Strain-responsive polyurethane/PEDOT:PSS elastomeric composite fibers with high electrical conductivity, Adv. Funct. Mater. 24 (20) (2014) 2957−2966.

[67] Y. Atwa, N. Maheshwari, I.A. Goldthorpe, Silver nanowire coated threads for electrically conductive textiles, J. Mater. Chem. C 3 (16) (2015) 3908−3912.

[68] S. Lee, et al., Ag nanowire reinforced highly stretchable conductive fibers for wearable electronics, Adv. Funct. Mater. 25 (21) (2015) 3114−3121.

[69] H. Cheng, et al., Textile electrodes woven by carbon nanotube-graphene hybrid fibers for flexible electrochemical capacitors, Nanoscale 5 (8) (2013) 3428−3434.

[70] T.-W. Lee, et al., Electrically conductive and strong cellulose-based composite fibers reinforced with multiwalled carbon nanotube containing multiple hydrogen bonding moiety, Compos. Sci. Technol. 123 (2016) 57−64.

[71] N. Behabtu, et al., Strong, light, multifunctional fibers of carbon nanotubes with ultrahigh conductivity, Science 339 (6116) (2013) 182.

[72] Y.A. Samad, et al., Non-destroyable graphene cladding on a range of textile and other fibers and fiber mats, RSC Adv. 4 (33) (2014) 16935−16938.

[73] Z. Xu, et al., Highly electrically conductive Ag-doped graphene fibers as stretchable conductors, Adv. Mater. 25 (23) (2013) 3249−3253.

[74] S. Zhu, et al., Ultrastretchable fibers with metallic conductivity using a liquid metal alloy core, Adv. Funct. Mater. 23 (18) (2013) 2308−2314.

[75] W. Xi, et al., Ultrathin and wearable microtubular epidermal sensor for real-time physiological pulse monitoring, Adv. Mater. Technol. 2 (5) (2017) 1700016.

[76] A. Kamyshny, S. Magdassi, Conductive nanomaterials for printed electronics, Small 10 (17) (2014) 3515−3535.

[77] Z. Stempien, et al., Inkjet-printing deposition of silver electro-conductive layers on textile substrates at low sintering temperature by using an aqueous silver ions-containing ink for textronic applications, Sens. Actuat. B: Chem. 224 (2016) 714−725.

[78] N. Matsuhisa, et al., Printable elastic conductors with a high conductivity for electronic textile applications, Nat. Commun. 6 (2015) 7461.

[79] E.B. Secor, et al., Gravure printing of graphene for large-area flexible electronics, Adv. Mater. 26 (26) (2014) 4533−4538.

[80] J.C. Yeo, Kenry, C.T. Lim, Emergence of microfluidic wearable technologies, Lab Chip 16 (21) (2016) 4082−4090.

[81] M.L. Hammock, et al., 25th anniversary article: the evolution of electronic skin (e-skin): a brief history, design considerations, and recent progress, Adv. Mater. 25 (42) (2013) 5997−6038.

[82] J.A. Rogers, R. Ghaffari, D.-H. Kim, Stretchable Bioelectronics for Medical Devices and Systems., Springer, Switzerland, 2016.

[83] H. Fang, et al., Ultrathin, transferred layers of thermally grown silicon dioxide as biofluid barriers for biointegrated flexible electronic systems, Proc. Natl. Acad. Sci. U.S.A. 113 (42) (2016) 11682−11687.

[84] M. Segev-Bar, et al., Tunable touch sensor and combined sensing platform: toward nanoparticle-based electronic skin, ACS Appl. Mater. Interfaces 5 (12) (2013) 5531−5541.

[85] J. Lee, et al., A stretchable strain sensor based on a metal nanoparticle thin film for human motion detection, Nanoscale 6 (20) (2014) 11932−11939.

[86] S. Jun, B.-K. Ju, J.-W. Kim, Ultra-facile fabrication of stretchable and transparent capacitive sensor employing photo-assisted patterning of silver nanowire networks, Adv. Mater. Technol. 1 (6) (2016) 1600062.

[87] A. Morteza, Y. Yong Jin, P. Inkyu, Ultra-stretchable and skin-mountable strain sensors using carbon nanotubes−Ecoflex nanocomposites, Nanotechnology 26 (37) (2015) 375501.

[88] S.-H. Bae, et al., Graphene-based transparent strain sensor, Carbon 51 (2013) 236−242.

[89] C.S. Boland, et al., Sensitive, high-strain, high-rate bodily motion sensors based on graphene−rubber composites, ACS Nano 8 (9) (2014) 8819−8830.

[90] H. Jang, et al., Graphene-based flexible and stretchable electronics, Adv. Mater. 28 (22) (2016) 4184−4202.

[91] H. Lee, et al., Moving beyond flexible to stretchable conductive electrodes using metal nanowires and graphenes, Nanoscale 8 (4) (2016) 1789−1822.

[92] Kenry, et al., Highly flexible graphene oxide nanosuspension liquid-based microfluidic tactile sensor, Small 12 (12) (2016) 1593−1604.

[93] D.-Y. Khang, et al., A stretchable form of single-crystal silicon for high-performance electronics on rubber substrates, Science 311 (5758) (2006) 208.

[94] W.M. Choi, et al., Biaxially stretchable "wavy" silicon nanomembranes, Nano Lett. 7 (6) (2007) 1655−1663.

[95] D.-H. Kim, et al., Stretchable and foldable silicon integrated circuits, Science 320 (5875) (2008) 507.

[96] D.-H. Kim, et al., Ultrathin silicon circuits with strain-isolation layers and mesh layouts for high-performance electronics on fabric, vinyl, leather, and paper, Adv. Mater. 21 (36) (2009) 3703−3707.

[97] D.-H. Kim, et al., Materials and noncoplanar mesh designs for integrated circuits with linear elastic responses to extreme mechanical deformations, Proc. Natl. Acad. Sci. U.S.A. 105 (48) (2008) 18675−18680.

[98] S. Xu, et al., Stretchable batteries with self-similar serpentine interconnects and integrated wireless recharging systems, Nat. Commun. 4 (2013) 1543.

[99] Y. Zhang, et al., Experimental and theoretical studies of serpentine microstructures bonded to prestrained elastomers for stretchable electronics, Adv. Funct. Mater. 24 (14) (2014) 2028−2037.

[100] W.-H. Yeo, et al., Multifunctional epidermal electronics printed directly onto the skin, Adv. Mater. 25 (20) (2013) 2773−2778.

[101] J.C. Yeo, et al., Triple-state liquid-based microfluidic tactile sensor with high flexibility, durability, and sensitivity, ACS Sens. 1 (5) (2016) 543−551.

[102] J.C. Yeo, et al., Wearable tactile sensor based on flexible microfluidics, Lab Chip 16 (17) (2016) 3244−3250.

[103] M.D. Dickey, Stretchable and soft electronics using liquid metals, Adv. Mater. (2017) 1606425.

[104] T. George, et al., Liquid metals as ultra-stretchable, soft, and shape reconfigurable conductors. 9467 (2015) 946708.

[105] J.R. Napier, The form and function of the carpo-metacarpal joint of the thumb, J. Anat. 89 (Pt 3) (1955) 362−369.

[106] C.-L. Tsai, et al., How kinematic disturbance in the deformed rheumatoid thumb impacts on hand function: a biomechanical and functional perspective, Disabil. Rehabil. 39 (4) (2017) 338−345.

[107] A.M. Wing, C. Fraser, The contribution of the thumb to reaching movements, Quart. J. Exp. Psychol. A 35 (2) (1983) 297−309.

[108] F.J. Leversedge, Anatomy and pathomechanics of the thumb, Hand Clin. 24 (3) (2008) 219−229.

[109] M.B. Trudeau, et al., Thumb motor performance varies by movement orientation, direction, and device size during single-handed mobile phone use, Hum. Fact. 54 (1) (2011) 52−59.

[110] M.G. Björksten, B. Almby, E.S. Jansson, Hand and shoulder ailments among laboratory technicians using modern plunger-operated pipettes, Appl. Ergon. 25 (2) (1994) 88−94.

[111] L.-C. Kuo, et al., Feasibility of using a video-based motion analysis system for measuring thumb kinematics, J. Biomech. 35 (11) (2002) 1499−1506.

[112] W. Park, et al., A soft sensor-based three-dimensional (3-D) finger motion measurement system, Sensors 17 (2017) 2.

[113] J. Guna, et al., An analysis of the precision and reliability of the leap motion sensor and its suitability for static and dynamic tracking, Sensors 14 (2) (2014).

[114] C.D. Metcalf, et al., Markerless motion capture and measurement of hand kinematics: validation and application to home-based upper limb rehabilitation, IEEE Trans. Biomed. Eng. 60 (8) (2013) 2184−2192.

[115] T. Dutta, Evaluation of the Kinect™ sensor for 3-D kinematic measurement in the workplace, Appl. Ergon. 43 (4) (2012) 645−649.

[116] G. Du, et al., Markerless kinect-based hand tracking for robot teleoperation, Int. J. Adv. Robot. Syst. 9 (2) (2012) 36.

[117] W. Ying, J. Lin, T.S. Huang, Analyzing and capturing articulated hand motion in image sequences, IEEE Trans. Pattern Anal. Mach. Intell. 27 (12) (2005) 1910−1922.

[118] T. Pham, et al., A non-contact measurement system for the range of motion of the hand, Sensors 15 (8) (2015).

[119] S. Ryu, et al., Extremely elastic wearable carbon nanotube fiber strain sensor for monitoring of human motion, ACS Nano 9 (6) (2015) 5929−5936.

[120] S. Yao, Y. Zhu, Wearable multifunctional sensors using printed stretchable conductors made of silver nanowires, Nanoscale 6 (4) (2014) 2345−2352.

[121] C. Yan, et al., Highly stretchable piezoresistive graphene−nanocellulose nanopaper for strain sensors, Adv. Mater. 26 (13) (2014) 2022−2027.

[122] Kramer, R.K., et al. Soft curvature sensors for joint angle proprioception, in: 2011 IEEE/RSJ International Conference on Intelligent Robots and Systems, 2011.

[123] Chossat, J.B., et al. Wearable soft artificial skin for hand motion detection with embedded microfluidic strain sensing, in: 2015 IEEE International Conference on Robotics and Automation (ICRA), 2015.

[124] S.G. Yoon, H.-J. Koo, S.T. Chang, Highly stretchable and transparent microfluidic strain sensors for monitoring human body motions, ACS Appl. Mater. Interf. 7 (49) (2015) 27562−27570.

[125] L. Wang, et al., PDMS/MWCNT-based tactile sensor array with coplanar electrodes for crosstalk suppression, Microsyst. Nanoeng. 2 (2016) 16065.

[126] Yeo, J.C., et al. Tactile sensorized glove for force and motion sensing, in: 2016 IEEE Sensors. 2016.

[127] V. Iyengar, et al., Grip force control in individuals with multiple sclerosis, Neurorehabil. Neural Repair 23 (8) (2009) 855−861.

[128] D.A. Nowak, J. Hermsdörfer, H. Topka, Deficits of predictive grip forcecontrol during object manipulation in acutestroke, J. Neurol. 250 (7) (2003) 850−860.

[129] S.E. Tomlinson, R. Lewis, M.J. Carré, The effect of normal force and roughness on friction in human finger contact, Wear 267 (5−8) (2009) 1311−1318.

[130] W.S. Yu, et al., Thumb and finger forces produced by motor units in the long flexor of the human thumb, J. Physiol. 583 (3) (2007) 1145−1154.

[131] R. Malhotra, et al., Normative values of hand grip strength for elderly Singaporeans aged 60 to 89 years: a cross-sectional study, J. Am. Med. Dir. Assoc. 17 (9) (2016). p. 864.e1-864.e7.

[132] R.W. Bohannon, Muscle strength: clinical and prognostic value of hand-grip dynamometry, Curr. Opin. Clin. Nutr. Metab. Care 18 (5) (2015) 465−470.

[133] R.W. Bohannon, Hand-grip dynamometry predicts future outcomes in aging adults, J. Geriatr. Phys. Ther. 31 (1) (2008) 3−10.

[134] A. Kerr, et al., Does admission grip strength predict length of stay in hospitalised older patients? Age Ageing 35 (1) (2006) 82−84.

[135] S.E. Reuter, N. Massy-Westropp, A.M. Evans, Reliability and validity of indices of hand-grip strength and endurance, Aust. Occup. Ther. J. 58 (2) (2011) 82−87.

[136] H.C. Roberts, et al., A review of the measurement of grip strength in clinical and epidemiological studies: towards a standardised approach, Age Ageing 40 (4) (2011) 423−429.

[137] I. Batzianoulis, et al., EMG-based decoding of grasp gestures in reaching-to-grasping motions, Robot. Auton. Syst. 91 (2017) 59−70.

[138] Z. Yang, et al., Surface EMG based handgrip force predictions using gene expression programming, Neurocomputing 207 (2016) 568−579.

[139] Z.G. Xiao, C. Menon, Performance of forearm FMG and sEMG for estimating elbow, forearm and wrist positions, J. Bionic Eng. 14 (2) (2017) 284−295.

[140] Makino, Y. and H. Shinoda. Comfortable wristband interface measuring myoelectric pattern, in: Second Joint EuroHaptics Conference and Symposium on Haptic Interfaces for Virtual Environment and Teleoperator Systems (WHC'07), 2007.

[141] L. Popović Maneski, et al., Stimulation map for control of functional grasp based on multi-channel EMG recordings, Med. Eng. Phys. 38 (11) (2016) 1251−1259.

[142] Amft, O., et al. Sensing muscle activities with body-worn sensors, in: International Workshop on Wearable and Implantable Body Sensor Networks (BSN'06), 2006.

[143] J.B.H. Tok, Z. Bao, Recent advances in flexible and stretchable electronics, sensors and power sources, Sci. China Chem. 55 (5) (2012) 718−725.

[144] Perlmutter, M. and S. Breit. The future of the MEMS inertial sensor performance, design and manufacturing, in: 2016 DGON Intertial Sensors and Systems (ISS), 2016.

[145] Benser, E.T. Trends in inertial sensors and applications, in: 2015 IEEE International Symposium on Inertial Sensors and Systems (ISISS) Proceedings, 2015.

[146] J. Allen, Photoplethysmography and its application in clinical physiological measurement, Physiol. Meas. 28 (3) (2007) R1.

[147] Haahr, R.G., et al. A wearable "electronic patch" for wireless continuous monitoring of chronically diseased patients. in: 2008 5th International Summer School and Symposium on Medical Devices and Biosensors, 2008.

[148] O. Pabst, et al., Comparison between the AC and DC measurement of electrodermal activity, Psychophysiology 54 (3) (2017) 374−385.

[149] A. Martínez-Rodrigo, et al., Arousal level classification of the aging adult from electro-dermal activity: from hardware development to software architecture, Pervasive Mob. Comput. 34 (2017) 46−59.

[150] E.P. Scilingo, G. Valenza, Recent Advances on Wearable Electronics and Embedded Computing Systems for Biomedical Applications, Multidisciplinary Digital Publishing Institute, Switzerland, 2017.

[151] J. Benito-Leon, E.D. Louis, Essential tremor: emerging views of a common disorder, Nat. Clin. Pract. Neuro 2 (12) (2006) 666−678.

[152] G. Deuschl, P. Bain, M. Brin, Consensus statement of the movement disorder society on tremor, Mov. Disord. 13 (S3) (1998) 2−23.

[153] A. Anouti, W.C. Koller, Tremor disorders. diagnosis and management, West. J. Med. 162 (6) (1995) 510−513.

[154] J. Timmer, et al., Characteristics of hand tremor time series, Biol. Cybernet. 70 (1) (1993) 75−80.

CHAPTER 7

Wearable Technologies and Force Myography for Healthcare

M.N. Victorino, X. Jiang and C. Menon

Menrva Research Group, Schools of Mechatronic Systems and Engineering Science, Simon Fraser University, Metro Vancouver, BC, Canada

In recent years, there has been a growing trend in wearable technologies for applications in personalized and clinical healthcare. In fact, the global wearable technology market is expected to grow $5.8 billion by the year 2018 from $750 million in 2012 [1,2]. The rapidly advancing market of wearable technologies has been supported by the growing amount of research being conducted in the field of innovative healthcare devices and the increasing interest of end-users for easier, more effective, and more convenient technologies to assist them in their daily lives and provide them with greater independence [1]. In fact, the major driving factor for the global wearable technology market has been the increase in awareness of healthcare and wellbeing among the general population.

Most applications for wearable healthcare technology range from monitoring of daily tasks to assess mobility, strength, and level of independence in individuals, to integrating small sensors into devices and clothing to enhance the function of technologies that assist patients in performing functional motor tasks they are unable to do independently, without the help of family or a healthcare professional. Specifically, these technologies have fallen within physical activity monitoring, fall detection, blood pressure and blood glucose monitoring, brain and heart activity measurement, and muscle force measurement. These collected data are often used for diagnostic purposes, in addition to utilizing the data to determine the rehabilitation progression of an individual recovering from an injury or disease, given the ability of these devices to provide feedback on health and physiological status for each person and subsequently allow healthcare professionals to optimize treatment options based on this feedback [3]. In particular, wearable technologies for healthcare are gaining momentum in applications targeted for stroke, Parkinson's disease, diabetes, prosthetics, and cardiovascular disorders [1,4,5], and have focused on enhancing usability, efficiency, and convenience for end-users. Additionally, despite the focus on developing wearables for clinical assessments and applications that can be implemented in hospital and formal healthcare settings, such as rehabilitation clinics [4], optimal designs showcasing the most potential will be wearables that allow for the same quality of recording accuracy when the user is in a home setting or outdoors. This provides a more realistic setting for daily activities that a user partakes in since they generally spend more time outside a clinic or hospital, giving insight on muscle activity and treatment progression when users are simply living their daily lives.

Many wearable technologies utilize a wide array of sensors that are integrated into the devices depending on the parameter of interest. Some examples include optical fibers and other fiber-like smart materials embedded in clothing to detect biosignals and respiration, gyroscopes, and accelerometers for fall detection and movement monitoring, pressure cuffs to monitor blood pressure, and electroencephalography (EEG) to detect brain activity for neurological and neuromuscular disorders and recovery techniques. This chapter will focus on force detection and movement monitoring wearable technologies, specifically an emerging technique called force myography (FMG) and its applications and potential in healthcare and rehabilitation.

7.1 Movement Monitoring

Tracking human body movement is a complex goal that can be achieved in various different ways. When an individual suffers from neuromuscular disorders or musculoskeletal injuries, it is imperative to recovery that motor performance can be monitored and feedback on functional movement and performance is given to the individual as this allows therapists and clinicians to better assess and improve upon rehabilitation progress for each patient and allow for a more personalized treatment approach [6]. The information on motor functions collected from these wearable technologies, such as grasping motions, hand gestures, gross arm movements, and gait patterns, and parameters, are also highly relevant to patients and end-users as the objective feedback received can serve as motivation and encouragement, potentially resulting in increased participation and proactive movement and use of muscle by the individual. As an individual increases the use of their muscles in an impaired limb, the more likely an improvement in the motor function will be seen [5].

Another advantage of using wearable technologies for movement monitoring in healthcare and rehabilitation is that it provides therapists an option to track patient progress when they are in daily environments, such as at home or outdoors in the general community while they are doing activities of daily living (ADL). This can help to improve clinical management as it can gather the differences and nuances between how individuals move during therapist-led rehabilitation sessions and average day-to-day tasks performed in at home and in more familiar settings, such as a grocery store or a community park. These nuances can then be utilized to personalize therapies and enhance treatments to include exercises and strategies outside the hospital or clinic so individuals can apply them even after they are no longer attending rehabilitation sessions and during long-term activities.

Below, Section 7.1 expands on a few common sensors and devices that have been developed for the purpose of human movement monitoring and tracking.

7.1.1 Accelerometers

Accelerometers are sensing devices that measure a moving object's acceleration and can detect frequency and intensity of human movement [7]. For decades, accelerometers have been widely used in monitoring functional motor movement, including studies on neuromuscular disorders like stroke and Parkinson's disease, and have focused on measuring gait, posture, and tremor parameters, in addition to detecting falls [7,8]. Body-worn accelerometers have also been utilized to classify activities in sitting, walking, standing, cycling, and lying positions, and have shown high accuracies in such studies [8]. Additionally, accelerometers have been used to measure the gross amount of arm movement for upper-extremity rehabilitation; however, it cannot track finger movement, such as grasping, making it unsuitable for tracking movements involved in ADL given that grasping is a gesture integral to actions, such as holding and lifting objects. Given that accelerometers are small and low-cost, they have been designed and integrated into many wearable technologies, typically fashioned into watches, bracelets, and bands, easy to wear clothing that can appeal to end-users for continuous and long-term use.

7.1.2 Inertial Measurement Units

Inertial Measurement Units (IMU), sensors which comprise accelerometers and gyroscopes that use microcontrollers to process its collected measurements and Bluetooth modules for system communication, are another technology growing in use for human motion tracking and analysis. IMUs, like accelerometers, are noninvasive and relatively easy to integrate into a wearable strap, making them appropriate sensors to utilize for wearable technologies aimed at healthcare and rehabilitation since they are capable of assessing and monitoring activities and symptoms of a particular movement disorder and subsequently enabling the evaluation of an individual's quality of life [9]. For example, in Parkinson's disease, a progressive neurodegenerative disorder that severely affects a person's motor function through symptoms, such as tremors, slow movements, and difficulties in walking, IMUs have been employed to monitor gait patterns and stride parameters to further understand the intricacies of the Parkinsonian gait and to determine the efficacy of medications on the gait and other motor symptoms of Parkinson's disease patients [10]. However, while IMUs have shown to be reliable in movement monitoring, IMUs often require the use of at least two sensors in different locations, typically, placed on the foot and calf, which can be bulky and susceptible to the sensors sliding down the leg, consequently, affecting the accuracy of the motion being tracked. IMUs can also be uncomfortable to wear and unable to continuously monitor movement over long periods of time, such as overnight [11].

7.1.3 Data Gloves

Data gloves, which sense and detect hand and finger movements, are wearable and can accurately track hand motion. However, data gloves are difficult to do when a person has a clenched hand, such as those who have suffered from stroke or for those with a limited range of motion in their fingers due to injuries, making the data gloves unideal. Data gloves also have the drawback of limiting the tactile sensation of the fingers, in addition to being obtrusive and unsuitable for many ADL, including bathing, washing, and dressing [12].

Others technologies, such as wearable robotics are generally expensive, not portable, or obtrusive to natural movement, and therefore are not the ideal type of wearable device to use for real-time motion tracking in healthcare and rehabilitation applications. There is still, therefore, a lack in the current wearable technologies available for healthcare that will provide users with the optimal benefits for motion tracking.

7.1.4 Conductive Fibers and Smart Clothing

Interactive electronic textiles (e-textiles) are another developing research topic in the field of wearable technologies due to its suitability for providing comfort and aesthetic to users. While e-textiles are utilized in the fashion, entertainment, and communication industries, this technology is increasingly becoming applied to healthcare matters [13]. Textile sensors typically come in the form of conductive fibers that are sensitive to the strain applied by the muscles they're in contact with. Gibbs and Asada [14] conducted preliminary experiments using a wearable sensing garment integrated with conductive fibers to continuously detect joint movement, specifically in the hip and knee, and found that it was effective for joint monitoring [14]. The current state of the art for smart clothing includes Google and Levi's Project Jacquard, which uses conductive yarns woven into clothing to allow users to interact with their environment and the technology through touch. However, in Project Jacquard simple sensors are only embedded in a very localized area of the wearable garment, such as the sleeve of a jacket, to provide limited functionality [15]. Sensing clothing is therefore an active area of research and development, considering that its optimal success would immerse users in a new world where information related to motion activities could be recorded, extracted, processed, and made available in real time while wearing comfortable day-to-day garments.

Below, we will discuss the state of the art of a technique called myography, and more importantly, FMG and its potential for real-time movement monitoring that could shift the landscape of wearable healthcare technologies.

7.2 Myography

To measure muscle force production during contraction, a technique called myography is often utilized. Like the previously mentioned sensors (inertial sensors, etc.), myography can be used to track motion patterns and characterize gestures. There are different categories of measurement under the general term of myography, namely mechanomyography, ultrasound myography, optical myography, and electromyography (EMG).

Mechanomyography, which is also sometimes referred to as acoustic myography, uses inertial sensors, condenser microphones, laser sensors, and piezoelectric crystals to measure the vibration of muscle fibers and the subsequent dimensional changes during activation, which creates pressure waves that can be detected on the skin surface [16]. However, mechanomyography requires a significant amount of signal processing and is susceptible to muscle artifact sensitivity, making it difficult to integrate into clothing or into other portable and wearable designs. Ultrasound myography or sonomyography, measures the velocity of muscle movement which is directly related to force production. Ultrasound myography, often used for diagnostic and therapeutic applications, has the advantage of being able to detect the activity of deep muscles, however, it is far too expensive and the system is too bulky to be able to integrate into a wearable device. Additionally, it is susceptible to heating and often requires trained professionals to operate the system to ensure safety [17]. Optical myography is unobtrusive given that it optically tracks muscle movement by observing the surface deformations of a limb, however, just like any system that mainly relies on vision, it is greatly susceptible to occlusions, which can affect data accuracy [18].

Among these systems, EMG is the most well known and established technique. EMG, measures the electrical activity in muscles during activation and can be done through intramuscular electrodes or surface electrodes (sEMG). EMG is used in various applications in rehabilitation, human–machine interactions, ergonomics, and prosthetic and exoskeleton control. In fact, EMG has been paired with accelerometers to capture movement and muscle activity patterns, and to provide a more comprehensive tracking accuracy during functional motor tasks. However, EMG is subject to various sources of noise, such as the inherent noise of the electrodes, movement artifacts, electromagnetic noise, muscle cross-talk, ECG artifacts, blood flow velocity, skin temperatures, tissue structure, and the selected measuring site on the body, that can affect the accuracy of the signals that are collected. This susceptibility to noise makes it difficult for EMG to be utilized and integrated into wearable devices that can track motion in real-time when individuals are performing ADL. EMG is also impractical for unsupervised use at home given its lengthy preparation procedures and the hardware can be expensive and heavy. While cheaper sEMG systems are available, these give very poor accuracy in estimating finger positions and are still susceptible to skin impedance, making it unreliable.

7.3 Force Myography

Many neuromuscular and musculoskeletal disorders, such as stroke, Parkinson's disease, arthritis, and cerebral palsy significantly affect an individual's capacity for movement and often result in disabilities that reduce their independence and prevent them from being able to perform ADL, such as walking from room to room, bathing, dressing, and grooming. There is a rich history in research and technology wherein scientists and engineers have attempted to generate innovative ideas and devices to compensate for these disabilities and assist individuals through their daily lives [4,19−22]. Devices such as defibrillators, rehabilitation robots, and machines, exoskeletons, and wearable sensors have been in existence for decades and are constantly being upgraded and improved upon in design and function by researchers, healthcare professionals, and engineers. However, many technologies continue to be inaccessible, whether due to factors associated to cost, aesthetic, or perceived usefulness and ease of use [1,23]. With healthcare costs showing an increasing trend, there is a greater need for affordable and accessible healthcare technology that can assist patients through recovery and provide them with the capability to perform daily tasks.

Force myography (FMG) is one such technology. FMG is a mechanomyographic [24] approach sometimes referred to as pressure myography (PMG) [25], topographic force mapping (TFM) [26], residual kinetic imaging (RKI) [27], and surface muscle pressure (SMP) [28]. Despite the various terms associated with it, the technique is similar. FMG is an emerging technique that uses force or strain sensors surrounding a limb to measure the volumetric changes in the underlying musculature to distinguish movements of tendons and muscles. An FMG signal is recorded by sensors measuring local radial expansions of a limb. Typically, FMG signals are detected from sensors wrapped around the forearm, the wrist, or both. For example, considering that movements of the fingers are primarily controlled by muscles/tendons of the forearm, arrays of strain or pressure sensors around the forearm can be associated with distinct actions of the fingers.

Compared to the previously mentioned myography technologies, FMG presents various advantages that suggests its potential as a highly innovative option to further pursue in the field of wearable technologies. FMG technology does not require skin preparation, the variation of electrical impedance does not affect the signal quality, meaning there is no effect from sweating unlike sEMG, and the sense of touch is not altered like it is the case with data gloves. Additionally, FMG provides a more promising and reliable option for accessible movement monitoring technologies given that it is lightweight, cost-effective, easy-to-use, and has signals that are stable over time. FMG can be worn on the wrist, arm, or ankle without being obtrusive to motion, providing comfort to users and a pleasing aesthetic that can make it more appealing to use in a daily basis. These factors combined to make FMG technology an attractive and highly advantageous option for human movement monitoring, especially for individuals with a significant amount of disability and may already spend large amounts of money on major clinical and healthcare costs, such as medications and rehabilitation expenditures. In addition, FMG can be seamlessly integrated into clothing, allowing for a more seamless and convenient wearable garment that can provide real-time feedback to users without the hassle of donning bulky devices. Further, while FMG has not been as extensively investigated as the other technologies in healthcare technology, it has very recently been considered one of the most promising emerging methods for noninvasive human—machine interfaces (HMIs) for the disabled [29]. In a study by Ravindra and Castellini [29], FMG was also found to be the most promising noninvasive sensor compared to sEMG and ultrasound, with FMG showing the potential to have the highest prediction accuracies.

Besides movement monitoring in clinical settings, FMG can also be utilized to control external devices by integrating the FMG technology with an HMI. FMG can be integrated into a band that can be worn around the arm and the sensor array within the band will pick up the force activity produced by the muscles involved in the area of the limb where the band is wrapped around. Then, FMG, along with the HMI system tasked to interpret the collected muscle activity data and implement command signals, can function to send voice commands, power a wheelchair, and control home appliances, such as refrigerators and audio systems for the disabled. In this way, FMG technology can provide assistance for ambient assisted living for seniors or individuals with severe neuromuscular deficiencies, such as those affected by cerebral palsy.

The accuracy in estimating finger and hand movements using an FMG band worn around the arm has been shown in research studies to be comparable to or better than sEMG performance acquired with highly sophisticated hardware [6,30]. Further, similar to forearm sEMG, studies have shown that FMG signals extracted from the forearm can detect some hand gestures, wrist positions, and elbow related movements [6]. Connan et al. [31] proposed and tested a wearable FMG and EMG device on 10 healthy subjects and found that FMG was able to generate signals which are more stable across time, with a lower signal variance, compared to EMG. It was also speculated that fusion of FMG and EMG may be promising and advantageous in improving myocontrol.

Despite the many advantages of FMG technology, one major drawback in its application is that it relies on the activity of the muscles, meaning the lack of musculature in individuals experiencing atrophy from muscle disuse or in those with amputated limbs leads to a more complicated scenario when attempting to detect FMG signals in that the accuracy of the recorded data cannot be as maximized as the accuracy that can be detected in musculature that is intact [25]. As such, majority of the early research work in FMG have been conducted in healthy and able-bodied participants for the upper extremities in which the arm was in a fixed position [32].

Overall, FMG's capability to track and monitor movement and muscle activity, especially in the arm, hand, and fingers, makes it a beneficial technology to utilize for healthcare applications, such as in rehabilitation and prosthetics. This allows the technology to potentially contribute to improving functional human movement and increase independence in those who are motor-impaired. To be able to measure the volumetric changes of a limb during functional movement, FMG technology uses force pressure sensors or strain gauge sensors typically integrated into a band for the arm and wrist, or integrated into other pieces of clothing to ensure that it is fully wearable. These sensors are discussed in the following section to give a better insight of the technology.

7.4 Force-Sensing Resistors

Force-sensing resistors (FSRs) are polymer thick film (PTF) devices. FSRs are comprised of conductive polymer films [33] whose resistance is based on the amount of force applied to its surface on the active area of a given limb, with its resistance decreasing when the force applied to it increases. FSRs have a conductive surface interdigitated with electrodes, with both surfaces facing each other. When the two surfaces are in contact and are pressed together, the pressure between the conductive layer and the printed electrodes reduces the electrical resistance.

FSRs are advantageous to use in designing and fabricating wearable devices and clothing as they are easy to implement, cost-effective (off-the-shelf FSRs typically cost less than $10), flexible, robust, and widely available, making them ideal to use and attach to body areas, such as the hands and fingers and measuring the forces produced by the respective musculature involved in a particular set of movements. Due to these reasons, much of FMG research has utilized FSRs for motion detection and gesture recognition. Various studies have reported high accuracies for predicting arm and finger gestures when using FSRs around the arm [3,32,34,35] and while the number of FSRs used vary from study to study, the resulting high prediction accuracies remain common among the current research. An example of customized FSR strap is shown in Fig. 7.1A, which has 8 FSRs (FSR 402, Interlink Electronics, Inc., Los Angeles, United States) in a circle and can be worn on either the wrist or the forearm. The FMG signal was extracted using a voltage divider circuit as shown Fig. 7.1B. The resistance of the FSR is more than 10 M Ohm with absence of pressure on the FSR, and it increases logarithmically as the pressure increases. A base resistor (R_{sense} in Fig. 7.1B) adjusts the output range for the muscle pressure sensing, which is usually empirically set to a value (12 k Ohm in this case). The voltage divider circuit is usually powered by a 5 V voltage source and its voltage output can be read by a data acquisition board (DAQ) transferred to a working PC.

During the performance of a hand movement or hand gesture, the muscle contraction resulting in the pressure variation is sensed by the FSR strap, forming distinctive signal patterns relating to each hand movement/gesture. Fig. 7.2 shows an example of FSR signals

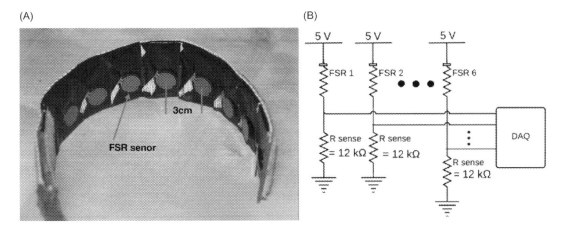

Figure 7.1
An example of FSR strap and voltage divider circuit. (A) the FSR strap with 8 FSR sensors, (B) the diagram of the voltage divider circuit. Part (A) is from Kadkhodayan et al. [35].

for hand movements performed by a subject during a drinking task [3]. We can see that the 8 FSR signals are similar between repetitions of the same type of hand movement, but distinctive between different types of hand movements (represented by different background colors). Using supervised machine learning algorithms such as linear discriminant analysis (LDA) [30], support vector machine (SVM) [36], or extreme learning machine (ELM) [3], the FSR data samples can be classified with the help of class labels [3,30,36].

However, there are certain drawbacks still associated with FSRs that could argue against this particular type of sensor being the most ideal to use in movement monitoring. FSRs have intrinsic limitations due to the type of sensor since wearable devices utilizing FSRs need close contact with the skin in a consistent manner and do not slide across the skin surface to ensure that the movement is detected and that the accuracy is acceptable. Taking into consideration, this factor, designs for wearable technology could end up constricting and uncomfortable, especially if FSRs are integrated into clothing.

7.5 Strain Gauges

Another type of sensor that can be implemented for FMG is a strain gauge sensor. This type of sensor measures the strain applied by the muscles during functional movement. Textiles have traditionally been integrated with strain gauges and then attached to clothing to monitor movement; however, the rigidity of an alloy strain gauge limits the static strain level to low-stress measurements [13]. Strain gauges transduce the elastic deformation of tendons, skin, and muscles during movement into electrical signals, which are then transmitted to a computer for data interpretation. Like FSRs, an FMG approach using strain

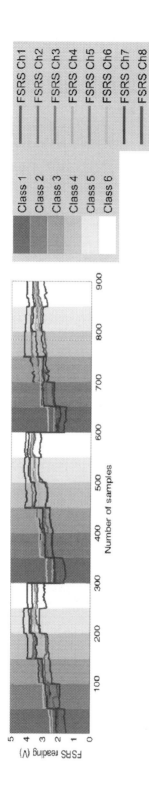

Figure 7.2

Example of FSR signals for hand movements performed by a subject during a drinking task. Classes 1–6 are the 6 hand movements, that is relax, 90 degrees Elbow Flexion, Fingers Extension, Soft Grasp, 120 degrees Elbow Flexion, and Wrist Pronation. *Adapted from Xiao et al. [3].*

gauges is cost-effective, easy to use, and has the capability to be integrated into clothing and other unobtrusive wearable devices. One of the main advantages that strain gauges have over FSRs is that strain gauges do not require the sensors to be in constant contact with the skin. This allows for more flexibility in design for a wearable device, such as an elastic wristband, while maintaining high reliability and data accuracy during movement monitoring. Since pressure sensors can be difficult to seamlessly integrate into clothing for optimal wearability, Ferrone et al. [37] investigated the capabilities of strain sensors and have published preliminary work showing that pressure sensors can be replaced by strain sensors. Specifically, strain sensors can show significant performance and accuracy via a band of cloth embedded with innovate sensing fibers that the Menon research lab have recently developed [37]. Further, the study examined the feasibility of using custom-fabricated strain gauge sensors to classify hand gestures and by employing cross-validation and cross-trial evaluations, the team achieved accuracies of 95% and 80% [38]. Fig. 7.3 shows wristband fabricated in Dr. Menon's lab, with strain gauges embedded.

Another study developed a novel strain gauge system that could be worn on the wrist to detect hand gestures was able to exhibit classification accuracies as high as 99% using machine learning algorithms [39]. In the aforementioned study, 14 hand gestures were tested on eight healthy participants and the sensor system, worn around the wrist via a band, was able to provide distinctive signal patterns corresponding to the different hand gestures.

It should be noted that compared to FSRs, strain gauges also exhibit high sensitivity to pressure variations [40] and can be utilized for high accuracy gesture recognition. FSRs have shown time lags and cannot react quickly enough to changes in pressure variation,

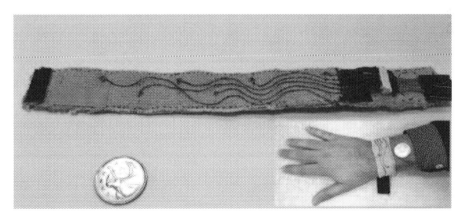

Figure 7.3
The strain gauge wristband and the band worn on the wrist. This figure is from Ferrone et al. [38].

resulting in unstable results, making strain gauges a more attractive option when recording on smaller and more complex musculature with high variations in pressure [39].

7.6 Applications

FMG technology has a wide range of biomedical applications. It has been utilized in prostheses control, rehabilitation, gait analysis, and gesture recognition. While FMG can potentially be used in additional approaches, this chapter will focus on the applications for the upper extremities, with the exception of gait analysis, which will focus on the lower limbs.

Gesture recognition: All the currently available systems for human gesture monitoring have relevant disadvantages and limitations. For example, systems based on 3D infrared cameras for recognizing finger gestures [41], in addition to other devices based on general camera recognition, suffer from occlusions and have heavy computation requirements, making it difficult to monitor and classify gestures, especially in large quantity. Other sensors are based on EMG, the disadvantages of which have been mentioned earlier in this chapter.

When detecting hand movements, FMG signals are generally extracted from the middle of the forearm given that this area can generate large radial displacements for data collection [3,32,34,42]. However, recent studies have explored the possibility of recording and processing FMG signals at the wrist to estimate hand movements. Morganti et al. [43] proposed a smart watch comprised of four FSRs for detecting wrist and hand gestures. Dementyev and Paradiso [44] developed and presented the WristFlex, a wearable interface that used an array of FSRs and could be worn on the wrist, which was reported to be capable of detecting six finger gestures at high accuracy. These works showed the potential of embedding FMG sensors inside a watch strap or wristband, which could make the technology more appealing for users, especially for those who take into high consideration the aesthetic of the device as a factor for long-term usage. Jiang et al. [31] systematically evaluated the performance of a customized FMG strap vs a commercially available medical quality sEMG system in distinguishing three sets of 48 hand gestures, on both the forearm and wrist, respectively. The authors found that the FMG band is comparable to the sEMG system in all three sets hand gestures classification; more interestingly, they found the FMG band achieved a similar classification accuracy on both the forearm and wrist.

FMG has been shown to provide accurate gesture recognition for healthy individuals in a study that Xiao and Menon [32] conducted, in which the ability of using FMG to count the number of grasping motions during a series of pick-and-place (PAP) actions was investigated. In this study, wireless FMG straps were prototyped and used on the wrist and the forearm of healthy individuals. The collected hand gesture data were evaluated by well-established machine learning algorithms, such as LDA and SVM and the study reported that

FMG can accurately monitor grasping motions during functional arm movements when in a controlled environment. The work of Xiao and Menon build a foundation for future work in which FMG can be tested to detect grasping during noncontrolled environments, such as performing ADL outdoors, and in which, FMG accuracy in grasp detection can be investigated for nonhealthy participants, particularly those with weak arms and hands, such as stroke individuals. Moreover, to further expand FMG's capabilities for gesture recognition, Kadkhodayan, Jiang, and Menon [35] have reported that FMG technology can act as a control interface and continuously predict finger movements in healthy participants by predicting the displacement of the fingertips in three different grasp types: the power grasp, tripod pinch grasp, and the index pinch grasp. An average squared correlation coefficient of 0.96 for the thumb and middle and index fingers were reported in this study, a much higher performance compared to the commonly-used sEMG, which achieves an average correlation coefficient within the range of 0.74−0.85 for finger joint movement estimation [35]. The study showcased the feasibility of FMG for continuous grasp prediction of the hand and is a step towards more extensive research on using FMG for full applications in robotic exoskeletons [35].

Gesture recognition and prediction are highly relevant in healthcare applications for the telemanipulation of assistive robotic devices for rehabilitation and for prosthetic control. It can also have applications in virtual reality, which in itself is an emerging technology in healthcare and rehabilitation. Gesture recognition through FMG technology can enhance virtual reality environments by allowing lightweight wearable bands or other types of clothing to be worn during the motion to track movement while an individual performs repetitive exercises or ADL in an immersive virtual setting. The combination of both FMG and virtual reality has the potential to augment therapy exercises and treatment protocols for motor-impaired populations.

Prosthesis Control: FMG has been widely applied to gesture recognition for prosthesis operation through the use of sophisticated machine learning technologies [34]. While the history of prosthesis devices date back to centuries [45] and the current prosthesis technology has developed significantly to assist in mitigating the loss of a limb, there are still limitations to the current state of the art. The main weakness presented by limb prosthesis is the lack of adequate and responsive control an individual has in using the artificial limb, with the devices unable to effectively mimic the human hand and arm [25,45]. Subsequently, the overall rate of upper extremity prosthesis use in amputees remains low [25] despite the commercial availability of prosthesis devices, such as Motion Control Inc.'s Utah Arm, Otto Bock's Michelangelo hand, Touch Bionics' i-Limb, and Steeper Group's Bebionic3. The difficulty in controlling prosthetic limbs can be attributed to a number of factors including the prosthesis weight, socket discomfort, associated pain, and lack of functionality [46]. In addition, the misclassification of user intentions frequently leads to unintentional and unplanned movements [25,47], which results in an unreliable and unsatisfactory prosthetic limb.

It is common for prosthetic devices to be myoelectrically controlled, with the external prosthesis utilizing the electric signals generated during muscle contraction to direct movement. To be able to detect the electrical activity during muscle activation, standard myoelectric prosthetic devices, including the major commercial products, have relied on EMG and surface electromyography (sEMG) [48]. EMG can be detected on the skin surface through electrodes, providing a noninvasive approach to detecting electrical muscle activity, and has shown the capability to track upper extremity movements and hand gestures. However, as previously mentioned, EMG are highly susceptible to noise, sweat and skin impedance, electrode shifts during movement, and signal cross-talking between adjacent muscles, resulting in signal inconsistencies that can significantly limit the accuracy and robustness of the prosthetic device.

FMG has been recently explored as a potential alternative to sEMG for prosthesis control. A study conducted by Cho et al. [25], determined the feasibility of the use of FMG to classify grip patterns in four transradially amputated individuals, with measurements done on both the sound and residual limbs to be able to compare classification accuracies. The team was able to determine an accuracy of above 70% in the residual limb, showing that it is possible to classify at least six primary grips important in ADL using FMG. In the same study, they found that the residual limb was outperformed by the sound limb, which is expected given the lack of significant musculatures, such as muscles, tendons, and ligaments, in addition to the atrophy caused by reduced muscle use in amputees, in the residual limb. While future work is still required to determine the extent of FMG's capabilities for prosthetic control in upper extremities and possibly for lower extremities as well, the preliminary work has pointed to the direction that FMG has the potential to be a cheaper, easier, and more accurate option.

Gait analysis: Majority of the studies in literature for FMG have focused on its use for upper extremities. However, FMG can also monitor and detect lower limb movement for healthcare applications as sports injury prevention, rehabilitation exercises, and senior activity monitoring. Ankle movements, such as plantarflexion, dorsiflexion, inversion, eversion, lateral axial rotation, and medial axial rotation are involved in most basic gait phases and lower limb activities, especially in those comprising ADL. In addition to these, simple ankle movements are often the target of rehabilitation exercises to strengthen muscles and potentially improve motor function for individuals with gait impairments, such as those who've suffered stroke [49]. Similarly to the already mentioned applications, FMG has also been recently investigated as an alternative technology for lower limb motion tracking given the success of the technology in detecting upper extremity activities. The feasibility of FMG to monitor and classify ankle movements has been explored in studies [50,51] with a wearable ankle strap consisting of an FSR array. The study showcased a high average prediction accuracy of 94% and 85% in cross-validation and cross-trial evaluation, respectively. Despite the study only being conducted in three healthy individuals, these results are promising and provide groundwork for future research using FMG technology for lower limb motion tracking in both healthy individuals and those who are motor-impaired.

Successfully acquiring information of ankle position through a wearable ankle strap can provide relevant information on an individual's gait patterns and assist in generating better exercises and treatments for those with gait abnormalities. This could prove to be useful to individuals with Parkinson's disease, stroke, and arthritis, who are known to suffer from a variety of gait issues that limit their independence and prevent them from moving freely.

Rehabilitation: Arm, hand, and finger movements are integral to day-to-day human activity. Muscle limb activity contributes to all types of arm movements, such as grasping, holding, and pinching, and contribute to more complex daily motions, such as lifting objects and drinking from a cup. These movements are frequently used in rehabilitation exercises with the goal to retrain the impaired limbs or strengthen atrophied muscles [30]. FMG's ability to monitor and classify movement, both in upper and lower limbs as discussed in previous sections, has contributed in its emergence as a potential rehabilitative technology. Early work on an FMG band designed to predict the upper limb postures involved in a drinking task, a movement sequence often used in rehabilitation, demonstrated results that confirmed FMGs capability to generate distinct patterns for some upper-extremity postures involved in a drinking task [3]. Some upper limb postures included a relaxed arm and hand, a 90-degree flexed elbow while grasping a cup, and raising the arm. The study was able to achieve an average overall accuracy of 92.33%, however, the test was only done in healthy male volunteers [3]. Nevertheless, this ability to monitor and predict human movement in real-time and provide feedback to users, especially to those undergoing rehabilitation, shows significant potential and can be highly beneficial in promoting an increase in the amount of muscle activity, exercise, patient motivation since exercise progress can be more easily collated and visualized, potentially facilitating a quicker recovery.

In conclusion, FMG is an emerging technique showing high potential in various healthcare applications, including gesture recognition, rehabilitation, prosthesis control, and gait analysis. It has shown significant recording advantages compared to other more commonly-used technologies, such as sEMG while being significantly more affordable. FMG is also more easily integrated into a lightweight unobtrusive wearable device given its sensors, FSRs or strain gauges, are neither bulky nor tethered to computers by wires. While more work is needed to determine FMG's full capabilities in healthcare, the early research work on the topic has proven to be promising, paving a road for innovation and improved healthcare.

References

[1] J. Casselman, N. Onopa, L. Khansa, Wearable healthcare: lessons from the past and a peek into the future, Telemat. Informatics 34 (2017) 1011−1023.

[2]   Transparency Market Research. *North America to Lead Global Wearable Technology Market, Healthcare Sector Dominates Demand.* (2015). https://www.transparencymarketresearch.com

[3]   Z.G. Xiao, C. Menon, Towards the development of a wearable feedback system for monitoring the activities of the upper-extremities, J. Neuroeng. Rehabil. 11 (2014) 2.

[4]   Bonato, P. Advances in wearable technology and its medical applications, in *32nd Annual International Conference of the IEEE Engineering in Medicine and Biology Society (EMBC)* 2021–2024 (2010). doi:10.1109/IEMBS.2010.5628037

[5]   P. Sungmee Park, et al., Enhancing the quality of life through wearable technology, IEEE Eng. Med. Biol. Mag. 22 (2003) 41–48.

[6]   Z.G. Xiao, C. Menon, Performance of forearm FMG and sEMG for estimating elbow, forearm and wrist positions, J. Bionic Eng. 14 (2017) 284–295.

[7]   C.-C. Yang, Y.-L. Hsu, A review of accelerometry-based wearable motion detectors for physical activity monitoring, Sensors 10 (2010) 7772–7788.

[8]   M.J. Mathie, B.G. Celler, N.H. Lovell, A.C.F. Coster, Classification of basic daily movements using a triaxial accelerometer, Med. Biol. Eng. Comput. 42 (2004) 679–687.

[9]   D. Rodríguez-Martín, C. Pérez-López, A. Samà, J. Cabestany, A. Català, A wearable inertial measurement unit for long-term monitoring in the dependency care area, Sensors 13 (2013) 14079–14104.

[10] S.T. Moore, H.G. MacDougall, J.-M. Gracies, H.S. Cohen, W.G. Ondo, Long-term monitoring of gait in Parkinson's disease, Gait Posture 26 (2007) 200–207.

[11] T. Seel, J. Raisch, T. Schauer, IMU-based joint angle measurement for gait analysis, Sensors 14 (2014) 6891–6909.

[12] L. Dipietro, A.M. Sabatini, P. Dario, A survey of glove-based systems and their applications, IEEE Trans. Syst. Man, Cybern. Part C Applications Rev. 38 (2008) 461–482.

[13] T.W. Shyr, J.W. Shie, C.H. Jiang, J.J. Li, A textile-based wearable sensing device designed for monitoring the flexion angle of elbow and knee movements, Sensors (Switzerland) 14 (2014) 4050–4059.

[14] P.T. Gibbs, H.H. Asada, Wearable conductive fiber sensors for multi-axis human joint angle measurements, J. Neuroeng. Rehabil. 2 (2005) 7.

[15] Poupyrev, I., Gong N.-W., Fukuhara, S., Karagozler, M.E., Schwesig, C. Robinson, K.E. Project Jacquard: Interactive Digital Textiles at Scale, in *Proceedings of the 2016 CHI Conference on Human Factors in Computing Systems—CHI '16* 4216–4227 (ACM Press, 2016). doi:10.1145/2858036.2858176

[16] M.O. Ibitoye, N.A. Hamzaid, J.M. Zuniga, A.K. Abdul Wahab, Mechanomyography and muscle function assessment: a review of current state and prospects, Clin. Biomech. 29 (2014) 691–704.

[17] D.L. Miller, N. Smith, M. Bailey, G. Czarnota, K. Hynynen, I. Makin, Overview of therapeutic ultrasound applications and safety considerations, J. Ultrasound Med. 31 (2012) 623–634.

[18] C. Nissler, N. Mouriki, C. Castellini, Optical myography: detecting finger movements by looking at the forearm, Front. Neurorobot. 10 (2016) 3.

[19] R. Looned, J. Webb, Z.G. Xiao, M. Menon, Assisting drinking with an affordable BCI-controlled wearable robot and electrical stimulation: a preliminary investigation, J. Neuroeng. Rehabil. 11 (2014) 51.

[20] A.M. Elnady, X. Zhang, Z.G. Xiao, X. Yong, B.K. Randhawa, L. Boyd, et al., A Single-Session Preliminary Evaluation of an Affordable BCI-Controlled Arm Exoskeleton and Motor-Proprioception Platform, Front. Hum. Neurosci. 9 (2015) 168.

[21] M. Bächlin, M. Plotnik, D. Roggen, I. Maidan, J.M. Hausdorff, N. Giladi, et al., Wearable assistant for Parkinsons disease patients with the freezing of gait symptom, IEEE Trans. Inf. Technol. Biomed. 14 (2010) 436–446.

[22] M. Capecci, L. Pepa, F. Verdini, M.G. Ceravolo, A smartphone-based architecture to detect and quantify freezing of gait in Parkinson's disease, Gait Posture 50 (2016) 28–33.

[23] R.J. Holden, B.-T. Karsh, The technology acceptance model: its past and its future in health care, J. Biomed. Inform. 43 (2010) 159–172.

[24] M.A. Islam, K. Sundaraj, R.B. Ahmad, N.U. Ahamed, M.A. Ali, Mechanomyography sensor development, related signal processing, and applications: a systematic review, IEEE Sens. J. 13 (2013) 2499–2516.

[25] E. Cho, R. Chen, L. Merhi, Z.G. Xiao, B. Pousett, Force myography to control robotic upper extremity prostheses: a feasibility study, Front. Bioeng. Biotechnol. 4 (2016) 1–12.

[26] C. Castellini, P. Artemiadis, M. Wininger, A. Ajoudani, M. Alimusaj, A. Bichii, et al., Proceedings of the first workshop on peripheral machine interfaces: going beyond traditional surface electromyography, Front. Neurorobot 8 (2014) 22.

[27] S.L. Phillips, W. Craelius, Residual kinetic imaging: a versatile interface for prosthetic control, Robotica 23 (2005) 277–282.

[28] D.A. Yungher, M.T. Wininger, J.B. Barr, W. Craelius, A.J. Threlkeld, Surface muscle pressure as a measure of active and passive behavior of muscles during gait, Med. Eng. Phys. 33 (2011) 464–471.

[29] V. Ravindra, C. Castellini, A comparative analysis of three non-invasive human-machine interfaces for the disabled, Front. Neurorobot. 8 (2014) 24.

[30] X. Jiang, L.K. Merhi, Z.G. Xiao, C. Menon, Exploration of force myography and surface electromyography in hand gesture classification, Med. Eng. Phys. 41 (2017) 63–73.

[31] M. Connan, E. Ruiz Ramírez, B. Vodermayer, C. Castellini, Assessment of a Wearable force- and electromyography device and comparison of the related signals for myocontrol, Front. Neurorobot. 10 (2016) 17.

[32] Z.G. Xiao, C. Menon, Counting grasping action using force myography: an exploratory study with healthy individuals, JMIR Rehabil. Assist. Technol. 4 (2017). e5.

[33] A. Hollinger and M.M. Wanderley, Evaluation of commercial force-sensing resistors, in: Proceedings of the International Conference on New Interfaces for Musical Expression, Paris, France, 2006.

[34] N. Li, D. Yang, L. Jiang, H. Liu, H. Cai, Combined use of FSR sensor array and SVM classifier for finger motion recognition based on pressure distribution map, J. Bionic Eng. 9 (2012) 39–47.

[35] A. Kadkhodayan, X. Jiang, C. Menon, Continuous prediction of finger movements using force myography, J. Med. Biol. Eng. 36 (2016) 594–604.

[36] G.P. Sadarangani, X. Jiang, L.A. Simpson, J.J. Eng, C. Menon, Force myography for monitoring grasping in individuals with stroke with mild to moderate upper-extremity impairments: a preliminary investigation in a controlled environment, Front. Bioeng. Biotechnol. 5 (2017) 42.

[37] A. Ferrone, X. Jiang, F. Maita, M. Arquilla, L. Maiolo, C. Menon, et al., Wearable Band for Hand Gesture Recognition based on Strain Sensors, in *IEEE RAA/EMBS International Conference on Biomedical Robotics and Biomechatronics (BioRob)* 4–7 UTown, Singapore (IEEE, 2016). doi:10.1109/BIOROB.2016.7523814

[38] A. Ferrone, X. Jiang, L. Maiolo, A. Pecora, L. Colace, C. Menon, A Fabric-based Wearable Band for Hand Gesture Recognition Based on Filament Strain Sensors: a preliminary investigation, in IEEE EMBS Special Topic Conference on Healthcare Innovations & Point-of-Care Technologies, Mexico, USA (2016).

[39] H.W. Ng, X. Jiang, L.K. Merhi, C. Menon, Investigation of the feasibility of strain gages as pressure sensors for force myography, in *Lecture Notes in Computer Science (including subseries Lecture Notes in Artificial Intelligence and Lecture Notes in Bioinformatics)* 10208 LNCS, 261–270 (Springer, Cham, 2017).

[40] K. Hoffmann, An introduction to measurements using strain gages, Hottinger Baldwin Messtechnik GmbH 273 (1989). Available from: https://doi.org/10.1111/j.1475-1305.2001.tb01242.x.

[41] D. Kim, O. Hilliges, S. Izadi, A. Butler, J. Chen, I. Oikonomidis, et al., Digits: Freehand 3D interactions anywhere using a wrist-worn gloveless sensor, in *Proceedings of the 25th annual ACM symposium on User interface software and technology—UIST '12* 167 (ACM Press, 2012). doi:10.1145/2380116.2380139

[42] C. Castellini, V. Ravindra, A wearable low-cost device based upon Force-Sensing Resistors to detect single-finger forces, in *5th IEEE RAS/EMBS International Conference on Biomedical Robotics and Biomechatronics* 199–203 (IEEE, 2014). doi:10.1109/BIOROB.2014.6913776

[43] E. Morganti, L. Angelini, A. Adami, D. Lalanne, L. Lorenzelli, E. Mugellini, A smart watch with embedded sensors to recognize objects, grasps and forearm gestures, Procedia Engineering 41 (2012) 1169–1175.

[44] Dementyev, Artem; Paradiso, J.A. WristFlex: low-power gesture input with wrist-worn pressure sensors, in *Proceedings of the 27th Annual ACM Symposium on User Interface Software and Technology, UIST'14* (2014).

[45] D.P. Ferris, B.R. Schlink, Robotic devices to enhance human movement performance, Kinesiol. Rev. 6 (2017) 70–77.

[46] D. Farina, S. Amsüss, Reflections on the present and future of upper limb prostheses, Expert Rev. Med. Devices 13 (2016) 321–324.

[47] E. Biddiss, D. Beaton, T. Chau, Consumer design priorities for upper limb prosthetics, Disabil. Rehabil. Assist. Technol. 2 (2007) 346–357.

[48] C. Behrend, W. Reizner, J.A. Marchessault, W.C. Hammert, Update on advances in upper extremity prosthetics, J. Hand Surg. Am. 36 (2011) 1711–1717.

[49] M.-J. Lee, S.L. Kilbreath, K.M. Refshauge, Movement detection at the ankle following stroke is poor, Aust. J. Physiother. 51 (2005) 19–24.

[50] K.H.T. Chu, X. Jiang, C. Menon, Wearable step counting using a force myography based ankle strap, J. Rehabil. Assist. Technol. Eng. 4 (2017) 1–11.

[51] X. Jiang, H.T. Chu, Z.G. Xiao, L.-K. Merhi, & C. Menon, Ankle positions classification using force myography: An exploratory investigation, in *2016 IEEE Healthcare Innovation Point-Of-Care Technologies Conference (HI-POCT)* 29–32 (IEEE, 2016). doi:10.1109/HIC.2016.7797689

CHAPTER 8

Fiber-Based Sensors: Enabling Next-Generation Ubiquitous Textile Systems

Michael McKnight[1], Talha Agcayazi[1], Tushar Ghosh[2] and Alper Bozkurt[1]

[1]*Electrical and Computer Engineering, North Carolina State University, Raleigh, NC, United States*
[2]*College of Textiles, North Carolina State University, Raleigh, NC, United States*

8.1 Introduction

Wearable electronic systems for health monitoring can take many different forms ranging from common commercial wrist-worn systems, to more advanced wireless bandages, socks, and fitness shirts. Some major advantages of integrating sensors and actuators directly into body-worn textiles rather than as separate wearable entities are enhanced comfort, and larger sensing area [1,2]. Textiles with integrated sensing capabilities could significantly improve healthcare monitoring in remote or harsh locations, where data sensed across the surface of the entire body could help to quickly identify problems. Such embedded ubiquitous sensors could provide novel insights and correlations between healthcare parameters measured from the textile and additional body-worn environmental or health sensors. Textiles used in spacesuits, extreme cold weather clothing [3], and military uniforms [4] can be outfitted with sensing fibers to simultaneously detect events, such as blunt force impact, excessive bleeding, and abnormal heart rates. The additional spatial information provided by arrays of sensing fibers could indicate not only the occurrence of such events, but also where they occur on the body and to what extent.

In this chapter, we examine the challenges to textile integration of sensors, the requirements for integrated sensors, and potential fiber-based sensing modalities. Although we emphasize the health applications as a motivator of this article, our discussions can be extended to other applications, such as environmental monitoring and energy harvesting. We also discuss briefly some of the methods of sensing fiber production from benchtop prototyping to large-scale production, and examine how fiber-based sensors can be integrated in textiles with corresponding sensing circuitry.

Wearable Technology in Medicine and Health Care.
DOI: https://doi.org/10.1016/B978-0-12-811810-8.00008-7

8.2 Conventional Textile Wearable Integration Techniques

Textile fabrics consist of a hierarchical structure, which incorporates building blocks (most commonly fibers/yarns) of different sizes and scales through combinations of many processing steps, resulting in structures that can exhibit unique physical and mechanical properties. Fig. 8.1 shows different hierarchical levels within textiles, which may be engineered to enable sensing. The fibers are made of polymers or macromolecules. During the process of fiber manufacture, these macromolecules are arranged in a fibrillar structure consisting of crystalline and noncrystalline phases, producing a structure of high aspect ratio and with diameters (or equivalent dimension) ranging from nanometers to microns [5]. Continuous and discrete lengths of these fibers are subsequently assembled into yarns with dimensions in the range of hundreds of microns. The yarns are then combined in a fabric structure (woven, knitted, etc.) to produce textile fabrics. Needless to say, fabrics can also be manufactured directly from fibers through the nonwovens processes. Desirable properties for a given application can be enhanced or engendered in a textile fabric at any of the hierarchical levels (molecule, fiber, yarn, etc.). Most of the work done to produce electronic textiles, or e-textiles, has focused on altering the electrical properties of the bulk textile fabrics or the yarns in order to create textiles capable of sensing and signal transmission [2,6]. While some of these methods are advantageous, they have not made the necessary transition into commercial applications, primarily because they require multiple additional processing steps [7]. Arguably, the next-generation e-textiles will consist of fibers which are designed to perform sensing and actuation functions, and which can be produced using standard textile production techniques [8]. The methods described here explain the more conventional methods of e-textile integration.

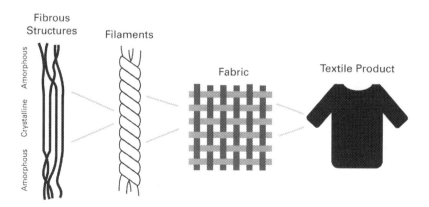

Figure 8.1
"Textile structures are enabled by hierarchical integration of components ranging in size from nanometer scale fiber structures, to macroscale textile products."

8.2.1 Rigid Component Integration

The earliest sensing textile systems consisted of rigid commercial off the shelf (COTS) components integrated into textiles using methods, such as sewing to affix the sensor and corresponding electronics within the textile [6]. In many examples of integrating rigid components, conductive fibers/yarns are used as interconnects to both sew components in place and provide an electrical connection to other rigid components. Similar sewing techniques have been used to incorporate sensors fabricated on flexible polymeric substrates into textiles [9,10]. For example, temperature and humidity sensors developed on polyimide have been sewn into woven textiles and shown to have a stable response [11]. While this form of integration functions well for prototyping and demonstrations, it does not translate well to commercial applications because it fails to address issues such as processability and comfort.

8.2.2 Printing

Printing methods, including screen and inkjet printing have been used to print necessary conductive material directly onto textile surfaces [12,13]. Because screen printing is already frequently used for textile production, it provides a potential avenue for commercial e-textile production. Printed conductors can enable flexible and conformal sensors and interconnects; however, these systems suffer from size limitations because smaller printed conductors produce higher electrical resistance. This is due, in part, to the porosity of the fabrics onto which these materials are generally printed. Stability of these printed conductive lines is also susceptible to repeated bending/stretching of the substrate fabric and long-term use may be limited [14]. Screen printed conductive pastes based on silver/silver chloride are most commonly used. These materials can be printed as both electrodes and interconnects, and can adequately bend and stretch along with the textile substrates they are printed on. Novel printable inks consisting of carbonaceous and silver nanomaterials may enable improvements in printed conductor/sensor quality. An advantage of producing printed e-textiles is that sensor sizes and geometries can be tailored for different applications. For example, subsequent layers can be printed onto the Ag/AgCl inks to enable potentiometric ion sensing [15,16] or to functionalize the electrode surface for biopotential sensing.

8.2.3 Conductive Yarns/Fibers

Highly conductive yarns also provide promise in enabling e-textiles, particularly as textile-embedded interconnects. Conductive yarns have been developed with conductivities ranging from 5 Ω/m to several kΩ/m [1]. The conductive filaments/yarns with the best electrical properties are comprised in whole or in part of metallic fibers, with stainless steel being the most commonly used material. These conductive yarns are generally not insulated, and instead must be packaged in textiles such that they are only exposed in the desired areas. These yarns have been woven to form large electrodes which can measure biopotentials [17], sense strain [18], or detect touch events [19]. Google's Project Jacquard is an example of a commercial effort aiming to produce such yarns for use in textile-embedded touch interfaces for controlling external devices.

8.3 Textile Requirements for Integrated Sensors

Many flexible, conformal sensors have been developed for textile integration; however, the planar geometry of these sensors often requires them to have a large active surface area. These sensors often have a low aspect ratio (length approximately equal to width), which is less suitable for fiber-based integration. Initial design of sensors with a geometry mimicking that of traditional textile systems, has produced promising results, indicating that multiple types of sensors can be scaled to have a high aspect ratio (length much greater than width) making them suitable candidates for fiber-based sensing. Fiber-based sensors designed to be embedded in textiles, will generally be incorporated in conjunction with conventional textile fibers. Conventional textile fibers can include natural fibers, such as cotton and wool, or synthetic fibers, such as polyesters and nylons. The mechanical properties of these fibers can vary significantly, and ideal fiber-based sensors should closely match the mechanical and physical properties of the textiles. Additionally, stretching and flexing should not cause changes in the sensor output (unless desired, as in strain sensors).

8.4 Spatial Sensing using Fiber-Based Sensors

The advantage of incorporating fiber-based sensors into a wearable system, is that these sensors can be easily routed to or integrated at different locations within the textile. Fiber-based sensors can either be independent, meaning that a single fiber can perform some sensing capability, or dependent, meaning that the fiber works in conjunction with another fiber or system to enable sensing. Of the sensing modes described, strain, biopotential, and temperature sensing generally represent independent fiber-based sensing modes, whereas pressure and wetness represent dependent sensing modes. The independent modes normally sense across the entire length of the fiber, such as in the strain sensing mode, where the resistive response of the entire length of the fiber will be used to determine strain levels. Such independent sensing fibers are advantageous because they require fewer signal transmission channels, and thus fewer interconnects, however, they are limited in their spatial sensing capabilities. For example, a fiber may only be strained along a small portion of the entire fiber length, but resistive sensing will not provide us with any information about the location of strain, or what percent of the total fiber length is being strained. This limitation should be accounted for when deciding how to integrate the fibers into a sensing fabric. Textile knitting and weaving methods could be used to enable or restrict fiber straining in certain portions of a larger sensing fabric as shown in Fig. 8.2. It is important to note that for most wearable applications of sensing fibers, it is not necessary to have extremely high-density of sensing fibers.

Additionally, array-based sensors are preferred to help increase the accuracy and redundancy of the sensed information if the information coming from multiple sensing points is averaged. To resolve data from such sensor configurations, a multiplexing circuit is most commonly used. This circuit allows for the circuitry to select each sensor individually in order to retrieve the data from the array one sensor at a time. A fast sweep through the array of all sensing points within the textile allows for the necessary sampling rate of the full array to be reached.

8.5 Fiber-Based Sensing Modalities

Multiple sensing modalities could be enabled in textile applications using fiber-based sensors. The sensing mechanisms described here exhibit methods that could be enabled by tuning conventional textile production methods. Most of the fiber sensors described here also incorporate a limited number of different materials into the fiber cross-section designs. Other more complex fiber-based sensing structures could be explored for biochemical sensing and gas sensing. Also, though fiber-based sensors are discussed here, other components that could be useful in wearable textile systems could also be built directly into fibrous structures. These include fiber-based energy harvesters, super capacitors, antennas, and batteries [8,20].

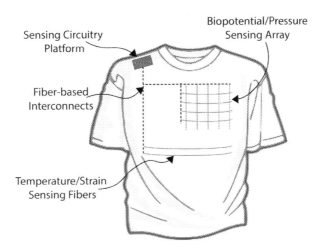

Figure 8.2
"Sensing textiles may include different component sensing fibers at different locations for highly customized healthcare sensing applications."

8.5.1 Pressure Sensing

Pressure sensing using fibers can be done in multiple configurations. If capacitive pressure sensing alone is desired from the fibers, then a conductor can be entirely encapsulated within the fiber itself, as in Fig. 8.3A. Using this configuration, the point where two fibers cross perpendicular to one another will act as the sensing location with the insulating material used to coat the fiber serving as the capacitor dielectric medium. The material properties of the insulating material, particularly its elastic modulus, will determine the amount of compression the fibers undergo when pressure is applied. As compression occurs, the proximity of the crossing conductors will be increased and the resultant capacitance will increase. Fibers for this type of sensing could also be designed such that the conducting segment of the fibers is exposed, with air acting as the dielectric [21]. If this configuration is used, the fibers may be simultaneously capable of other impedance-based sensing modalities. Highly-sensitive fiber-based capacitive pressure sensors have already been developed for textile systems [22]. These systems show promise for monitoring breathing rates, taking ballistocardiogram measurements, detecting impact, and as a user interface using touch-based input.

8.5.2 Wetness Detection

By either fully exposing portions of the fiber conductive segments or by using a porous insulating material, the presence of certain fluids within a textile could be detected via impedance measurements [23,24]. Many biological fluids of relevant interest for health monitoring contain relatively high concentrations of ions, which enables fluid detection using impedance-based methods. When an ionic fluid is present in the space where two sensing fibers crossover one another, the electrical impedance will be orders of

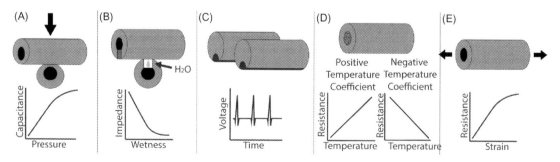

Figure 8.3

Potential embodiments of fiber-based sensors/sensing modalities using multicomponent fibers. (A) Pressure sensing; (B) wetness detection; (C) biopotential recording; (D) temperature sensing; and (E) strain sensing.

magnitude lower, compared to the impedance when no fluid is present (due to the high impedance of air). An example of wetness detection using multicomponent fibers as shown in Fig. 8.3B.

Fibers capable of detecting wetness could be used to detect the location and extent of bleeding, sweating, and urination. More specific information about the salinity of the fluid that is penetrating the sensor could potentially be extracted if the fibers are designed such that a controlled amount of fluid enters each sensing crossover point. Choice of fiber materials and geometric designs can either aid or inhibit the uptake of fluid into the fiber crossover point. For example, more hydrophilic and/or wettable fiber material surfaces will more readily carry the fluid into the crossover point. The high aspect ratio of fibers enables them to be designed to promote capillary action, which also increases fluid uptake into the sensing crossover point. Fibers designed for multimodal sensing, cannot simultaneously perform both capacitive and impedance-based sensing functions when the fluid is present between the two conductors because the fluid will render the capacitive sensing mode inaccurate.

8.5.3 Biopotential Monitoring

Biopotential recordings using textile electrodes have been a major focus of research and development. The use of conductive yarns to produce macroelectrodes capable of monitoring heart rate and generating electrocardiogram waveforms has been reported by numerous researchers. In order to obtain adequate biopotential signals, intimate skin-electrode contact is required [25]. Appropriate textile integration techniques that help bring the electrode fibers close to the skin surface can help ensure functional sensing if the fibers are to be used as dry electrodes. Dry electrodes are desired for textile sensing systems, since repeated application of electrode gel or moisturizer is not practical for long-term and repeated recording [26]. Poor electrode interfacing can lead to an increase in contact impedance at the skin-electrode interface, leading to poor signal recordings. Biopotential recordings require at least two conductive fiber segments to be in contact with the skin, in a configuration as shown in Fig. 8.3C. A major challenge to recording biopotentials using fiber-based sensors is the inherent impedance of small fibers. As fiber size is reduced to produce sensors that mimic common textile fibers, the reduced conductor cross-section leads to an increase in the conducting fiber resistivity. This challenge is exacerbated because of the need for conductors with flexible, textile-like properties. The incorporation of conductive materials with polymeric materials to produce conductive fiber segments, results in fibers that are inherently more resistive than metallic conductors. Fibers with increased resistivity are more susceptible to electrical noise, especially when attempting to record and amplify biopotential signals with amplitudes on the order of millivolts.

8.5.4 Temperature Sensing

Body temperature is a clinically relevant healthcare parameter frequently used to aid diagnosis of both short-term and chronic illnesses. Carbonaceous polymer composites can be tailored to have thermistive properties, which enables them to be used for temperature sensing applications. Temperature sensing fibers using thermistive polymers, can comprise either positive temperature coefficient materials (resistance increases due to rising temperature) or negative temperature coefficient materials (resistance decreases due to rising temperature) as shown in Fig. 8.3D [27]. Temperature sensing fibers function in the independent sensing mode, because they do not require interaction between multiple fibers. If the entire length of the fiber is comprised of the thermistive material, the length of the fiber will sense changes in temperature. However, if temperature sensing is desired at a particular location, the fiber could be functionalized with thermistive material only at that location, with other nonthermistive conductive polymers incorporated for interconnection throughout the length of the polymer. A challenge to fabric integration of temperature sensing fibers is that impedance-based sensing using independent fibers requires electrical connections at both ends of the fiber. In applications, where multiple fiber sensing types are to be incorporated into a textile fabric, dedicated temperature sensing fibers may also be incorporated to compensate the measured impedances of other sensing fibers to changes in temperature.

8.5.5 Strain Sensing

Another parameter that can be measured using fibers is strain. Wearable strain sensors can be used to monitor breathing and motion. Conductive polymers can be tuned to exhibit piezoresistive properties, such that in-plane strains along the length of the conductor induce an increase in impedance. Both integrated yarns [28,29], as well as monofilament fibers [30] incorporating carbonaceous conductors have been developed for textile integrated piezoresistive strain sensing. Methods of coating elastomeric fibers with carbon, or combining carbon with thermoelastic polymers have been used to produce the conductive portions of yarns and fibers [31]. Strain sensors for wearable applications should function for lower strain regimes, and few wearable applications require fibers capable of being strained more than 30%. Carbon-based conductors experience a degradation of mechanical strength as carbon is added, so lower levels of carbon are also generally preferred [32]. Polymer-based conductive composites containing a percolating network of carbonaceous materials, such as carbon-black and carbon nanotubes have been used to create piezoresistive strain sensing fibers. Fibers for strain sensing ideally exhibit a linear resistive response to increasing strain, are not susceptible to changes in environmental temperature and humidity, and should exhibit little hysteresis following mechanical cycling [33]. As such, it is best to encapsulate the conductive portion of the fiber using an insulating polymer for both sensor stability and mechanical support as shown in Fig. 8.3E. Much strain sensing fiber research has found sensing yarns and fibers to suffer from nonlinearities due to the geometric irregularities, and sensor degradation due to repeated washing and folding. Further research is needed using new materials and fabrication methods, to address some of these problems.

8.5.6 Fiber-Optic Sensors

Fiber-optic sensors have found some utility in textile sensing systems for sensing strain and fiber bending. Small-scale fiber-optic systems are inherently suitable for textile integration because they can have outer fiber diameters of 300 μm or less. They also have advantages over the aforementioned sensing fiber types because of their electromagnetic immunity [34,35]. Fiber-optic sensors consist of a core material and a cladding material with differing refractive indices which enable sensing based on analysis of the light that is either reflected back to the emitting end of the fiber or transmitted to the end of the fiber. There are two primary types of fiber-optic sensors, which are commonly used in textile applications. Fiber Bragg grating sensors are optical sensors, which consist of a short section of periodic alternating core segments of differing refractive indices. These differences in refractive indices cause a shift in the frequencies of the reflected light in response to changes in either temperature or pressure [34]. Because the reflected light is measured at the emitter, these fibers do not require connections at both fiber ends in a textile application. The second common type of fiber-optic sensors are intensiometric sensing fibers, which measure the intensity of light at the optical fiber outlet. These fibers function by modulating the intensity of input light, and examining output intensity. For fibers where microbending is present, some of the light will be dispersed before reaching the end of the fiber so light intensity will be diminished [34]. Intensiometric sensing also enables sensing of both pressure and temperature. Fiber-optic sensors can be used to measure important physiological parameters such as heartbeat and respiratory rate [36].

8.6 Fiber Sensor Prototypes

8.6.1 Microfabricated Fibers

Highly flexible, microfabricated sensors with a high aspect ratio have been demonstrated to mimic textile fibers [37]. These microfabricated sensors were fabricated as planar devices, each consisting of an H-shaped cross-section with the conductive components located in the middle of the cross-section. When a second set of these devices is placed on top of and perpendicular to the first set, the points where the fibers crossover one another, enable multiple modes of sensing. These microfabricated sensors have been fabricated using gold as the conductive segment, and a flexible polyimide material as the insulating component. The production cost per device of these microfabricated fiber sensors is much higher than other fiber sensors, but may be suitable for high technology applications where sensor precision and stability is of great importance. Microfabricated fibers can contain biocompatible precious metal conductors such as gold, platinum, and silver as the conducting component, thus enabling low fiber resistance and lower sensor noise susceptibility.

8.6.2 Paper-Based Fibers

Low-cost sensing alternatives are desired for some clinical healthcare settings, such as remote health clinics in rural locations, or for at home short-term monitoring. In these settings, disposable, yet readily available materials could be used to produce sensors. Materials such as paper can serve as a low-cost sensing substrate, which can be processed using inexpensive rapid prototyping techniques. Paper also can be used for chemical and biological sensing, where contact with bodily fluids requires subsequent incineration of sensor materials [38,39]. By combining printing and cutting techniques, low-cost sensors which mimic fibers have been produced for multimodal sensing. An advantage of using this method is that paper-folding techniques (such as folding/origami) can be used to quickly enable novel sensing modalities [40]. A foldable, three-layer fiber sensing patch was produced using conventional screen printing and laser cutting techniques, which was capable of pressure sensing, wetness detection, and biopotential recordings [41]. For these sensors, a silver/silver chloride paste was printed on filter paper to form the conductive portion of the sensor, with subsequent layers of a flexible insulating material screen printed on the substrate, leaving only a small segment of the sensing traces exposed. In addition to printing the conductive lines to form fiber sensors, interconnects were also printed on the paper to connect these fibers to external sensing circuitry. The printed interconnects are encapsulated by the printed insulating paste, so that they are less susceptible to interference. Because paper substrates are capable of quickly absorbing moisture, the substrate should be well insulated in all places except where sensing is desired. Interference could occur if moisture penetrates the paper, enabling potential electrical short circuits between neighboring or overlapping conductive traces. Additionally, the conductive interconnects should be well insulated to prevent electrical short circuits between stacked layers when the sensor is folded. Laser cutting and folding lines were used to enable perpendicular alignment of the paper-based sensing fibers when the sensor is folded. The location where two sensing fibers crossover was evaluated for capacitive response to application of force in the range of general human touch. This disposable sensor exhibited pressure sensing capabilities, wetness detection, and biopotential recordings.

8.6.3 Printed Fibers

Though 3D printing is not a conventional textile fabrication method, it has gained great interest in recent years for its versatility in printing objects of all shapes, sizes, and materials including conductive materials [42]. Conductive polymeric materials for 3D printing have been developed and commercialized. Both liquid and solid conductive filaments have been used to 3D print conductive fiber structures. Using liquid deposition modeling, conductive nanoparticles have been dispersed in common 3D printable plastics such as poly(lactic acid) (PLA) using high volatility solvents [43]. Liquid deposition methods might allow the integration of other functional nanofillers (i.e., graphene [44]) into printable filaments, which could enable sensing modalities beyond those afforded by conducting filaments alone. Challenges when using liquid-phase filament materials arise from the rheological properties of the material, which can be compensated for by the extrusion nozzle geometry and the shear rate at the nozzle during extrusion. These parameters must be characterized in order to tune the print flow settings to achieve the desired printed material dimensions, such that printing multiple materials simultaneously may be challenging. While solid filament materials can also be embedded with conducting particles, they are generally used to produce more rigid printed materials, making them less suitable for textile applications where soft, flexible fibers are desired. Printing of soft, flexible sensing fibers with an appropriately shaped cross-section has been demonstrated to produce multimodal sensing fibers [21]. A challenge to producing printed fiber sensors is the printable resolution of each component in multicomponent fibers, but this may be improved as new filament materials and new printing systems are developed. Using multifilament 3D printing techniques, both conductive polymers and insulating polymers can be simultaneously printed to form complex structures. Printing techniques can be used to print macroscale polymer-based textiles or fabrics, with sensing components/fibers printed simultaneously within the textile itself. A 3D printed woven mesh structure that mimics textiles comprised entirely of conducting material has been printed using liquid deposition techniques [43]. Methods for producing similar printed fabrics have also been developed using conventional inkjet printing techniques [45]. These examples of printing textile substrates could enable novel woven sensor geometries for multiple applications.

8.7 Large-Scale Fiber Production

One of the major barriers to large-scale commercial production of sensing textile fibers is the fabrication of such fibers using conventional textile fiber production methods. Polymeric fibers for textile applications are generally produced using fiber spinning methods such as melt spinning, solution spinning, and reaction spinning. Of these, melt spinning is the most promising method of fabricating multifilament sensing fibers with unique cross-sectional geometries or multiple component filaments. Solution spinning and reaction spinning may be used to produce monofilament material sensing fibers using certain polymers depending on the desired sensing application.

8.7.1 Melt Spinning

Melt spinning is most commonly used to produce polymeric fibers because it enables rapid fiber production without the use of solvents. In melt spinning, the molten thermoplastic polymer is extruded through a spinneret (shaped holes), as shown in Fig. 8.4A. Liquid-phase polymer is quenched to the solid-phase upon exiting the spinneret using air, liquid, or a combination of the two. The spinneret design contributes significantly to the size and shape of the final fiber. The fibers are drawn (or stretched) to tune the fiber dimensions, fine structure (e.g., molecular orientation, etc.) and mechanical properties. This well-established fiber production technique presents unique challenges for production of sensing fibers. Most sensing fibers will require integration of two or more materials, including at least one insulator and one conductor. Though techniques exist to simultaneously extrude multiple component materials through the spinnerets, different materials may respond differently to the extrusion, quenching and drawing processes. In melt extruded carbonaceous conducting polymers, for example, drawing speed significantly affects both the viscosity and storage modulus of the extruded material [46]. Bi-component melt extrusion is also affected by shear rates of the flowing polymers as they pass through the spinneret [47]. Conventional spinnerets are designed to extrude fibers with circular cross-sections, and exhibit differences in flow shear rates from the outer edge of the fiber, where they are highest, to the center of the fiber, where they are lowest. When multiple immiscible polymers are extruded, the differences in shear flow contribute to differences in interfacial forces between materials. Even polymeric materials with similar viscosities may

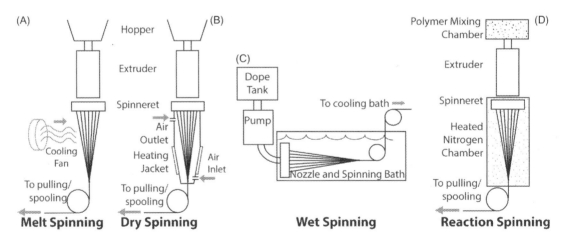

Figure 8.4

Potential methods of multicomponent fiber sensor extrusion. (A) Melt spinning uses cooling fans to solidify extruded fibers. (B) Dry spinning uses heated gases at the extrusion outlet to solidify fibers. (C) Wet spinning uses a submerged spinneret system to cool and solidify fibers. (D) Reaction spinning can be used to induce fiber forming of reactive polymer materials.

suffer from elastic layer rearrangement which causes alterations in the desired cross-sectional geometry. The effects of elastic layer rearrangement are greater for extruded cross-sections of square or teardrop shape than for circular shape, when the colaminar flow path is longer, and when more elastic materials are used [48].

8.7.2 Solution Spinning—Dry spinning/Wet spinning

Solution spinning is generally used to form fibers from polymers that do not form stable melts, but that dissolve in solvents [49]. In dry spinning (Fig. 8.4B), as the polymer solution exits the fiber spinneret, it enters a drying tower where it is exposed to hot gas (generally air) which rapidly removes nearly all of the solvent. The fiber may be further stretched or processed following solvent evaporation. In wet spinning (Fig. 8.4C), the spinneret itself is submerged in a liquid nonsolvent coagulation bath. As the polymer exits the spinneret, the miscible solvent is removed by the liquid bath, causing the skin of the filament to be solidified first. The fiber is stretched while the solvent diffuses out of the polymer making room for the nonsolvent, effecting coagulation. The production of sensing fibers using solution spinning methods presents unique challenges due to the effects of mass transfer that occur as the solvent is removed either in air or in the liquid bath. Because of this mass transfer, the fiber shaping dynamics are more difficult to control, and this process does not lend itself well to producing sensing fibers with unconventional geometries or multicomponent fibers.

8.7.3 Reaction Spinning

Reaction spinning can be used to form polymeric fibers from monomers or prepolymers, as shown in Fig. 8.4D [49]. These components are mixed in solution with additional additives before being extruded through the spinneret. As they exit the spinneret, the filaments are heated and exposed to nitrogen to initiate polymer forming reactions. This method can be used to create polymers with different reactive components providing multiple advantageous properties. Some elastomeric fibers are produced using this method, and this method could potentially be used to produce fibers with unique mechanical properties for strain sensing or biopotential recording.

8.8 Interconnects

An important challenge when making sensors with fiber and yarns is interfacing them with other components using the same ubiquitous materials that they are made from. At the systems level, a sensor that is fully wearable needs to also have wearable interconnects that route the transduced information to a transmission/processing center. On normal printed circuit boards (PCB) this is achieved using metallic pads that sensors can be soldered to. However, when the substrate changes to flexible textiles, new materials and methods are needed to both route and connect conductive textiles to components that have metallic leads.

One of the simplest methods to route circuits in textiles is embroidery. Traditionally, the embroidery process is used for decorating textile fabrics; however, in this application, its main purpose is to route a circuit using conductive fibers or yarns [50]. While this method may seem like the simplest for prototyping, it is difficult to automate due to the limitations of commercially available conductive fibers that can be used for sewing. Another challenge with embroidering is insulating the connections with other materials to avoid electrical shorts when the fabric is deformed. An alternative method to route circuits is through weaving of conductive yarns inside the textile substrate [51]. Weaving allows for conductive fibers to be integrated during fabric production in a roll-to-roll process. More complicated multilayer weave designs can also be used to make vias with the warp and weft fibers traversing from one circuit layer to another. Knitting is a method of fabric production where loops of yarns are interlaced [52]. This method can be used to incorporate conductive yarns in a textile substrate. Knitted fabrics may also be more suitable for printing circuits on textiles when intimate skin contact is needed [53]. A comparison of the advantages and disadvantages of these routing methods can be found in Table 8.1A.

Once these circuits are routed in a textile substrate, they will need to be connected to common integrated circuits with metallic leads. While some metallic yarns can withstand soldering temperatures, this kind of a connection is not flexible and is easy to break if mechanical movement is present. Mechanical gripping is a simple way to connect a fiber to a metallic lead [50,54]. Sewing the conductive fiber through a metallic connection or crimping it in a metallic connector are two examples of the mechanical gripping technique. Although this connection is simple to implement, its susceptibility to damage due to mechanical movement and humidity make it unreliable to use on textile substrates. Another way of connecting a conductive fiber to a metallic lead is by applying a conductive epoxy which secures the connection physically and reduces the effect that movement and humidity have on the electrical connection quality [55]. Conductive epoxies are made with conductive particles that provide electrical connection. A semi-flexible connection can be achieved if the conductive epoxy material is made with a flexible polymeric material. Table 8.1B offers a comparison of the commonly used textile-to-circuit interconnect methods. A need still exists for research and development of more reliable, highly conductive interconnects for interfacing sensing textile fibers with corresponding circuitry.

8.9 Challenges of Fiber-Based Sensing

Though fiber-based sensors show promise for enabling next-generation electronic textiles, there are still major hurdles to overcome for these sensors to be realized in commercial

Table 8.1: (A) Textile Routing Methods. (B) Textile-to-Circuit Interconnect Methods.

	Advantages	Disadvantages
A		
Embroidery	— Simple to prototype by hand — Can be performed on a final product — Independent of fabric formation processes — Routing can be done along the shortest path	— Difficult to automate — Needs to be sealed to prevent electrical shorting
Woven conductive yarns	— Roll-to-roll, highly automated process — Wider choice of conducting yarns — Stable and robust structure — Vias possible	— Crossover point interconnect required — May not be suitable for next-to-skin products — Needs to be sealed to prevent electrical shorting — Relatively fragile
Screen printing	— Can be performed on a final product — Independent of fabric formation processes — Stretchable and flexible — Routing can be done along the shortest path	
Knitting	— Roll to roll, highly automated process — Fully fashioned and whole garment knitting possible — More suitable for next-to-skin garments	— Limited choice of conducting yarns
B		
Soldering	— Common and easy process — Strong mechanical and electrical connection	— Connection not flexible — Requires the use of a temperature resistant metallic fiber/yarn
Mechanical gripping	— Connection is based on tension (i.e., gluing material not required)	— Humidity and mechanical movement affect the electrical connection quality — Brittle connection
Conductive epoxy	— Flexible connection if flexible epoxies are used — Robust electrical connection	

applications. The need for fiber-based sensors that produce a stable, repeatable sensor response demands precise fabrication methods and highly efficient interconnect methods. All materials must be able to withstand repeated mechanical cycling, as well as washing and drying without degradation in sensor response and stability. For improved fiber-based sensors, novel materials are needed for insulating segments, conducting segments, and sensing elements. In order to be used for healthcare applications, fiber-based sensors must be benchmarked against highly accurate and precise healthcare sensors that are currently used in clinical settings. Fiber-based sensors are more likely to find use in applications of remote health/wellness monitoring, where sensors may not provide specific diagnoses but can indicate the need for further health evaluations.

8.10 Conclusion

As a natural interface between humans and their environment, textiles offer tremendous surface area to functionalize and deploy sensors, actuators, and other devices ubiquitously and with relatively lower production costs. In this chapter, we discussed the need in e-textile-based sensors for the ability to (i) integrate various electronic functionalities into textiles in a truly unobtrusive manner, and (ii) preserve the unique and essential "textile" characteristics of fibrous structures, such as strength, flexibility, texture, softness, porosity, comfort, and stability. We presented a systematic report on sensory characteristics of a strategically designed textile structure, assembled from coextruded multicomponent fibers that are capable of generating useful electrical response under various stimuli. These fabrics utilize the unique structural and material characteristics of coextruded multicomponent fibers to create sensing element crossover points for concurrent detection of multiple physical parameters.

8.11 Conflicts of Interest

The authors acknowledge that they had no involvement in the development of fiber sensors or related sensing systems for commercial gain at the time of writing of this book chapter. The authors are engaged in publicly funded research on the topic of fiber-based sensors and were involved in the publication of certain references used throughout the chapter.

Acknowledgments

The authors would like to acknowledge the support of the North Carolina State University Chancellor's Innovation Fund, as well as the National Science Foundation (NSF) for grants ECCS-1509043, IIS-1622451, and DGE-1252376 (NSF Graduate Research Fellowship). We would also like to thank Kony Chatterjee, Ashish Kapoor, Hannah Kausche, and Jordan Tabor for providing insight during the preparation of this book chapter.

References

[1] K. Cherenack, L. Van Pieterson, Smart textiles: challenges and opportunities, J. Appl. Phys. 112 (2012).
[2] L.M. Castano, A.B. Flatau, Smart fabric sensors and e-textile technologies: a review, Smart Mater. Struct. 23 (2014) 53001.
[3] J. Rantanen, J. Impiö, T. Karinsalo, M. Malmivaara, A. Reho, M. Tasanen, et al., Smart clothing prototype for the arctic environment, Pers. Ubiquitous Comput. 6 (2002) 3–16.

[4] O. Sahin, O. Kayacan, E. Yazgan-Bulgun, Smart textiles for soldier of the future, Defence Sci. J. 55 (2005) 195–205.

[5] S. Eichhorn, J. Hearle, M. Jaffe, T. Kikutani, Handbook of Textile Fibre Structure, Volume 1-Fundamentals and Manufactured Polymer Fibres (2009). ISBN: 978-1-84569-380-0

[7] E. Devaux, V. Koncar, B. Kim, C. Campagne, C. Roux, M. Rochery, et al., Processing and characterization of conductive yarns by coating or bulk treatment for smart textile applications, Trans. of the Inst. of Measurement and Control 29 (2007) 355–376.

[6] M. Stoppa, A. Chiolerio, Wearable electronics and smart textiles: a critical review, Sensors (Switzerland). 14 (2014) 11957–11992.

[8] W. Zeng, L. Shu, Q. Li, S. Chen, F. Wang, X.M. Tao, Fiber-based wearable electronics: A review of materials, fabrication, devices, and applications, Adv. Mater. 26 (2014) 5310–5336.

[9] C. Ataman, et al., Humidity and temperature sensors on plastic foil for textile integration, Proc. Eurosensors XXV (2011) 136–139.

[10] R. Rahimi, M. Ochoa, W. Yu, B. Ziaie, Highly stretchable and sensitive unidirectional strain sensor via laser carbonization, Appl. Mat. & Int. 7 (2015) 4463–4470.

[11] G. Mattana, T. Kinkeldei, D. Leuenberger, C. Ataman, J.J. Ruan, F. Molina-Lopez, et al., Woven temperature and humidity sensors on flexible plastic substrates for e-textile applications, IEEE Sens. J. 13 (2013) 3901–3909.

[12] J. Suikkola, et al., Screen printing fabrication and characterization of stretchable electronics, Sci. Reports 6 (2016) 25784.

[13] M. Yokus, R. Foote, J. Jur, Printed stretchable interconnects for smart garments: design, fabrication, and characterization, IEEE Sens. J. 16 (2016) 7967–7976.

[14] M. de Kok, H. de Vries, K. Pacheco, G. van Heck, Failure modes of conducting yarns in electronic-textile applications, Text. Res. J. 85 (2015) 1749–1760.

[15] A. Cranny, N.R. Harris, M. Nie, J.A. Wharton, R.J.K. Wood, K.R. Stokes, Screen-printed potentiometric Ag/AgCl chloride sensors: lifetime performance and their use in soil salt measurements, Sensors Actuators, A Phys. 169 (2011) 288–294.

[16] W.J. Lan, X.U. Zou, M.M. Hamedi, J. Hu, C. Parolo, E.J. Maxwell, et al., Paper-based potentiometric ion sensing, Anal. Chem. 86 (2014) 9548–9553.

[17] M. Pacelli, G. Loriga, N. Taccini, R. Paradiso, Sensing fabrics for monitoring physiological and biomechanical variables: E-textile solutions, Proc. 3rd IEEE-EMBS Int. Summer Sch. Symp. Med. Devices Biosensors, ISSS-MDBS 2006. (2006) 1–4.

[18] H. Zhang, X. Tao, T. Yu, S. Wang, Conductive knitted fabric as large-strain gauge under high temperature, Sensors Actuators, A Phys. 126 (2006) 129–140.

[19] J.-S. Roh, Textile touch sensors for wearable and ubiquitous interfaces, Text. Res. J. 84 (2014) 739–750.

[20] H. Qu, O. Seminikhin, M. Skorobogatiy, Flexible fiber batteries for applications in smart textiles, Smart Mater, And Struct. 24 (2014) 025012.

[21] A. Kapoor, M. McKnight, K. Chatterjee, T. Agcayazi, H. Kausche, T. Ghosh, et al., Soft, flexible 3D printed fibers for capacitive tactile sensing, IEEE Sensors Conf (2016) 1535–1537.

[22] J. Lee, H. Kwon, J. Seo, S. Shin, J.H. Koo, C. Pang, et al., Conductive fiber-based ultrasensitive textile pressure sensor for wearable electronics, Adv. Mater. 27 (2015) 2433–2439.

[23] T. Pereira, P. Silva, H. Carvalho, M. Carvalho, Textile moisture sensor matrix for monitoring of disabled and bed-rest patients, EUROCON—IEEE Int. Conf. on Computer as a Tool (2011).

[24] M. McKnight, T. Agcayazi, H. Kausche, T. Ghosh, A. Bozkurt, Sensing textile seam-line for wearable multimodal physiological sensing, IEEE Int. Conf of the Engineering in Medicine and Biology Society (EMBC) (2016) 311–314.

[25] L. Vojtech, R. Bortel, M. Neruda, M. Kozak, Wearable textile electrodes for ECG measurement, Adv. in Elec. And Elect. Eng. 11 (2013) 410–414.

[26] J. Marquez, F. Seoane, E. Valimaki, K. Lindecrantz, Comparison of dry-textile electrodes for electrical bioimpedance spectroscopy measurements, J. Physics: Conference Series 224 (2010) 012140.

[27] S. Bielska, M. Sibinski, A. Lukasik, Polymer temperature sensor for textronic applications, Mater. Sci. Eng. B Solid-State Mater. Adv. Technol. 165 (2009) 50–52.

[28] H. Zhao, Y. Zhang, P.D. Bradford, Q. Zhou, Q. Jia, F.-G. Yuan, et al., Carbon nanotube yarn strain sensors, Nanotechnology 21 (2010) 305502.

[29] C.T. Huang, C.F. Tang, M.C. Lee, S.H. Chang, Parametric design of yarn-based piezoresistive sensors for smart textiles, Sensors Actuators: A Phys. 148 (2008) 10–15.

[30] M. Melnykowycz, B. Koll, D. Scharf, F. Clemens, Comparison of piezoresistive monofilament polymer sensors, Sensors (Switzerland) 14 (2014) 1278–1294.

[31] R. Alagirusamy, J. Eichhoff, T. Gries, S. Jockenhoevel, Coating of conductive yarns for electro-textile applications, J. of Textile Institute 104 (2013) 270–277.

[32] J. Hwang, J. Muth, T. Ghosh, Electrical and mechanical properties of carbon-black-filled, electrospun nanocomposite fiber webs, J. App. Poly. Sci. 104 (2007) 2410–2417.

[33] C. Yang, X. Wang, Y. Jiao, Y. Ding, Y. Zhang, Z. Wu, Linear strain sensing performance of continuous high strength carbon fibre reinforced polymer composites, Composites Part B.: Eng. 102 (2016) 86–93.

[34] C. Massaroni, P. Saccomandi, E. Schena, Medical smart textiles based on fiber optic technology: an overview, J. Funct. Biomater. 6 (2015) 204–221.

[35] X.M. Tao, Integration of fibre-optic sensors in smart textile composites: design and fabrication, J. Text. Inst. 91 (2000) 448–459.

[36] X. Yang, Z. Chen, C.S.M. Elvin, L.H.Y. Janice, S.H. Ng, J.T. Teo, et al., Textile fiber optic microbend sensor used for heartbeat and respiration monitoring, IEEE Sens. J. 15 (2015) 757–761.

[37] F. Lin, M. McKnight, J. Dieffenderfer, E. Whitmire, T. Ghosh, A. Bozkurt, Microfabricated impedance sensors for concurrent tactile, biopotential, and wetness detection, IEEE Int. Conf of the Engineering in Medicine and Biology Society (EMBC) (2014) 1312–1315.

[38] D.D. Liana, B. Raguse, J. Justin Gooding, E. Chow, Recent advances in paper-based sensors, Sensors (Switzerland) 12 (2012) 11505–11526.

[39] A.W. Martinez, S.T. Phillips, G.M. Whitesides, E. Carrilho, Diagnostics for the developing world: Microfluidic paper-based analytical devices, Anal. Chem. 82 (2010) 3–10.

[40] E.W. Nery, L.T. Kubota, Sensing approaches on paper-based devices: a review, Anal. Bioanal. Chem. 405 (2013) 7573–7595.

[41] M. McKnight, F. Lin, H. Kausche, T. Ghosh, A. Bozkurt, Towards paper based diaper sensors, IEEE Biomed. Circuits Syst. Conf (BioCAS), 2015 (2015).

[42] D. Espalin, D.W. Muse, E. MacDonald, R.B. Wicker, 3D Printing multifunctionality: Structures with electronics, Int. J. Adv. Manuf. Technol. 72 (2014) 963–978.

[43] G. Postiglione, G. Natale, G. Griffini, M. Levi, S. Turri, Conductive 3D microstructures by direct 3D printing of polymer/carbon nanotube nanocomposites via liquid deposition modeling, Compos, Part A Appl. Sci. Manuf. 76 (2015) 110–114.

[44] D. Zhang, B. Chi, B. Li, Z. Gao, Y. Du, J. Guo, et al., Fabrication of highly conductive graphene flexible circuits by 3D printing, Synth. Met. 217 (2016) 79–86.

[45] M.N. Karim, S. Afroj, M. Rigout, S.G. Yeates, C. Carr, Towards UV-curable inkjet printing of biodegradable poly (lactic acid) fabrics, J. Mater. Sci. 50 (2015) 4576–4585.

[46] C. Guo, L. Zhou, J. Lv, Effects of expandable graphite and modified ammonium polyphosphate on the flame-retardant and mechanical properties of wood flour-polypropylene composites, Polym. Polym. Compos. 21 (2013) 449–456.

[47] E. Ayad, A. Cayla, F. Rault, A. Gonthier, T. LeBlan, C. Campagne, et al., Influence of rheological and thermal properties of polymers during melt spinning on bicomponent fiber morphology, J. Mater. Eng. Perform. 25 (2016) 3296–3302.

[48] H.F. Giles Jr, J.R. Wagner Jr, E.M. Mount, Extrusion: The Definitive Processing Guide and Handbook, 2005. www.williamandrew.com\nwww.knovel.com.

[49] X. Zhang, Fundamentals of Fiber Science, 2014. https://books.google.com/books?id = -36gAgAAQBAJ& pgis = 1.

[50] T. Linz, New Interconnection Technologies for the Integration of Electronics on Textile Substrates - Torsten Linz - Fraunhofer IZM, (2005).

[51] I. Locher, Technologies for System-on-TextileIntegration, Phd Thesis, ETH. (2006) 164. doi:10.3929/ethz-a-005135763.

[52] Li Li, Wai Man Au, Kam Man Wan, Sai Ho Wan, Wai Yee Chung, Kwok Shing Wong, et al., Network model for conductive knitting stitches, Text. Res. J. 80 (2010) 935−947. Available from: https://doi.org/10.1177/0040517509349789.

[53] H. Kim, Y. Kim, B. Kim, H.J. Yoo, A wearable fabric computer by planar-fashionable circuit board technique, Proc.—2009 6th Int. Work. Wearable Implant. Body Sens. Networks, BSN 2009. (2009) 282−285. doi:10.1109/BSN.2009.51.

[54] T. Linz, C. Kallmayer, R. Aschenbrenner, H. Reichl, Embroidering electrical interconnects with conductive yarn for the integration of flexible electronic modules into fabric, Proc. Int. Symp. Wearable Comput. ISWC. 2005 (2005) 86−89. doi:10.1109/ISWC.2005.19.

[55] T. Linz, Analysis of Failure Mechanisms of Machine Embroidered Electrical Contacts and Solutions for Improved Reliability, Ghent University Faculty of Engineering and Architecture, 2011.

CHAPTER 9

WearUp: Wearable Smart Textiles for Telemedicine Intervention of Movement Disorders

Mohammadreza Abtahi[1], Nicholas P. Constant[1], Joshua V. Gyllinsky[2], Brandon Paesang[3], Susan E. D'Andrea[4], Umer Akbar[5] and Kunal Mankodiya[1]

[1]*Electrical, Computer, and Biomedical Engineering, University of Rhode Island, Kingston, RI, United States* [2]*Computer Science and Statistics, University of Rhode Island, Kingston, RI, United States* [3]*Textiles, Fashion Merchandising, and Design, University of Rhode Island, Kingston, RI, United States* [4]*Providence Veteran Affairs Medical Center, Providence, RI, United States* [5]*Department of Neurology, Rhode Island Hospital, Providence, RI, United States*

9.1 Introduction

Today, healthcare challenges are escalating at an unprecedented rate across the globe due to the increased prevalence of chronic conditions in conjunction with the increasing elderly population, and rising costs of healthcare. For example, by 2050, the population with the age of 65 year and older is projected to increase to 83.7 million, almost doubled compared to 43.1 million in 2012 [1]. In addition, as of 2012, 117 million people in the U.S.—half of nation's adult population—suffer from chronic diseases, such as heart and brain disorders, cancer, and diabetes [2]. Two major questions remain unanswered:

- How do we address the problem of healthcare for an increasing number of individuals with chronic conditions?
- How will we provide scalable quality care to these individuals remotely?

Telemedicine is thought to be one of the most transformative methods in healthcare today. American Telemedicine Association defines telemedicine/telehealth as "the use of medical information exchanged from one site to another via electronic communications to improve patients' health status" [3]. Telemedicine includes not only actual patient–physician interactions, but also education and information services. Technologies such as wearable sensors, telecommunications, cloud storage, and data analytics are vital in remote healthcare applications. In this way, telemedicine increases the awareness of diagnoses and medical conditions, treatments, and good health practices.

Wearable Technology in Medicine and Health Care.
DOI: https://doi.org/10.1016/B978-0-12-811810-8.00009-9

Especially, wearable systems that are designed for telemedicine applications in which patients are asked to wear them on a daily basis in unsupervised environments such as patients' homes and offices. The wearable devices could provide assistance with both diagnosis and treatment to a vast number of individuals with chronic and progressive disorders. There is still a significant gap between the technology and its clinical implementation. Although smart watches and fitbits are helpful in increasing general awareness of one's wellness and health, but are far from providing detailed information of activity expected in a telemedicine application. Wearable technology is not a "one-size-fits-all", rather, it demands a unique combination of human-centered design and technology matching the user's needs. One of the wearable technologies that is advancing rapidly is smart textiles. Textiles and fabrics are a part of our daily lives and, when integrated with sensors and computational ability, have the potential for improving people's lives via telehealth systems. Textiles such as clothes, gloves, socks, and hats make easy coupling with the body. This could provide an expansive opportunity to fuse together wearable design, engineering, and data science for improved telemedicine (Fig. 9.1).

We have designed WearUP that is a glove made of smart electronic textiles with an objective to practice telemedicine and to collect the diagnostic and treatment specific information of patients with Parkinson's disease (PD). WearUP consists of two flexible sensors woven into a base fabric, an embedded processor with Bluetooth Low Energy for wireless communication of the sensor data, and a power management module. The two flex sensors are woven onto the index finger and thumb. They measure the angular deformation when patients perform the finger tapping task that is commonly used in the clinic to screen PD patients. The wireless data are transmitted to a smartphone/tablet app for further analysis and interpretation. WearUP is a highly interdisciplinary research conducted among domain experts from Biomedical Engineering, Electrical Engineering, Computer Science, Biomechanics, Neurology, and Textile Design. The chapter provides a deepened

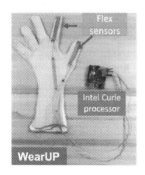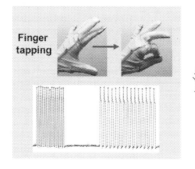

Figure 9.1
An overview of WearUP, a smart glove technology for Parkinson's disease (PD).

understanding of how smart textiles are fabricated and how they perform in controlled and natural environments. The rest of this chapter is organized as following:

- Section II discusses PD; its interventions, and existing wearable technologies for movement disorders and the advancements of smart textiles.
- Section III is dedicated to the design aspects of WearUP, including the textile design, wearable systems design, and the experimental design for validating the involved technology.
- Section IV describes the development of algorithms required to process the sensor data recorded from WearUP.
- Section V focuses on the experimental results involving the WearUP glove to quantify the repeatability and accuracy of finger tapping in time and in amplitude. Our pilot study involves nine healthy human participants. We have also used a robotic hand to calculate the standard deviation of error when the known movement was measured by WearUP.
- Section VI provides our insights about smart textiles, lessons learned from this research and our vision for the future of this technology.

9.2 Background and Related Work

9.2.1 Parkinson's Disease and its Assessment

PD is one of the most prevalent hypokinetic movement disorders and is the second most common neurodegenerative disorder [4]. PD currently affects 4 million people worldwide and over 9 million PD cases are projected by 2030 [5]. PD incurs $23 billion in annual medical costs in the U.S. alone and increasing as our population ages [6], thus there is an urgent need to improve the lives of those afflicted with PD. The characteristic motor features of PD are slowness of movement (bradykinesia), stiffness (rigidity), tremor at rest, and walking instability [7,8].

Patients with PD make periodic visits to clinics every 3–6 months for routine management and medication adjustments. In the clinic, physicians conduct motor screening that involves various upper- and lower-body exercises, enabling doctors to estimate disease severity and medication effects. The Unified Parkinson's Disease Rating Scale (UPDRS) [9] is designed to score patients from 0 to 4 (0 for no symptoms and 4 for severe symptoms) based on finger tapping performance and hand opening–closing tasks. UPDRS scoring allows physicians compare task performance to previous assessments and evaluate disease progression and medication effects, and make necessary adjustments. The evaluation is based on a short 15–20 minutes visit during which the clinician witnesses only a small part of the patient's daily function.

The problem arises when PD patients return to their natural environment wherein physicians are uninformed about the conditions of their patients. For example, physicians cannot monitor how the patients respond to prescribed medications because patient reporting can often times be unreliable, especially if patients suffer from dementia, a comorbidity often associated with PD. Poor reporting can make medication adjustments difficult for clinicians and this can lead to overdosing, which could cause potential harm to the patients with various side effects. At the same time, the under-dosing may have no or little effect on reducing the symptoms. Physicians need to rely on patients' next clinical visit to assess the effect of medication. Physicians have to undertake this "trial-and-error" approach to treat their patients. Medication efficacy (or effectiveness) monitoring is reported to be one huge challenge for physicians [10–13] since it is difficult to optimize the medication dosage for each patient without performing quantified assessments. Another issue with PD management is medication adherence (or compliance) to the correct dosage and at the correct intervals [14–16].

9.2.2 A Brief Survey on Existing Wearable Technologies for Parkinson's Disease

Given their potential for improving healthcare, wearable devices have been proposed and used in the treatment of PD. As listed in Table 9.1, there is a large body of research works dedicated to wearable technology and its applications in PD or other movement disorders. The research works can be divided into two categories: (1) wearable systems design that aims to make the sensing approach more specific to motor symptoms, and (2) algorithms that use off-the-shelf wearable systems to improve the detection, quantification, and classification of the symptoms. A growing number of commercial products are emerging in the market to fill the gap between the clinical practice and telemedicine.

The research work presented in this chapter explores both aspects: (1) the design and development of wearable system made of smart textiles (more details in Section III) and (2) the development and evaluation of algorithms that could detect and classify motor behaviors in the upper limb extremities (more details in Sections IV and V).

9.2.3 Smart Textiles: State of the Art

Nowadays, the research and developments on smart textile is growing at a rapid rate. A flexible and sensitive textile-based pressure sensor has been developed [30] by using highly conductive fibers coated with dielectric rubber materials. Compared to the previous textile-based pressure sensors, this new developed pressure sensor provides higher sensitivity, fast response time, and high stability. The pressure sensor can be used to make the smart gloves or clothes, by using a custom weaving method.

Nishiyama et al. [31] developed a wearable sensing glove with embedded hetero-core fiber-optic nerve sensors. The glove can detect finger flexion to achieve unconstrained hand motion monitoring. Due to the capability of optical intensity-based measurements, the hetero-core fiber sensor has been suited to the wearable sensing glove.

Table 9.1: A list of selected publications presenting wearable technologies for Parkinson's disease (PD).

Domains	Devices/Algorithms	Descriptions
Research and developments	Tremor prediction algorithms [17–19]	Prediction algorithms were developed to detect ON-OFF periods of either medication or deep brain stimulation (DBS) in patients with PD or essential tremors. Researchers achieved a high accuracy for predicting tremors in PD patients.
	Mercury [20]	A sensor node called "Throttle Gyro" was developed to provide automated activity detection at low-energy consumption.
	eGaIT—Smart Shoes [21,22]	eGaIT is an automated gait (walking) analysis system, consisting of a shoes pair with embedded inertia sensors. The shoes pair helps analyze and classify abnormal strides during UPDRS gait task.
	Various other Smart Gloves [23–26]	A number of smart gloves have been proposed and tested to monitor dyskinesia or tremors in PD. Some of the projects have adopted flex sensors to measure tremors on individual fingers.
	SPARK, a smart watch-based system [27]	Previously, our research group developed a smart watch-driven solution to quantify the motor symptoms of PD. We studied this approach on 24 patients with PD in a clinical focus group study. Although it produced a higher accuracy, it was not reliable for fine-motor symptoms which are generally observed in the finger tapping task.
Products	Kinesia Products [28] (Great Lakes NeuroTechnologies Inc.)	Kinesia One is a product consisting of a ring sensor that streams the finger motion data to a smartphone. Kinesia 360 is a multisensor system for patients to wear at home. Physicians have access to the tremor information for making informed decision about medication.
	PKG Data Logger [29] (Global Kinetics Co.)	The logger looks like a wristwatch and collects movement data remotely from patients with movement disorders to assist doctors with the diagnosis and treatment.

E-textiles have also earned new respect in the last decade through a series of advancements in conductive threads [32], embroidered electronics [33], knitted antennas [34], electric screen-printing [35] and conductive paints [36]. E-textile refers to a broad range of research activities to integrate advanced functionalities such as sensing, actuation, computing, energy harvesting, communication, and human—computer interaction into the textile fabrics [37–42]. Based on the functionality, e-textiles can be classified into two subgroups: (1) Passive e-textiles, which have only sensing capabilities for users or environment conditions. For example, passive (battery-less) UHF RFID (radio frequency ID) tags are woven into regular clothing for indoor human motion tracking [43] and fetal monitoring [44]. (2) Active e-textiles, which have both sensing and actuation capabilities with a power source on board. This chapter targets active textiles that are more complex and demand sophisticated integrative approaches for design, fabrication and application.

Active e-textiles offer varieties of complex functionalities such as smart dancing shoes [45] that can perform digital drawings in real-time and Smart Gloves [46] that can work as a multimeter for industrial workers. Electronic components can be knitted or woven in a double-face fabric so that the outer part of the electronics does not contain any conductive yarn while insulating the electronics [35,47].

9.3 WearUP Design

In this section, we will explore WearUP's design from the standpoint of textile, hardware, and initial validation. Specifically in the case of gloves, finger tapping is a standard and well-established method for assessing the state of disease progression or treatment in patients with PD. The movement requires fine motor control, the ability to keep track of time and tempo, and the ability to clasp and tap [48−53]. While this test is often performed in front of the clinician and interpreted only through visual inspection, various data gloves have been proposed. Moreover, some of these systems have been shown to perform better than through eyesight alone, while being able to record a precise model of the hand. WearUp is a smart glove that is developed to record and detect patient's hand movements. The intent is to have the patients wear the glove in their natural environment and send the data to the physicians in order to provide a more frequent monitoring.

9.3.1 WearUP: Textile Design

9.3.1.1 Layout considerations

The previous design of our smart glove (Fig. 9.2) consisted of two flex sensors stitched onto the cotton over the phalanges of the thumb and index fingers and a microcontroller

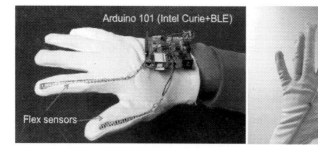

Figure 9.2
The evolution of wearable smart gloves in our own research works: (1) the old version of smart glove in the left [54] and (2) the improved design of the glove, WearUP in the right.

attached over the carpus [54]. All electronic components were attached to the right-handed glove on the dorsum of the hand and fastened with either cross-stitches or loops through holes in the printed circuit boards (PCBs). The cotton glove extended only to the wrist and included space for onboard battery power and recharging circuit. The components were cross-stitch sewn together loosely for easy disassembly. Three main fabric pieces were used, accounting for the top, bottom, and thumb. Although slight sliding of the sensor may be possible during the flexion phase of hand clasping, this was observed to be minimal. Once the appropriate placement of sensors and components were determined, we improved stitching to minimize potential sliding.

Redesigning the glove was necessary for more effective integration of the sensors. This addressed issues such as the stability of the wired sensor connector over the metacarpophalangeal joint (MCP) and the body of the flex sensors over the proximal interphalangeal joint (PIP) to the distal interphalangeal joint (DIP). To accomplish this, a pattern design for a well-fitted glove was drawn based on a healthy volunteer's left hand with fingers spread. Note that the design can also be used for the creation of a right-handed glove, with mirroring modifications.

9.3.1.2 Fabric considerations

In order to make the smart glove, the material needs to be form-fitting but also elastic, so it stays on the hand without slipping, during the 15−20 minutes time to complete a telemedicine screening protocol. After empirical review of several different fabrics, a thin polyester spandex blend was selected for the glove foundation as it performed well on the hand and was also washable. Rather than integrating the sensors directly into this fabric layer, sensors were fastened between two layers of fabric with the top layer being a thin mesh (∼1 mm holes) with matching elasticity. This setup protected the sensors while keeping them visible and accessible for maintenance inside the glove. For convenience and backward compatibility, electronic components remained in securely the same locations (Fig. 9.3).

9.3.1.3 Assembly

The design lines were trued and a one-inch seam allowance was added to increase error tolerance for the stitching line. The pattern was treated as a "cut 2" for the fashion fabric and a "cut 1" was used for mesh material for only one hand. Since the fabric stretch was the same in all four directions in both fabrics, the grainline was put through the center of the glove's hem where the patient would insert the hand.

To increase error tolerance for the stitching line when attaching the bottom layer of the glove, the glove's mesh cutout was affixed to the fashion fabric of the top of the glove by stitching all the way on the outside, increasing tolerance for error when stitching the fingers. A regular 2.5 mm stitch was used to attach the bottom layer while backstitching at

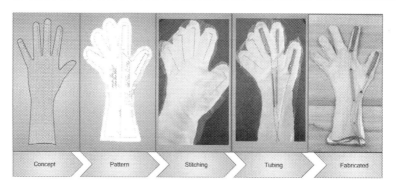

Figure 9.3
WearUP textile design steps.

the base where the fingers meet. This finishing touch provided decreased likelihood of tearing when putting on or taking off the glove.

Before attaching the bottom layer to the top layer, an L stitch was made on the top layer where the fingers are the length of the sensor, providing a stable place to press the sensor against within the mesh polyester layers so it stayed on the finger in the appropriate place. The L stitch was color coded to aid in later visual inspection during experiments with patients. Each flex sensor was encased in a sleeve made of the mesh material using an overlocked stitch with the connecting leads protruding through the mesh at the base.

After slipping the encased sensors against the L stitch on the fingers, a basting stitch along the overlocked seam of the encasement of the sensor was used to keep it stable and in place while allowing it to be easily removed if necessary. A button hole was added at the bottom of the sensors on the mesh fabric to more easily remove the sensors if necessary at a later time. A long thin, flexible wire was soldered to the flexor leads and heat shrink was used to ensure insulation. The other end of these wires were connected to the microcontroller as described previously.

The special tubing was affixed to the mesh such that when the sensor's wire was threaded through the tube, the resulting movement of the wire caused by the bending of the corresponding finger could easily glide without impediment from the knuckle. The tubing also protected the wire from unwanted crimping. This sheath design was inspired by the carpal tunnel of the hand. A matching color scheme for each finger was used, such that the corresponding sets of wires, tubes, and L stitches were more identifiable from a color theme. Fig. 9.4 shows the stages of the WearUp development.

9.3.2 WearUP: Wearable Sensing & Computing System Design

Sensing, communicating, and processing in real-time requires orchestrating information between the glove and an mobile device (e.g., smartphone or tablet) that is less constrained and has more processing capacity.

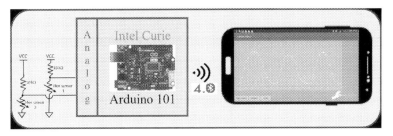

Figure 9.4
Components used in the smart glove and the voltage divider circuit schematic.

9.3.2.1 Flex sensors

Assessment of hand movements requires the device to sense and compute the data reliably and accurately. To assess the hand's position and movement requires relative spatial orientation of the fingers and palm. WearUP focuses on the use of single angle flex sensors. The flex sensors being used are in essence variable resistors, which use conductive ink with an electrical characteristic that causes them to increase in resistance as their surface area increases. The flex sensors we have used are Spectra Symbol flex sensor that has a thickness of 6.35 mm, with 84.86% of the part length designated as active length. These are placed between two flexible sensors, but conductive traces housed in a clear insulative material. The resistance when flat is roughly 20 kΩ and when bent to a 90 degree angle increases to roughly 70 kΩ.

Measuring this physical state requires a static state for relativity. We accomplish this static state by introducing a standard resistor. To measure the relative change in state between the standard resistor and flex sensor, we design a simple voltage divider. This allows us to extrapolate the desired information using Ohm's Law. One last challenge on this setup is maximizing the voltage range to increase the signal to noise ratio. We chose to use a standard 10 kΩ resistor, which provides a potential voltage swing of approximately 1 V. Later, we mapped this range from 0 to 90 as the flex sensors resistances reaches its maximum at a 90 degree angle.

9.3.2.2 An Intel Curie-based, low-power embedded system

The microcontroller unit (MCU) must be small enough to fit on the back of the hand, consume minimal power such that the battery life of the glove extends at least beyond one practice session, and have the processing power to compute the simple mapping of the analog to angular values at close to real-time speeds. Currently, the MCU being used, Arduino 101, meets these requirements. The Arduino 101 is a learning and development board which contains an Intel Curie Module. This board is designed to integrate the core's low power-consumption and high performance with the Arduino's ease-of-use. The Arduino

101 also provides 19 channel 12-bit ADCs, Bluetooth Low Energy capabilities, and power management circuitry to ensure stability in reference to analog readings.

9.3.2.3 Companion Android app

All the sensors are connected to the MCU, which collects the data from the flex sensors. In order to provide the movement freedom for the subject wearing the WearUp glove, we have designed an application that can collect the data wirelessly. The data from the Arduino board is sent to the WearUp application on the smartphone. WearUp application is specifically designed to run on any Bluetooth Smart or Bluetooth Smart Ready-enabled device running Android 4.4 or higher. This Android device utilizes the Bluetooth Low Energy (BLE) technology available on both the Arduino 101 and the device itself to receive updates every 100 milliseconds from the glove. This update comes in the form of a byte array which contains all the essential information from the WearUp Glove including raw sensor data from the Flex Sensors. The Android device then parses this data and saves all of it into a comma separated values (CSV) file which can be used for further processing.

For testing purposes, a simplified wired setup was also built. Serial output over a USB connection was relayed at 9600 baud with 8-bits and no parity. This provided raw sensor values at incremental steps sampled based on the CLK, separated by a tab character. Each line ended in a line-feed to ensure proper handling of the incoming data stream.

9.3.3 WearUP: Experimental Design

This section presents the procedure of the preliminary experiment to evaluate the performance of the WearUp glove (Fig. 9.5).

9.3.3.1 Robotic hand platform for testing the WearUP Glove

In the first set of the experiment, we tested the performance of the glove with a standard, specific measurement and removing the variability of participants. To achieve this goal, an equivalent right-handed WearUP glove was created and was placed on a 3D printed robotized robotic right-hand based on the InMoov design. The robotic hand was set to perform finger tapping task at two different speeds. For fast finger tapping, the goal was to finish a cycle of finger tapping in 1.5 seconds. For slow finger tapping, the goal was to finish a cycle of finger tapping in 3 seconds. Since the robotic hand was set for a specific task with a fixed angle of bending the fingers, we anticipated that the results of the robotic finger tapping experiment should not have much variation. Therefore, the main goal of this experiment was to validate the performance of the WearUp glove.

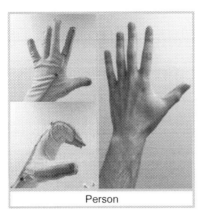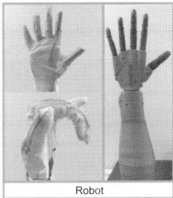

Figure 9.5
Experimental WearUp glove setup; WearUp glove worn by robotic hand (right) and person (left). The above demonstrates the range of motion available to the test robot versus human participants.

Figure 9.6
Preliminary experiment protocol shown in a block diagram.

9.3.3.2 A pilot study on human participants

In the second set of experiment, we recruited nine healthy human participants, after receiving IRB approval (#1020863-1). As shown in Fig. 9.6, the participants were asked to perform the finger tapping task that involves tapping the pointer finger to the thumb for 15 times at a relatively fast speed. The finger tapping was followed by resting their hands in the fully opened position for 5 seconds. The participants were then asked to perform the finger tapping for 15 times at a slower speed. The data was collected from each participant for three rounds of the experiment. The participants were instructed to perform the task with their own determination of speed to evaluate the performance of the WearUp glove in detecting variations in finger tapping velocity.

9.4 WearUP: Algorithms and Methods

9.4.1 Algorithm

This section presents the analysis of algorithms that have been developed to evaluate the performance of the WearUP glove in the preliminary experiment. We used WearUp Glove on both robotic hand and healthy human participants. The sensors used in the WearUp glove are flexible resistive sensors which change their resistance when any bending or angular deformation is applied to them. In other words, by placing the sensors on the fingers, and bending the fingers, an increase in the amount of resistance of the sensors can be detected.

The protocol of the experiment included the task of finger tapping, which required the participants to bend their index finger to complete the task. Therefore, we sought to detect peaks in the collected data during finger tapping.

Due to the micro-movements of the hand and fingers, and also due to the environmental noises, the raw data collected by the sensors is noisy. To reduce the noise from the data, a 4th order lowpass Butterworth filter was applied. The Butterworth filter, also called maximally flat magnitude filter, is designed to have a flat frequency response in the passband. The 4th order Butterworth filter has the roll off of -24 dB per octave in the stopband frequencies. Fig. 9.7 shows a sample of the raw data, filtered data and depicts the differences regarding the smoothness of data.

9.4.2 Finger Tapping Detection

Since the resistance of the flex sensors increases due to finger tapping movements, we observed periodic peaks in the data. We developed a peak detection algorithm capable of detecting the peak angles in finger tapping. A threshold was set such that if the resistance of the sensor exceeds this value, a potential peak related to the finger tapping is noted. The resistance of the flex sensor is equal to 20 kΩ in the straight (nonbending) position and above 70 kΩ at the bending of 90 degrees. Therefore, we have chosen the threshold to be 20% bigger than the straight value (1.2 $\times$ 20 k) to enhance the sensitivity of peak detection. Due to the small changes in the sensor value which are not related to the finger tapping, we also set a temporal threshold. This means that if the amplitude of the peaks are very close to each other, they are most likely not two different peaks of interest, and represent an "interruption" or freezing of movement which is seen and rated on the clinical exam, UPDRS. Therefore, we have chosen the temporal threshold to be 0.2 seconds. Setting this threshold is reasonable in the sense that healthy humans cannot volitionally tap the fingers more than 5 times in a second. Thus, our algorithms avoid missing any peaks related to the actual finger tapping.

9.4.3 Tapping Rate/Frequency

The velocity and amplitude of the finger tapping task are important in the assessment of PD patients. This information refers to the symptom severity and treatment efficacy. In order to provide this information from the WearUP glove, it is important to find the frequency of

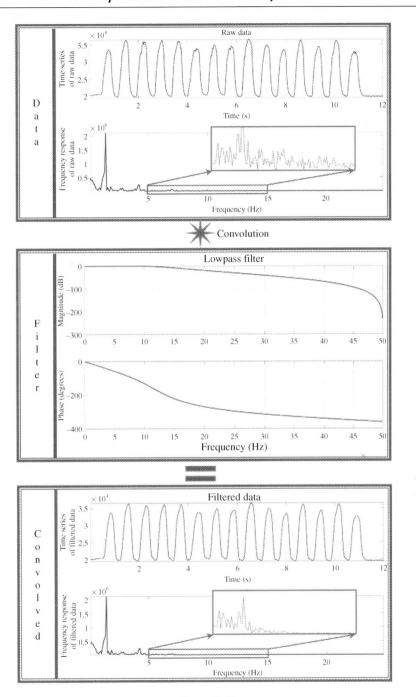

Figure 9.7
Illustrations of the raw data from the glove (top panel), Lowpass filter characteristics (middle panel), and the filtered data (bottom panel).

finger tapping (time interval between each finger tap). To achieve this, we tested the algorithm for detecting the peaks. Due to the characteristics of the sensors and the protocol of the experiment, we anticipate to see peaks in the data at the time of finger tapping. By using the algorithm, we are able to detect the peaks, and by finding the number of samples between two nearby peaks, we can find the time in seconds that has taken for the subject to tap their thumb. Knowing the time between each finger tap, allow us to calculate the instantaneous frequency of finger tapping.

9.5 Experimental Results and Discussions

This section presents the results of preliminary analysis of the data collected when using the robotic hand and from healthy participants. Fig. 9.8 shows the results of the peak detection algorithm on the data from the robotic hand. The red markers are the peaks detected by the algorithm. Since we are operating a robotic machine with defined angles and flexions, we observed that the amplitude (see Fig. 9.8) and frequency (see Fig. 9.10) of the finger tapping activity were almost identical at each consecutive tap.

Fig. 9.9 shows the results of the algorithm applied to the data recorded from a healthy participant, with the detected peaks marked in red. It is apparent that each finger tap is represented with one peak, and the variation of human movements can be seen by the magnitude of the peaks and the width (duration) of each finger tap.

After applying the peak detection algorithm, and finding the location of the peaks, we calculated the frequency of finger tapping by finding the number of samples between each two adjacent peaks and dividing the sampling rate with that difference as follow:

$$f = \frac{f_s}{N}$$

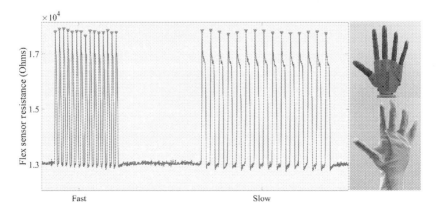

Figure 9.8

Results of the peak detection algorithm on the data from the robotic hand.

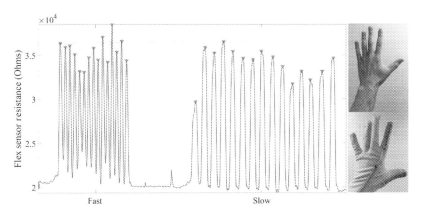

Figure 9.9
Results of the peak detection algorithm on the data from a healthy participant.

where f is the frequency in Hz, f_s is the sampling rate, and N is the number of samples between two adjacent peaks:

$$N = peak(i) - peak(i-1)$$

Consistency of the performance can be revealed by the results of peak detection and frequency calculation. Looking at the resistance value of the sensor at the peaks (the magnitude) reveals whether or not the subject has performed the finger tapping task in a similar way each time or if the position and the angle of the finger was different at each single finger tap. As an example, we can see that the first finger tap of the slow speed has a significantly lower peak compared to the other finger taps. On the other hand, by looking at the frequency of finger tapping for fast or slow speed task, we can understand how consistently the subject pursued the task. A good measure of this consistency is the mean and variance of the finger tapping frequency. The mean of the frequencies reveals how fast or slow the subject has been performing the finger tapping task, and the variance reveals the magnitude of fluctuation in frequencies. The smaller variance means more similarities of the frequencies or more consistency in the speed of finger tapping.

Table 9.2 shows the mean and variance (σ^2) of the finger tapping for the three rounds. Due to the fact that the robotized hand is operated by a computer with specific parameters, we should observe a very small variation in the frequencies compared to the healthy subjects that show higher error because of hand micro-movements or human error.

Fig. 9.10 shows the boxplot of the frequencies of the fast finger tapping from five participants and the robotic hand. All of the participants have been very consistent in performing the task. On the other hand, the frequency of finger tapping by the robotic hand is consistent and there is no variation in performing the task, thus these results provide the validation of the WearUp glove's hardware design, textile design, and algorithmic model.

Table 9.2: Mean and variance of finger tapping frequency. Data collected in 3 rounds from 5 healthy participants.

		Subject 1		Subject 2		Subject 3		Subject 4		Subject 5		Robot	
		Mean	σ^2	Mean	σ^2	Mean	σ^2	Mean	σ^2	Mean	σ^2	Mean	σ^2 ****
Round 1	Fast	3.76	0.052	2.78	0.149	2.90	0.029	2.81	0.049	3.01	0.022	0.654	6×10^{-5}
	Slow	1.54	0.022	1.59	0.027	0.94	0.013	1.31	0.008	1.39	0.001	0.327	1.9×10^{-6}
Round 2	Fast	3.05	0.018	2.49	0.034	2.58	0.010	3.12	0.042	2.62	0.009	0.653	1.8×10^{-5}
	Slow	1.49	0.025	1.39	0.012	1.00	0.003	1.36	0.008	1.10	0.002	0.327	1.2×10^{-6}
Round 3	Fast	3.04	0.054	2.42	0.087	2.20	0.007	2.74	0.017	3.20	0.028	0.654	2.9×10^{-5}
	Slow	1.58	0.014	1.51	0.010	1.01	0.004	1.34	0.009	1.29	0.009	0.327	1.2×10^{-6}

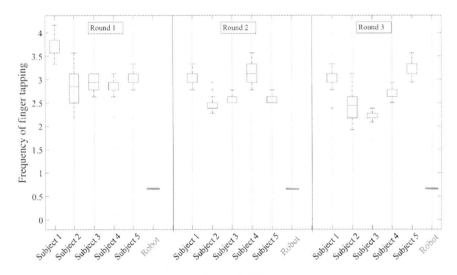

Figure 9.10

Fast finger tapping frequencies of five subjects and the robotic hand for three different rounds. It is clear that the finger tapping with robotic hand is very precise without any variations.

Table 9.2 shows that we can understand the average speed of the finger tapping task for each individual participant, and understand how much variability each participant has in performing the task. The average speed of finger tapping is observed from the average of finger tapping frequency provided in Table 9.2. As an example, Subject 1 has performed the fast finger tapping task at a higher speed in the first round (average frequency of 3.76 Hz) compared to the second and third round which the average frequency is 3.05 Hz and 3.04 Hz, respectively. Regarding the variability of the participants, the variance (σ^2) provides the information. As an example, Subject 2 in the fast finger tapping of round 1 was not very consistent, and had variability in performing the task (σ^2 is 0.149 which is slightly high). On the other hand, a good example of a consistent with small variability is Subject 5 in the slow finger tapping of round 1, which the variance is 0.001. The robotized hand, as discussed, should have a steady speed in performing the task. Table 9.2 shows that the average frequency of finger tapping is consistent for different rounds as 0.654 and 0.327 for fast and slow finger tapping respectively. Also, the variation of the task is negligible in the range of 10^{-5} for robotized hand.

The preliminary results of this study are promising but need further validation. The system was able to consistently detect finger taps and calculate the frequency of movements. The amplitude of these movements can also be assessed with this technology, but future work will evaluate the additional metrics.

9.6 Conclusions and Future Works

Today, neurologists face challenges of screening a growing number of patients with PD. The screening tests require patients to make frequent visits to the clinics. In response to this challenge, WearUP is aimed at offering a remote assessment of PD and its progress. Although UPDRS screening is complex, physicians still can estimate the severity of the disease from the symptoms present in upper extremity. Therefore, WearUP, a smart glove is designed such that it is possible to quantify subtle motor symptoms derived from a subset of UPDRS tests that involves repeated movements of hands and fingers. This chapter provides the design of WearUP that was tested on healthy humans and a robotic arm to measure sensitivity and measurement ranges.

The chapter provides an in-depth understanding of the most challenging part that was the design of the glove because WearUP was designed from raw textile materials from scratch without the use of off-the-shelf gloves. The process demanded many design iterations. Particularly, One difficult design issue involved the identification of a stable stitching line which kept the sensors in the appropriate places on the finger joints while fitting well on the hand during movement. The material selection was an iterative process that we expect to explore further in our future work. For every-day use, breathability (that refers to an air-flow within the fabric) will need to be improved as it might be an issue among some participants. Future generations of the glove need to improve the design to produce a robust, washable smart textile device which can be patient-friendly.

The first version of WearUP was tested on nine healthy subjects who performed three rounds of finger tapping task at two different tapping speed. We developed a peak detection algorithm based on amplitude and temporal thresholds which differentiates the finger tapping strength and frequency. Our experiments on healthy subjects demonstrated the ability of WearUP to detect the subtle intrapersonal variation in finger tapping task. In addition, we also tested the accuracy, precision, and repeatability of WearUP on robotized hand which was programmed for finger tapping task. The error rate is negligible in the range of 10^{-5} for robotized hand. The results of our experiment are promising and motivate us to proceed for the next steps involving clinical testing of WearUP on patients with PD. In the future, we will compare the ability of WearUP to produce a clinical-grade scoring. More advanced algorithms will be explored and validated in future trials on participants with PD.

Acknowledgment

We are very thankful to the study participants. This research work was supported by NSF research grants (#1565962 and #1652538). Any opinions, findings, and conclusions or recommendations expressed in this material are those of the author(s) and do not necessarily reflect the views of the National Science Foundation. We appreciate Andrew Peltier's help with the Android app development.

References

[1] J.M. Ortman, V.A. Velkoff, H. Hogan, An Aging Nation: The Older Population in the United States, US Census Bureau, Washington, DC, 2014, pp. 25−1140.

[2] B.W. Ward, J.S. Schiller, R.A. Goodman, Peer reviewed: Multiple chronic conditions among us adults: a 2012 update, Prev. Chronic Dis. 11 (2014).

[3] NTT Data White Paper: "Trends in Telehealth" published in 2014.

[4] J. Jankovic, E. Tolosa (Eds.), Parkinson's Disease and Movement Disorders, Lippincott Williams & Wilkins, Philadelphia, PA, 2007.

[5] E. Dorsey, R. Constantinescu, J.P. Thompson, K.M. Biglan, R.G. Holloway, K. Kieburtz, et al., Projected number of people with Parkinson disease in the most populous nations, 2005 through 2030, Neurology 68 (5) (2007) 384−386.

[6] D. Weintraub, C.L. Comella, S. Horn, Parkinson's disease—part 1: pathophysiology, symptoms, burden, diagnosis, and assessment, Am J Manag Care 14 (2 Suppl.) (2008) S40−S48.

[7] D.B. Calne, B.J. Snow, C. Lee, Criteria for diagnosing Parkinson's disease, Ann. Neurol. 32 (S1) (1992) S125−S127.

[8] A.J. Hughes, S.E. Daniel, L. Kilford, A.J. Lees, Accuracy of clinical diagnosis of idiopathic Parkinson's disease: a clinico-pathological study of 100 cases, J. Neurol. Neurosurg. Psychiatr. 55 (3) (1992) 181−184.

[9] Christopher G. Goetz, et al., Movement Disorder Society-sponsored revision of the Unified Parkinson's Disease Rating Scale (MDS-UPDRS): Scale presentation and clinimetric testing results, Mov. Disord. 23 (15) (2008) 2129−2170.

[10] L. Elmer, M. Asgharnejad, B. Boroojerdi, F. Grieger, L. Bauer, 24-hour efficacy profile of rotigotine in patients with advanced Parkinson's disease: a post-hoc Analysis (P6. 057), Neurology 84 (14 Suppl.) (2015) P6−057.

[11] T.A. Finseth, J.L. Hedeman, R.P. Brown, K.I. Johnson, M.S. Binder, B.M. Kluger, Self-reported efficacy of cannabis and other complementary medicine modalities by Parkinson's disease patients in Colorado, Evid.-Based Complement. Altern. Med. 2015 (2015).

[12] D.B. Miller, J.P. O'Callaghan, Biomarkers of Parkinson's disease: present and future, Metabolism 64 (3) (2015) S40−S46.

[13] L. Collins, G. Cummins, R.A. Barker, Parkinson's disease: diagnosis and current management, Prescriber 26 (5) (2015) 16−23.

[14] N. Malek, D.G. Grosset, Medication adherence in patients with Parkinson's disease, CNS Drugs 29 (1) (2015) 47−53.

[15] J.Y. Shin, B. Habermann, I. Pretzer-Aboff, Challenges and strategies of medication adherence in Parkinson's disease: a qualitative study, Geriatr. Nurs. 36 (3) (2015) 192−196.

[16] A. Puschmann, L. Brighina, K. Markopoulou, J. Aasly, S.J. Chung, R. Frigerio, et al., Clinically meaningful parameters of progression and long-term outcome of Parkinson disease: an international consensus statement, Parkinsonism Relat. Disord. 21 (7) (2015) 675−682.

[17] I. Basu, D. Graupe, D. Tuninetti, P. Shukla, K.V. Slavin, L.V. Metman, et al., Pathological tremor prediction using surface electromyogram and acceleration: potential use in 'ON−OFF' demand driven deep brain stimulator design, J. Neural Eng. 10 (3) (2013) 036019.

[18] S. Patel, H. Park, P. Bonato, L. Chan, M. Rodgers, A review of wearable sensors and systems with application in rehabilitation, J. Neuroeng. Rehabil. 9 (1) (2012) 21.

[19] C. Ossig, A. Antonini, C. Buhmann, J. Classen, I. Csoti, B. Falkenburger, et al., Wearable sensor-based objective assessment of motor symptoms in Parkinson's disease, J. Neural Transm. 123 (1) (2016) 57−64.

[20] K. Lorincz, B.R. Chen, G.W. Challen, A.R. Chowdhury, S. Patel, P. Bonato, et al., Mercury: a wearable sensor network platform for high-fidelity motion analysis, SenSys'09, 2009, pp. 183−196.

[21] J. Klucken, J. Barth, P. Kugler, J. Schlachetzki, T. Henze, F. Marxreiter, et al., Unbiased and mobile gait analysis detects motor impairment in Parkinson's disease, PLoS One 8 (2) (2013) e56956.

[22] B.M. Eskofier, M. Kraus, J.T. Worobets, D.J. Stefanyshyn, B.M. Nigg, Pattern classification of kinematic and kinetic running data to distinguish gender, shod/barefoot and injury groups with feature ranking, Comput. Methods Biomech. Biomed. Eng. 15 (2012) 467−474. Available from: https://doi.org/10.1080/10255842.2010.542153.

[23] K. Niazmand, K. Tonn, A. Kalaras, U.M. Fietzek, J.H. Mehrkens, T.C. Lueth, Quantitative evaluation of Parkinson's disease using sensor based smart glove, in: 2011 IEEE 24th International Symposium on Computer-Based Medical Systems (CBMS) (2011, June) pp. 1−8.

[24] Y. Su, C.R. Allen, D. Geng, D. Burn, U. Brechany, G.D. Bell, et al., 3-D motion system ("data-gloves"): application for Parkinson's disease, IEEE Trans. Instrum. Meas. 52 (3) (2003) 662−674.

[25] H. Dai, B. Otten, J.H. Mehrkens, L.T. D'Angelo, T.C. Lueth, A novel glove monitoring system used to quantify neurological symptoms during deep-brain stimulation surgery, IEEE Sens. J. 13 (9) (2013) 3193−3202.

[26] S. Kazi, A. As' Arry, M.M. Zain, M. Mailah, M. Hussein, Experimental implementation of smart glove incorporating piezoelectric actuator for hand tremor control, WSEAS Trans. Syst. Control 5 (6) (2010) 443−453.

[27] Vinod Sharma, Kunal Mankodiya, Fernando De. La Torre, Ada Zhang, Neal Ryan, Thanh G.N. Ton, et al., SPARK: personalized parkinson disease interventions through synergy between a smartphone and a smartwatch, International Conference of Design, User Experience, and Usability, Springer International Publishing, 2014.

[28] Kinesia Products, <http://glneurotech.com/kinesia/products/ > , 2017 (accessed February 2017).

[29] PKG Data Logger, <http://www.globalkineticscorporation.com/ > , 2017 (accessed February 2017).

[30] J. Lee, H. Kwon, J. Seo, S. Shin, J.H. Koo, C. Pang, et al., Conductive fiber-based ultrasensitive textile pressure sensor for wearable electronics, Adv. Mater. 27 (15) (2015) 2433−2439.

[31] M. Nishiyama, K. Watanabe, Wearable sensing glove with embedded hetero-core fiber-optic nerves for unconstrained hand motion capture, IEEE Trans. Instrum. Meas. 58 (12) (2009) 3995−4000.

[32] P. Gould, Textiles gain intelligence, Mater. Today 6 (10) (2003) 38−43.

[33] Linz, T., Kallmayer, C., Aschenbrenner, R., & Reichl, H. (2006), Fully untegrated EKG shirt based on embroidered electrical interconnections with conductive yarn and miniaturized flexible electronics, in: IEEE International Workshop on Wearable and Implantable Body Sensor Networks, 2006. BSN 2006, pp. 4.

[34] Zhang, S., Chauraya, A., Whittow, W., Seager, R., Acti, T., Dias, T., et al. (2012, November). Embroidered wearable antennas using conductive threads with different stitch spacings. in: IEEE Antennas and Propagation Conference (LAPC), 2012 Loughborough, pp. 1−4.

[35] I. Kazani, C. Hertleer, G. De Mey, A. Schwarz, G. Guxho, L. Van Langenhove, Electrical conductive textiles obtained by screen printing, Fibres Text. East. Eur. 20 (1) (2012) 57−63.

[36] L.M. Castano, A.B. Flatau, Smart fabric sensors and e-textile technologies: a review, Smart Mater. Struct. 23 (5) (2014) 053001.

[37] K. Jost, G. Dion, Y. Gogotsi, Textile energy storage in perspective, J. Mater. Chem. A 2 (28) (2014) 10776–10787.

[38] L. Van Langenhove (Ed.), Smart Textiles for Medicine and Healthcare: Materials, Systems and Applications, Elsevier, Amsterdam, Netherlands, 2007.

[39] S. Park, S. Jayaraman, Smart textiles: wearable electronic systems, MRS Bull. 2 8 (08) (2003) 585–591.

[40] S. Wagner, E. Bonderover, W.B. Jordan, J.C. Sturm, Electrotextiles: concepts and challenges, Int. J. High Speed Electron. Syst. 12 (02) (2002) 391–399.

[41] A.G. Avila, J.P. Hinestroza, Smart textiles: tough cotton, Nat. Nanotechnol. 3 (8) (2008) 458–459.

[42] J. McCann, D. Bryson (Eds.), Smart Clothes and Wearable Technology, Elsevier, Amsterdam, Netherlands, 2009.

[43] D. Zhang, J. Zhou, M. Guo, J. Cao, T. Li, TASA: tag-free activity sensing using RFID tag arrays, IEEE Trans. Parallel Distrib. Syst. 22 (4) (2011) 558–570.

[44] D. Patron, T. Kurzweg, A. Fontecchio, G. Dion, K.R. Dandekar, Wireless strain sensor through a flexible tag antenna employing inductively-coupled RFID microchip, 2014 IEEE 15th Annual Wireless and Microwave Technology Conference (WAMICON), 2014, June, pp. 1–3.

[45] E-Traces, Ballet Slippers That Make Drawings from the Dancer's Movements, 2017, <http://makezine. com/2014/11/06/e-traces-ballet-slippers-that-make-drawings-from-the-dancersmovements/ > (accessed February 2017).

[46] ProGlove product, 2017, <http://www.proglove.de/> (accessed February 2017).

[47] O. Sahin, O. Kayacan, E.Y. Bulgun, Smart textiles for soldier of the future, Def. Sci. J. 55 (2) (2005) 195.

[48] Á. Jobbágy, P. Harcos, R. Karoly, G. Fazekas, Analysis of finger-tapping movement, J. Neurosci. Methods 141 (1) (2005) 29–39.

[49] A. Kandori, M. Yokoe, S. Sakoda, K. Abe, T. Miyashita, H. Oe, et al., Quantitative magnetic detection of finger movements in patients with Parkinson's disease, Neurosci. Res. 49 (2) (2004) 253–260.

[50] S.R. Muir, R.D. Jones, J.H. Andreae, I.M. Donaldson, Measurement and analysis of single and multiple finger tapping in normal and Parkinsonian subjects, Parkinsonism Relat. Disord. 1 (2) (1995) 89–96.

[51] P.K. Pal, C.S. Lee, A. Samii, M. Schulzer, A.J. Stoessl, E.K. Mak, et al., Alternating two finger tapping with contralateral activation is an objective measure of clinical severity in Parkinson's disease and correlates with PET [18 F]-DOPA Ki, Parkinsonism Relat. Disord. 7 (4) (2001) 305–309.

[52] I. Shimoyama, T. Ninchoji, K. Uemura, The finger-tapping test: a quantitative analysis, Arch. Neurol. 47 (6) (1990) 681–684.

[53] M. Yokoe, R. Okuno, T. Hamasaki, Y. Kurachi, K. Akazawa, S. Sakoda, Opening velocity, a novel parameter, for finger tapping test in patients with Parkinson's disease, Parkinsonism Relat. Disord. 15 (6) (2009) 440–444.

[54] B. Farahani, F. Firouzi, V. Chang, M. Badaroglu, N. Constant, K. Mankodiya, Towards fog-driven IoT eHealth: promises and challenges of IoT in medicine and healthcare, Future Gener. Comput. Syst 78 (Part 2) (2018) 659–676. Available from: https://doi.org/10.1016/j.future.2017.04.036. ISSN 0167-739X.

CHAPTER 10

A Soft Wearable Elbow Exosuit: Design Considerations

M. Xiloyannis[1], K.B. Dhinh[2], L. Cappello[3], C.W Antuvan[2] and L. Masia[4]

[1]PhD Student, Program for Research in Future Healthcare in the Interdisciplinary Graduate School (IGS), Nanyang Technological University, Singapore [2]PhD Student, School of Mechanical and Aerospace Engineering, Nanyang Technological University, Singapore [3]Postdoctoral Fellow, Biorobotics Institute, Scuola Superiore Sant'Anna, Pisa, Italy [4]Assistant Professor, School of Mechanical & Aerospace Engineering, Nanyang Technological University, Singapore

10.1 Introduction

Disorders of the nervous system account for approximately 28% of all years of life lived with disability worldwide [1]. These include, among others, disorders related to age, spinal cord injury, multiple sclerosis, traumatic brain injury, and cerebral palsy, but stroke outweighs all others combined in terms of disability-adjusted life years (DALYs) [2]. Although epidemiological studies report controversial results, 30%−60% of stroke survivors experience chronic movement dysfunction in the upper limbs 6 months after the stroke [3]. The distal parts of the limbs are known to be the most affected by persistent deficits, with weakness of voluntary contraction and hypertonus of the flexor muscles strongly reducing the ability to extend the joints [4]. Losing the remarkable and unmatched dexterity of our arms in manipulating and communicating severely hampers one's ability to accomplish simple, yet fundamental, activities of daily living (ADLs) such as eating, drinking, and getting dressed.

The incidence of stroke increases significantly with age [5] and, as the lifetime expectancy rises, its prevalence and impact on society are expected to grow. Various strategies have been explored to improve motor recovery after stroke, and, so far, repetitively performing isolated [6] and functional [7,8] movements in the acute phase of recovery has produced the best outcomes. These findings support the hypothesis that practice is the main leading factor in promoting synaptogenesis and brain plasticity after a stroke [9].

Wearable Technology in Medicine and Health Care.
DOI: https://doi.org/10.1016/B978-0-12-811810-8.00010-5

Robotic machines have thus been introduced to increase the intensity of practice and relieve therapists from the demanding task of manually assisting the patient. Robot-assisted therapy has shown encouraging results, comparable to the ones achieved with traditional therapy, while allowing greater patient compliance—patients can perform the tasks by playing a video game—and a quantitative, more accurate monitoring of the subject's performance. Yet, only 5%−20% of patients fully recover their lost motor function [10] and the great majority do not regain enough dexterity to be independent when discharged from physical therapy. Moreover, in hemiparetic patients—where only one side of the body is impaired—interruption of assistance to the impaired side may result in a compensation strategy, known as maladaptive plasticity [11], that could even worsen the affected side's functionality.

Despite this, the current contribution of robotic devices to aid stroke survivors is limited to the clinical environment: very little effort has been devoted to design a wearable robotic device that is simple and portable enough to be used at home and in common daily activities. A wealth of exoskeletons and orthoses has been engineered to assist movements of the upper limbs [12], but most of them are made of a heavy and bulky frame. Their structural complexity makes them ideal to perform accurate, repetitive movements at each individual joint and for quantifying the outcomes of physical therapy in clinical environments, but these same features cause them to be very poor candidates for daily use, where portability, simplicity and low profile are fundamental.

An innovative solution to transmit forces to the human body while meeting the criteria mentioned above was only very recently proposed by Asbeck et al. [13], and like most ingenious solutions, it is extremely simple: clothes. Replacing rigid frames with fabric reduces the weight and power requirements while increasing comfort and cosmetic acceptance of the device. The result is what has been named as *exosuit*, i.e., a device that looks more like a suit than a traditional exoskeleton, but delivers forces to the wearer's joints either by means of cables, routed in the fabric and pulled by electric motors [14−19], or using pneumatic soft actuators [20,21]. In both cases the device works in parallel with the wearer's own muscles to provide extra strength and improve the efficiency of movement [18,19,22].

In this chapter we present the design, control, and evaluation of a cable-driven soft suit for assisting elbow movements in ADLs. Our device specifically targets stroke survivors with moderate retained motor capacity in the upper limbs (Action Research Arm Test score between 10 and 35, and a Stroke Upper Limb Capacity Scale score of 4-757) and mild spasticity of the arm's flexor muscles (Modified Ashworth Scale (MAS) of 0−2). By using fabric instead of rigid links, we aim at designing a low-profile, lightweight, and power-efficient device that a patient can use to regain independence in tasks performed on a daily basis. The system, shown in Fig. 10.1, comprises a control unit that the wearer can carry in a backpack, and a fabric-based frame worn around the limb.

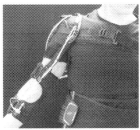

Figure 10.1

The soft, cable-driven exosuit for the elbow comprises a wearable component, fully made of fabric, and a proximally located actuator, mounted on a harness that loads its weight on the wearer's shoulder.

In the following sections we identify the technical design requirements of the actuation unit by combining prior studies on the force and motion characteristics of human movement in ADLs with a mathematical model of the tendon routing in the suit. We then propose our design of the device, comprising an actuation unit and a textile frame. Section 4 describes a novel control paradigm to detect the subject's intention and modulate the assistive force accordingly. Lastly, we test the design and control paradigm on healthy subjects, quantifying the level of assistance of the suit in terms of reduction in muscular effort, and evaluating its ability to tune the assistance level according to the wearer's ability to move.

10.2 Requirements

Defining the requirements of soft wearable devices for clinical use is still an open question. So far, exosuits have been proven to be effective increasing the efficiency of movement in unimpaired subjects [19,22] and have shown to yield improvements in key gait metrics in stroke patients [23,24], but complications such as severe spasticity and/or disuse osteoporosis—a localized bone loss condition due to reduction in mechanical stress, common in stroke patients—might cause such devices to perform rather poorly.

In this study we thus assume that our elbow exosuit will be used for assisting people suffering from muscle weakness but having no major spasticity or contractures (MAS 0−2). Our design objective is based on the average dynamic and kinematic requirements necessary to perform ADLs alongside reasonable practical considerations on the weight, size, and power consumption of the system.

This is done by combining prior studies on the average force/velocities of the elbow during ADLs with a simple mathematical model of the tendon routing in the suit. The requirements are summarized in Table 10.1 and further detailed in the following subsections.

Table 10.1: System requirements.

Characteristics	Requirements
Force/Motion: [25–27]	Elbow
Range of motion [degree]	146
Joint torque [Nm]	4.45
Bandwidth [Hz]	1.2
Practical Considerations: [28]	
Distal frame weight [kg]	0.7
Proximal pack weight [kg]	≤ 2.5

10.2.1 Force and Motion Characteristics

It is, first of all, fundamental for the device to span the whole range of motion (RoM) of the human joints. Magermans et al. [25] analyzed the RoM of the elbow and shoulder in nonimpaired subjects, finding a mean of 146 degree (0 degree corresponding to the fully extended configuration) for the elbow. Similarly, a wealth of studies has evaluated torques and average speeds of human joints in ADLs. Elbow flexion can require up to 4.45 Nm, with a mean of 1 Nm [26], and velocities can reach 331 degree per second [27]. Assuming a sinusoidal motion with a peak to peak movement equal to the RoM, these correspond to a frequency of movement of 1.2 Hz.

10.2.2 Practical Considerations

Being portability one of our main goals, we require the total weight mounted distally to be negligible when compared to the limb's weight (on average, 1.52 kg for males and 2.56 kg for females [28]); an acceptable value would be 0.3 kg. This can be easily achieved if the motors, controller and battery are located proximally (i.e., in a backpack or on a belt at the waist) and transmit forces to the joint via Bowden cables. A reasonable upper-limit for the proximally located part of the system, comprising the motor, controllers and power supply, is 2.5 kg.

10.2.3 Dynamics

In order to evaluate the actuator's power requirements we use a bi-dimensional model of the elbow on the sagittal plane. The elbow is modeled as a revolute joint, with the forearm thus being a simple pendulum, as shown in Fig. 10.2A, and assuming the arm to be aligned with the direction of gravity.

Starting from the Lagrangian form one can derive the dynamics of the elbow in the joint space:

$$M\ddot{\theta} + C\dot{\theta} + \mathrm{N}(\theta) = \tau \qquad (10.1)$$

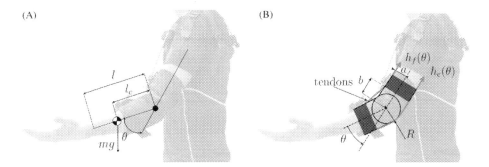

Figure 10.2
Planar model of the elbow joint and of the tendon routing. (A) The elbow joint is modeled as a revolute joint on the sagittal plane. Its dynamics in the joint space θ are described by Eq. (10.1). (B) From the geometry of the tendon-routing we can derive the extension functions (Eqs. (10.2) and (10.3)), used to map forces in the tendons of the suit to joint torques.

where M is the *Inertia* of the forearm, C is a matrix describing velocity-dependent torques and $N(\theta)$ is a column vector that takes into account gravitational forces. The vector τ expresses the external torques applied on the joints which, in our case, results from the force applied by the artificial tendons routed in the suit as shown in Fig. 10.2B. Although the suit is meant to work in parallel with its wearer's muscles, we hereby neglect the muscular torque and analyze the worst-case scenario.

A mapping from the tension in the tendons to the torque on the joints can be derived using geometrical considerations on the tendon's routing; referring to Fig. 10.2B, we can derive the *extension function* $h(\theta)$, projecting the elbow's joint angle to a corresponding displacement of the tendons:

$$h_f(\theta) = 2\sqrt{a^2 + b^2}\cos\left(\phi + \frac{\theta}{2}\right) - h_{f0} \tag{10.2}$$

$$h_e(\theta) = R\theta. \tag{10.3}$$

where h_f is the extension function for the flexor and h_e for the extensor tendon, a is half the width of the arm, b is the distance of the anchor point from the joint's center of rotation, $\phi = \arctan(a/b)$ and R is the radius of the joint. The constant h_{f0} assures that the extension functions are null when the arm is fully extended.

Using the principle of conservation of energy the functions $h(\theta)$ can be used to find a Jacobian-like matrix $P(\theta)$, which we shall call *coupling matrix*, mapping tension in the tendons to torques on the joint:

$$\tau = P(\theta)f \tag{10.4}$$

$$P(\theta) = \frac{\partial h^T}{\partial \theta}(\theta) \tag{10.5}$$

where f is the force applied on the tendons. By substituting Eq. (10.4) in Eq. (10.5) we obtain:

$$M\ddot{\theta} + C\dot{\theta} + N = Pf, \tag{10.6}$$

where we have removed the dependency on θ for the sake of conciseness.

Finally, we include a term modeling losses due to friction, not negligible in Bowden cable systems. The main parameter affecting the entity of such losses is the bending angle of the outer housing. Assuming that the contact between the inner cable and the outer sheet can be modeled as the sliding of a cable over a fixed cylinder, the force-transmission efficiency becomes:

$$\frac{f_{in}}{f_{out}} \approx e^{-\mu\phi}, \tag{10.7}$$

with μ being the friction coefficient between the cable and the outer sheet, ϕ the total wrap angle of the outer sheath of the Bowden cable, and f_{in}, f_{out} the tension in the cable before and after the transmission respectively. Assuming the actuator to be carried in a backpack, a reasonable and abundant estimate of the wrap angle is π. Using a Teflon-steel static friction coefficient yields to a loss in efficiency of $\approx 40\%$.

A numerical solution to the dynamic Eq. (10.6), accounting for the loss in efficiency expressed in Eq. (10.7), implemented in Simulink (The MathWorks, Inc., Natick, MA, United States) is shown in Fig. 10.3 for varying velocity of movement of the elbow. Joint motions, following a minimum-jerk trajectory, are shown in Fig. 10.3A. Torques on the joint can go up to ≈ 7 Nm for fast movements, requiring a cable tension above 200 N and a motor power of 25 W (Figs. 10.3B−D, respectively).

These, combined with the requirements listed in Table 10.1, served as guidelines to design the actuation unit and choose the tendon's material's for the soft exosuit.

10.3 Actuator's and Suit's Design

The exosuit comprises an actuation stage, driving a pair of tendons, and a wearable component. The actuation stage is located proximally, mounted on a harness on the wearer's back, and transmits power to the suit via Bowden cables.

The tendon-driving unit (shown in Fig. 10.4) comprises the following components: a brushless motor (Maxon EC-max, Ø 22 mm, 50 W) coupled to a gearhead (reduction of

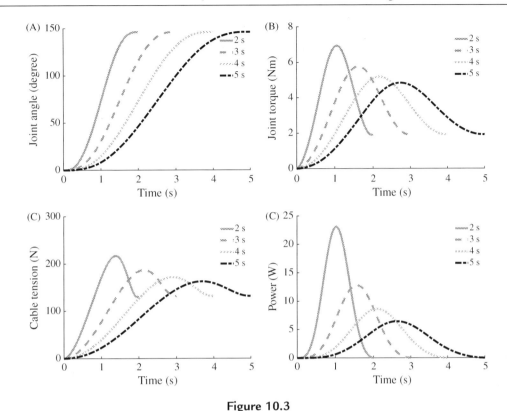

Figure 10.3
Dynamic model of human arm assisted by the exosuit with tendons routed as shown in
Fig. 10.2B. (A) Eq. (10.6) was solved for minimum-jerk trajectories requiring 5, 4, 3, and
2 seconds to cover the whole range of motion of the elbow joint. (B) Torques on the elbow joint
required to follow the corresponding trajectory of the joint. (C) Tension in the flexor tendon of
the soft exosuit for varying velocity of the elbow joint. (D) Motor power consumption for varying
velocities of the elbow joint, accounting for the losses in the transmission due to friction, as
described by Eq. (10.7).

33:1), whose position is sensed by a quadrature encoder (Maxon Encoder Mr, 512 CP) and
a spool around which two cables are coiled in opposite directions, so that rotation of the
motor in one direction causes retraction of one tendon and releases the other, in an
antagonistic fashion. A feeder mechanism, whose working principle is thoroughly described
in [29], confines the slack of the tendons outside of the actuation unit, ensuring that the
tendons are tightly wrapped around the spool at all times.

The two tendons, made of superelastic NiTi wire, are routed from the actuator unit on the
harness to the elbow joint through an outer cable housing (Nokon, Sava Industries). The
whole mechanism is enclosed in a 3D-printed case in ABS plastic.

The exosuit for the elbow (shown in Fig. 10.5B) was designed by modifying a
commercially available passive orthosis (MASTER-03, Reh4mat). The substrate of the suit,

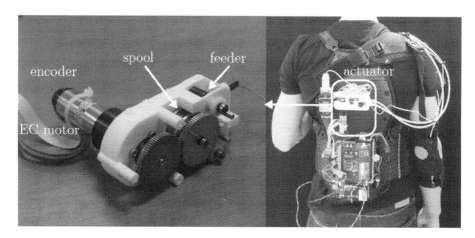

Figure 10.4

Actuator of the elbow exosuit. The tendon-driving unit is empowered by a 50 W EC-motor, equipped with an incremental encoder and planetary gearhead with a reduction of 33:1 that drives a spool around which the two tendons are wrapped in opposite directions. Rotation of the spool causes retraction of one tendon and release of the other one, in an antagonistic fashion. A feeder mechanism keeps the tendons tight around the spool. The actuator weighs 800 g and can be mounted on a harness that loads its weight on the shoulders of the wearer.

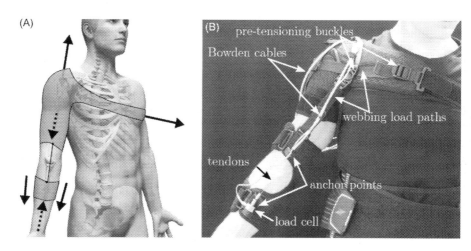

Figure 10.5

Transmission of forces in the exosuit when assisting the elbow and labeled photo of the fabric frame. (A) When either of the two tendons is in tension, compression forces are applied on the wearer's elbow joint through the humerus, radius and ulna (large dotted arrows), while the suit tensions along the load paths (small solid arrows), transmitting part of the forces to the shoulder and ribcage. Large solid arrows represent the reaction forces of the human body applied on the suit. Notice that the ones on the shoulder and armpit are normal to the body, whilst the distal anchor point only relies on shear forces on the skin. (B) The soft exosuit transmits forces from the actuator to the joint via Bowden cables, secured on the anchor points on each side of the joint. Load paths, made of webbing, ensure efficient transmission of forces while buckles allow to pretension the suit around the user's body.

having the function of adhering to the body of the user and keeping it in place, is made of a 3-layered fabric: an external layer used to attach hard components (buckles and webbing strips), an intermediate Ethylene-vinyl acetate (EVA) foam cushions high loads and avoids peaks of pressure and an internal 3D polyamid structure provides high air permeability and moisture absorption. Additionally the arm bands are lined with a silicone pattern at the interface with the skin to prevent slipping.

Fig. 10.5A highlights the directions along which forces are applied through the suit (small red arrows), which we shall refer to as *load paths*. Load paths need to be as stiff as possible to maximize force-transmission efficiency. They are thus made of webbing, i.e., nylon fibers woven in a flat strip, which is virtually inextensible and able to support high loads. Buckles attached on the webbing strips, moreover, allow to adjust the suit to different sizes and to pretension it around the user's body, so as to reduce backlash phenomena.

The suit covers the shoulder and encircles the chest, this allows it to rely on forces perpendicular to the body to prevent the fabric frame from slipping when the tendons are in tension. Shear forces, as a matter of fact, should be avoided as they are not only source of discomfort for the user but also of unreliable transmission, due to the large deformation of the underlying soft tissue.

Finally, to route the tendons along the load paths we sewed 3D-printed components on the webbing network on each sides of the joint. These serve as artificial ligaments, anchoring the tendons to the body; the distal anchor points house a load-cell to sense the tension applied by the actuation unit on the tendons.

In Fig. 10.4 the subject is also wearing a harness designed to carry the actuation unit on his torso. The harness, that can be tightened through a set of buckles, loads the weight of the device (2.2 kg including actuation, electronics, and power supply) on the wearer's shoulders.

10.4 Controller's Design

While the low profile, lightweight and compliance of soft wearable devices make them very appealing solutions for assisting human movements on a daily basis, their intrinsic soft nature poses unquestionable control challenges: deformation of the stretchable materials, friction in the Bowden cables and the viscoelastic properties of human soft tissues make a simple feedback control inadequate for achieving a reasonable tracking accuracy. Moreover, understanding the intentions of the wearer is a key but challenging task.

We propose a hierarchical cascade controller, schematized in Fig. 10.6, to achieve accurate and intuitive control of the soft exosuit using only a load cell on the tendons and a flex sensor to measure the elbow joint's position.

The proposed framework does so using a three-layered architecture: the highest layer decodes the user's intention and tunes the level of assistance accordingly; the middle layer

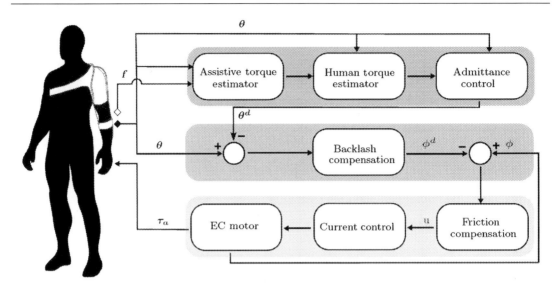

Figure 10.6
Schematics of the controller's architecture. The high level controller (blue) uses a load cell on the tendons and a flex sensor on the elbow joint to modulate the impedance perceived by the user and tune the level of assistance. The middle layer (red) continuously compensates for backlash hysteresis phenomena using an adaptive model. Similarly, the low level (green) deals with friction in the Bowden cable transmission.

compensated for backlash phenomena deriving from shifts of the fabric and Bowden cables upon the application of forces; the lowest layer accounts for friction in the transmission. Each level of the controller is further detailed in the following subsections.

10.4.1 High Level Controller

The high level controller tunes the impedance of the human–suit complex so as to make it smaller than the impedance of the arm alone. Moreover, it adapts the level of assistance to the subject's ability to move.

To estimate the torque exerted on the elbow joint by the subject we can start from the dynamic model of the human arm introduced in Section 2:

$$M\ddot{\theta} + C\dot{\theta} + N(\theta) = \tau \tag{10.8}$$

where τ is modeled as the sum of an assistive torque, exerted by the exosuit on the human joint, and the torque applied by the human muscles:

$$\tau = \tau_a + \tau_h. \tag{10.9}$$

The assistive torque can be estimated using Eq. (10.4), knowing the tension f and the elbow joint's angle θ, acquired via a load cell on the tendons and a flex sensor on the joint (Fig. 10.7), respectively. An estimate of the torque deriving from the human muscles $\hat{\tau}_h$ can then obtained from the inverse dynamic model:

$$\hat{\tau}_h = M\ddot{\theta} + C\dot{\theta} + N(\theta) - P(\theta)f \tag{10.10}$$

The subsequent step of the high level controller is to generate a reference trajectory of the elbow joint θ^d. The estimated human torque is thus used to solve the forward dynamic problem:

$$M^d\ddot{\theta}^d + C^d\dot{\theta}^d + K^d \sin \theta^d = \hat{\tau}_h \tag{10.11}$$

where M^d, C^d, and $K^d \sin \theta^d$ denote the desired inertial, viscous damping and gravitational torque acting at the elbow joint. For the device to be assistive, the impedance resulting from Eq. (10.11) must be smaller than the impedance of the human arm alone, i.e., $M^d \leq M$, $C^d \leq C$ and $K \sin \theta^d \leq N(\theta)$.

The gains of the admittance controller were thus chosen according to the following equation:

$$\begin{cases} M^d = M \\ C^d = C \\ K^d = \alpha m g l_c \\ \alpha = \alpha_0 \tanh\left(\dfrac{\dot{\theta}}{\epsilon_\alpha}\right) + \alpha_1 \end{cases} \tag{10.12}$$

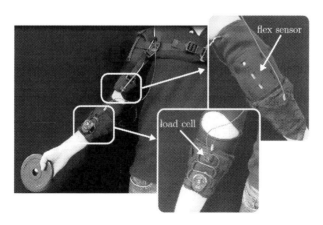

Figure 10.7

The elbow suit is equipped with a miniature load cell that senses the tension in the tendons and a flex sensor, sewn in the fabric, to monitor and control the joint's position.

where α_0 and α_1 are two constants experimentally chosen to bound the intervention of the assistance and ϵ_α denotes the sensitivity coefficient of the hyperbolic function $\tanh(\cdot)$. The factor α increases with the measured joint velocity $\dot{\theta}$, meaning that the level of assistance is strictly dependent on the user's residual motion capacity: a high motion speed from the user (i.e., high motor ability) corresponds to a low assistive torque provided by the exosuit and vice versa.

The output of the high level controller is a desired trajectory, θ^d, of the elbow, estimated so as to reduce the perceived impedance of the elbow-suit interaction and tuned according to the subject's ability of movement.

10.4.2 Mid-level Controller

The purpose of the mid-level control layer is to continuously compensate for the nonlinear backlash phenomenon resulting from deformation of the fabric and unmeasurable changes in the configuration of the Bowden cables (Fig. 10.8). The desired joint angular motion, θ^d, from the admittance control block must be converted into a motion of the actuator, ϕ, accounting for the loss of motion caused by backlash. We do so by adopting a model-based approach with an adaptive framework that fits the model's parameters at each sampling step so as to minimize the trajectory-following error.

The relationship between the desired motion θ^d and the motion the actuation unit ϕ can be defined adopting the Bouc−Wen hysteresis model [30], such that the desired elbow motion is a function of the actuator's rotation and a term representing the backlash uncertainties:

$$\beta\theta^d = \phi^d + \beta D \tag{10.13}$$

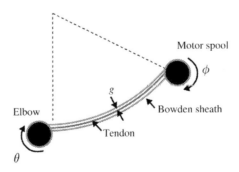

Figure 10.8

A significant amount of backlash in the transmission is caused by position hysteresis in the Bowden cables. Parameters such as the gap between the inner tendon and the Bowden sheath and the wrap angle of the Bowden cable strongly affect the entity of such phenomenon, causing loss of transmission between the motor spool rotation ϕ and the elbow joint θ.

where β represents the positive ratio of ϕ^d to θ^d and D represents the model uncertainties due to the Bowden sheath's variable geometry during operation.

The adaptive framework defines a reference motion θ^r and sliding surface s as in:

$$
\begin{cases}
e = \theta - \theta^d \\
s = \lambda e + \int_0^t e\,d\tau \to \dot{s} = \lambda\dot{e} + e \\
\theta^r = \theta^d - \lambda\dot{e},
\end{cases}
\tag{10.14}
$$

where e represents the tracking error between the desired elbow joint motion, θ^d, and the measured one, θ; and λ is an arbitrarily positive constant. Since the elbow joint is supposed to follow a given trajectory θ^d, the desired actuator state, ϕ^d, can be chosen as:

$$
\phi^d = \hat{\beta}\left(\theta^r - \widehat{D_m}\tanh\left(\frac{s}{\epsilon}\right)\right) - \kappa s
\tag{10.15}
$$

where $\hat{\beta}$ and $\widehat{D_m}$ are the estimated value of β and D_m, respectively (such notation will be used for all variables from now on). κ and ϵ are positive constants.

Replacing ϕ^d from Eq. (10.15) to Eq. (10.13) leads to the dynamics of the sliding surface s as:

$$
\beta\dot{s} + \kappa s = \widetilde{\beta}\bar{\theta} - \beta\widetilde{D_m}\tanh\left(\frac{s}{\epsilon}\right) - \beta\left(D_m\tanh\left(\frac{s}{\epsilon}\right) - D\right)
\tag{10.16}
$$

where $\bar{\theta} = \theta^r - \widehat{D_m}\tanh(s/\epsilon)$; $\widetilde{\beta} = \hat{\beta} - \beta$ is the estimated error of β; and $\widetilde{D_m} = \widehat{D_m} - D_m$ is the estimated error of D_m.

Therefore the adaptation law for backlash model parameters $\hat{\beta}$ and $\widehat{D_m}$ is:

$$
\begin{cases}
\dot{\hat{\beta}} = -\delta_1\bar{\theta}s \\
\dot{\widehat{D_m}} = \delta_2\beta\tanh\left(\frac{s}{\epsilon}\right)s
\end{cases}
\tag{10.17}
$$

where δ_1, δ_2 are positive adaptation gains. The initial values for $\hat{\beta}$ and $\widehat{D_m}$ are set to be zero.

At each sampling time, Eq. (10.17) thus updates the Bouc−Wen model's parameters so as to minimize the difference between the actual and desired elbow position. This allows continuous, online compensation of the exosuit's transmission backlash despite of the changes in the fabric and suit's configuration. The output of the mid-level controller is a desired actuator's position, ϕ^d, which is handled by the low level controller to compensate for friction phenomena.

10.4.3 Low Level Controller

The low level control layer is intended to drive the actuation stage by compensating for the nonlinear friction occurring because of the tendons sliding along the Bowden sheath.

The friction continuously and unpredictably changes according to the curvature of the sheath which varies as its wearer moves the arm. If not accounted for, the torque transmitted at the elbow joint would be significantly smaller than the one applied by the EC motor.

The actuator and the Bowden cable transmission can be modeled as follows:

$$J\ddot{\phi} + B\dot{\phi} + \tau_f = u \tag{10.18}$$

where J and B represent the inertia and damping coefficient of the actuation stage, τ_f the resisting torque on the motor axis caused by friction, and u the torque exerted by the EC motor. The dynamic parameters of the actuation stage comprising the EC motor and the Bowden cable are unknown.

We can adopt the LuGre model [31] to express the resisting torque caused by friction, τ_f, as:

$$\tau_f = B_v\dot{\phi} + \tau_z(\phi, \dot{\phi}, z) \tag{10.19}$$

where z is a variable introduced in the LuGre model; $B_v\dot{\phi}$ represents the viscous friction; and $\tau_z(\ddot{\phi}_a, \dot{\phi}_a, z)$ represents a dynamic term depending on z.

Similarly to what was done in Section 4.2, we can define a framework for continuously adapting the model's parameters. We can thus define the tracking error e_1, reference motion $\dot{\phi}^r$ and sliding surface s_1 for the actuator as:

$$\begin{cases} e_1 = \phi + \phi^d \\ s_1 = \dot{e}_1 + \lambda_1 e_1 \\ \dot{\phi}^r = \dot{\phi}^d - \lambda_1 e_1 \end{cases} \tag{10.20}$$

where ϕ and ϕ^d denote the measured (using the incremental encoder on the motor's axis) and desired rotation of the EC motor, respectively; and λ_1 is an arbitrary positive constant.

The control signal u for the EC motor thus becomes:

$$u = \hat{J}\ddot{\phi}^r + \left(\hat{B} + \hat{B}_v\right)\dot{\phi} - \widehat{\tau_{zm}}\tanh\left(\frac{s_1}{\epsilon_1}\right) - \kappa_1 s_1 \tag{10.21}$$

where ϵ_1 and κ_1 are two positive constants. Substituting Eqs. (10.19) and (10.21) in Eq. (10.18), and replacing $B_t = B + B_v$ result in the dynamics of the sliding surface s_1 as:

$$J\dot{s}_1 + \kappa_1 s_1 = \tilde{J}\ddot{\phi}^d + \tilde{B}_t\dot{\phi} - \widetilde{\tau_{zm}}\tanh\left(\frac{s_1}{\epsilon_1}\right) = \tau_{zm}\tanh\left(\frac{s_1}{\epsilon_1}\right) - \tau_z(\phi, \dot{\phi}, z) \tag{10.22}$$

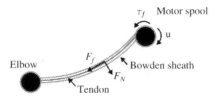

Figure 10.9
Friction in the Bowden sheath can be modeled as a resistive torque on the motor axis. Its magnitude is dependent on the wrap angle of the Bowden sheath between the motor spool and the elbow joint, which changes during operation and is difficult to accurately monitor.

where $\widetilde{J} = \widehat{J} - J$ is the estimated error of J; $\widetilde{B}_t = \widehat{B}_t - B_t$ is the estimated error of B_t; and $\widetilde{\tau}_{zm} = \widehat{\tau}_{zm} - \tau_{zm}$ is the estimated error of τ_{zm}. Therefore the unknown parameters $\widehat{J}$, $\widehat{B}_t$, and $\widehat{\tau}_{zm}$ are updated at each sampling time by:

$$\begin{cases} \dot{\widehat{J}} = -\delta_3 \ddot{\phi}^r s_1 \\ \dot{\widehat{B}}_t = -\delta_4 \dot{\phi} s_1 \\ \dot{\widehat{\tau}}_{zm} = \delta_5 \ \mathrm{anh}\,(s_1/\epsilon_1) s_1 \end{cases} \tag{10.23}$$

where δ_3, δ_4, and δ_5 are positive adaptation gains.

At each sampling step, Eq. (10.23) thus updates the LuGre model's parameters so as to minimize the difference between the actual and desired actuator's position. This allows continuous, online compensation of friction in the Bowden cables despite of changes in its configuration. The control signal u is then sent to a motor controller using a simple PID in current mode (Fig. 10.9).

10.5 Validation

The elbow exosuit was evaluated on three healthy subjects (age: 26.6 ± 1.5), with the control framework described above. We tested the controller in two conditions, both with the purpose of assessing its accuracy and stability and for examining its ability to provide a smooth intervention when assistance was requested by the user. Furthermore, the controller should be able to modulate the amount of torque at the elbow depending on the capacity of motion of the subject and decrease the muscular effort in lifting a weight.

The experimental set-up is illustrated in Fig. 10.10: in all conditions subjects were asked to move their elbow between 0 degree and 90 while holding a 1 kg weight in their hand. A surface electromyographic (EMG) electrode was placed on the biceps brachii (responsible for flexing the elbow) of the arm to monitor their muscular activity, while a flex sensor monitored the joint's position. To ensure that the movements were performed at the same, instructed velocity across repetitions, subjects were provided with a visual feedback of the position of their joint on a monitor.

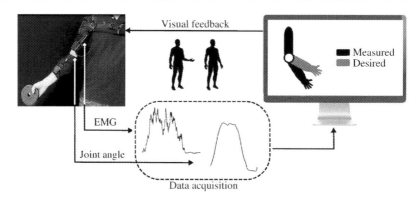

Figure 10.10

Experimental set-up. Subjects were asked to perform flexion/extension movements of the elbow while holding a 1 kg weight in their hand. We used the electromyographic (EMG) activity of their biceps brachii muscle to monitor their muscular effort and a flex sensor, sewn in the fabric, and to track their joint's angle. To standardize the velocity of movement, subjects were given a visual feedback of their elbow's position, compared to the movement of a virtual avatar moving at the desired velocity.

In the first condition subjects performed the movements at a velocity of 18 degree per second, 10 repetitions with and 10 without the exosuit. In the second condition subjects performed 10 repetitions moving at 18 degree per second and 10 repetitions moving at 36 degree per second, wearing the exosuit at all times. The higher velocity condition was chosen to replicate a comfortable daily action, requiring approximately 2 seconds for completion, and the lower velocity one was set to be its half, corresponding to a slow, unnatural movement of the elbow.

It is known that the kinematics of movement in stroke subjects are dramatically jeopardized [32,33] and a lower elbow speed is associated to a reduced voluntarily capacity of motion. For this reason, asking subjects to move at a lower and higher speed, was aimed as a preliminary examination of the ability of the proposed controller to tune the level of assistive torque according to the subject's need. Muscular effort was estimated from the root mean square (RMS) of the EMG activity [34] of the biceps brachii. The raw EMG was acquired using Trigno wireless EMG sensors (Delsys Inc.) and was preprocessed in Matlab Simulink using a full-wave rectification, followed by a low-pass second-order Butterworth filter with an 8 Hz cut-off frequency. The elbow joint angle was acquired during the experiment and used for control purposes using a resistive flex sensor (Spectrasymbol, USA) and the tendons' tension was recorded using a load cell (Futek, LCM300, 250 lb). Data acquisition and motor control were performed using the Quanser Quarc real-time workstation running at 1 kHz refresh rate.

Results of the first task comparing the EMG activity of one subject with and without the assistance are shown in Fig. 10.11A, showing a clear descrease in the amplitude of the

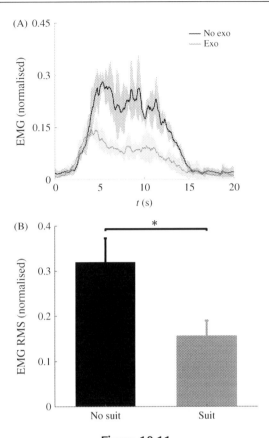

Figure 10.11

Comparison of muscular activity required to lift the elbow at 18 degree per second between the suit and no-suit cases. (A) Average EMG activity (solid line) and standard deviation (shaded area) of the biceps brachii muscle across 10 flexion/extension repetitions of the elbow for one subject; wearing the suit reduces the muscular effort required to perform the movement. (B) Mean and standard deviation over subjects of the root mean square (RMS) of the muscular activity, normalized by the maximum voluntary contraction of each subject. Wearing the suit causes a significant ($P = 0.01$) 48.3% drop in muscular effort.

EMG activity when the exosuit was assisting movement. Analysis of the RMS value of the EMG signal, normalized by the maximum voluntary contraction and averaged over repetitions and subjects, is shown in Fig. 10.11B. The latter shows an average significant ($P = 0.01$) drop in RMS of 48.3% between the nonassisted and assisted case. Fig. 10.12A and B display a comparison between the desired and measure tracking accuracy of the elbow and motor position. Both show a good tracking accuracy and demonstrate that the backlash and friction layers of the controller work reliably.

The second experiment examines the efficacy of the controller in adapting its contribution to the user's capacity of motion. Fig. 10.13A and B shows the EMG activity and the

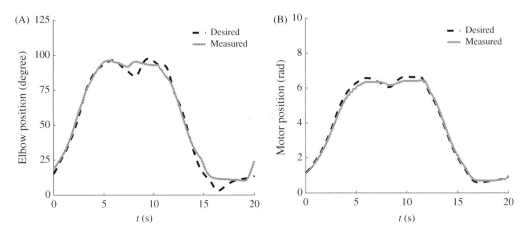

Figure 10.12

Joint and motor position-tracking accuracy of the proposed controller. (A) Desired and measured elbow joint position for one flexion/extension repetitions. The desired elbow position is the output of the high level controller (see Section 4.1). (B) Desired and measured motor angular position for one flexion/extension movement. The desired motor position is the output of the middle layer of the controller (see Section 4.2).

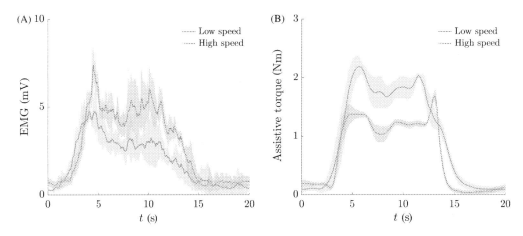

Figure 10.13

Comparison of muscular activity and assistive torque between elbow movements at 18 degree per second (low speed) and 36 degree per second (high speed). (A) Average EMG activity (solid line) and standard deviations (shaded area) across 10 flexion/extension repetitions of the elbow for one subject, at two different velocities of movement while wearing the suit. As expected, movements with higher velocity require higher muscular effort. (B) Average assistive torque (solid line) and standard deviations (shaded area) across 10 flexion/extension repetitions of the elbow for one subject, at two different velocities of movement while wearing the suit. For lower speed, corresponding to a higher level of impairment, the suit adapts by delivering a higher assistive torque.

delivered assistive torques respectively for one subject at the two execution speeds: it can be seen that for a lower speed (lower capacity of motion), requiring a smaller muscular effort, the controller delivers a higher assistive torque than when the subject moves at the higher speed (higher capacity of motion). The same applies to all subjects: taking the RMS of the time series and averaging over subjects, there is a 28.4% drop (nonsignificant) in the delivered assistive torque between the two cases, shown in Fig. 10.14.

10.6 Discussion

Disorders of the nervous system, and stroke most of all, can cause chronic movement dysfunctions that hamper the accomplishment of simple tasks, significantly reducing the independence and quality of life of those affected. Robotics has shown numerous advantages when applied to the rehabilitation phase, but a different approach is needed for daily assistance, where simplicity, portability, comfort, and cosmetic impact of the device play a fundamental role.

The introduction of soft material to transmit forces and torques to the human body allows to design wearable devices whose intrinsic compliance, kinematic transparency and lightweight are ideal for daily applications. The very same soft intrinsic nature, on the other hand, introduces non trivial technical challenges. How do we design a suit that is both compliant, to be comfortable, and stiff enough to transmit forces efficiently? How do we make the suit move in synchrony with its wearer's intentions, while tackling non linearities such as friction losses and backlash hysteresis in the transmission?

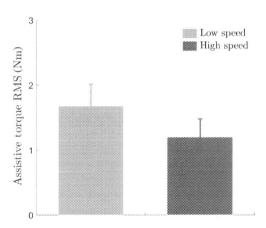

Figure 10.14
Mean and standard deviation across subjects of the RMS of the assistive torque, delivered by the exosuit. When the subject moves its joint at a higher velocity (36 degree per second), corresponding to a higher ability of movement, the exosuit lowers its assistance level by 28.4%. Vice versa, a higher level of assistance can be seen when the subjects shows a lower ability of movement (18 degree per second).

In this chapter we tried to answer these questions by presenting the design principles and control framework developed for an exosuit to assist elbow movements. A key feature of the system is the Bowden cable transmission, which allows to relocate the actuation unit close to the wearer's center of mass, removing unneeded weight from the limbs. A simple dynamic model, and geometrical considerations on the tendon's routing can be used to dimension the actuation's characteristics. The suit combines different materials, from webbing to elastic fabric, to transmit forces efficiently and cushion high loads on the skin. Its conformation allows it to avoid relying only on shear forces on the skin to keep the frame in place; buckles and velcro straps allow to fit it to different body sizes and pretension it before usage.

We further showed a controller that tunes the level of assistance depending on the voluntary motion capacity of the subjects. This strategy somewhat resembles the "assist-as-needed" paradigm [35]: the user is still in charge to control his/her actions and the exosuit is a complementary support to replace the lost functional movement ability. In this framework the concept of shared control authority between the machine and the neurogically damaged human counterpart covers another important aspect of motor rehabilitation, where the recovery process is optimized when the interaction between the physical therapist (human or robotic) and the patient results in a bidirectional exchange of dynamics, which is the essential substrate of the motor recovery process. The lower levels of the controller compensate for backlash and friction phenomena, resulting both from deformation of the fabric in the suit and from the Bowden cable transmission, thus ensuring an accurate position-tracking accuracy.

A limitation of our work is represented by its same strength: the soft nature of the suit limits its ability to apply high forces without loading its wearer's joints. Lacking of an external structure, the suit uses its wearer's own skeleton as the frame upon which to apply forces and torques. It is thus likely that the device would serve poorly patients suffering from disuse osteoporosis—a localized bone loss condition due to reduction in mechanical stress, common in stroke patients, or severe spasticity, which can generate position and velocity-dependent torques at the elbow joint of up to 2 Nm [36,37]. The target population of such a device may, thus, be restricted by this limitation. Lastly, although tests on healthy subjects are encouraging, clinical studies are needed to validate the device on its target population.

References

[1] N.J. Kassebaum, M. Arora, R.M. Barber, Global, regional, and national disability-adjusted life-years (DALYs) for 315 diseases and injuries and healthy life expectancy (HALE), 1990−2015: a systematic analysis for the Global Burden of Disease Study 2015, Lancet 388 (10053) (2016) 1603−1658. Available from: https://doi.org/10.1016/S0140-6736(16)31460-X.

[2] V.L. Feigin, G.A. Roth, M. Naghavi, P. Parmar, R. Krishnamurthi, S. Chugh, et al., Global burden of stroke and risk factors in 188 countries, during 1990−2013: a systematic analysis for the Global Burden of Disease Study 2013, Lancet Neurol 15 (9) (2016) 913−924. Available from: https://doi.org/10.1016/S1474-4422(16)30073-4.

[3] G. Kwakkel, B.J. Kollen, J. van der Grond, A.J. Prevo, Probability of regaining dexterity in the flaccid upper limb in acute stroke, Stroke 34 (9) (2003) 2181−2186. Available from: https://doi.org/10.1161/01. STR.0000087172.16305.CD.

[4] C.A. Trombly, Occupational Therapy for Physical Dysfunction (2004). Available from: https://doi.org/10.1111/j.1440-1630.2004.00470.x.

[5] D. Mozaffarian, E.J. Benjamin, A.S. Go, D.K. Arnett, M.J. Blaha, M. Cushman, et al., Heart disease and stroke statistics—2016 update: a report from the American Heart Association, Circulation 133 (4) (2016) e38−e60.

[6] C. Bütefisch, H. Hummelsheim, P. Denzler, K.H. Mauritz, Repetitive training of isolated movements improves the outcome of motor rehabilitation of the centrally paretic hand, J. Neurol. Sci. 130 (1) (1995) 59−68. Available from: https://doi.org/10.1016/0022-510X(95)00003-K.

[7] G. Kwakkel, B. Kollen, E. Lindeman, Understanding the pattern of functional recovery after stroke: facts and theories, Restor. Neurol. Neurosci 22 (3−5) (2004) 281−299. Available from: https://doi.org/10.1177/1545968308317972.

[8] R. Van Peppen, G. Kwakkel, S. Wood-Dauphinee, H. Hendriks, The impact of physical therapy on functional outcomes after stroke: what's the evidence? Clin. Rehabil. 18 (2004) 833−862.

[9] J.A. Kleim, T.A. Jones, T. Schallert, Motor enrichment and the induction of plasticity before or after, Brain Injury (2003). Available from: https://doi.org/10.1023/A:1026025408742.

[10] G. Kwakkel, B.J. Kollen, H.I. Krebs, Effects of robot-assisted therapy on upper limb recovery after stroke: a systematic review, Neurorehabil. Neural Repair 22 (2) (2007) 111−121. Available from: https://doi.org/10.1177/1545968307305457.

[11] N. Takeuchi, S.-I. Izumi, Maladaptive plasticity for motor recovery after stroke: mechanisms and approaches, Neural Plast. 2012 (2012) 1−9. Available from: https://doi.org/10.1155/2012/359728.

[12] H.S. Lo, S.Q. Xie, Exoskeleton robots for upper-limb rehabilitation: State of the art and future prospects, Med. Eng. Phys. 34 (3) (2012) 261−268. Available from: https://doi.org/10.1016/j.medengphy.2011.10.004.

[13] A.T. Asbeck, S.M.M. De Rossi, I. Galiana, Y. Ding, C.J. Walsh, Stronger, smarter, softer: next-generation wearable robots, IEEE Robot. Autom. Mag. 21 (4) (2014) 22−33. Available from: https://doi.org/10.1109/MRA.2014.2360283.

[14] A.T. Asbeck, S.M.M. De Rossi, K.G. Holt, C.J. Walsh, A biologically inspired soft exosuit for walking assistance, Int. J. Rob. Res. 34 (6) (2015) 744−762. Available from: https://doi.org/10.1177/0278364914562476.

[15] H. In, K.-j Cho, Exo-Glove: soft wearable robot for the hand using soft tendon routing system, IEEE Robot. Autom 22 (2015, March) (2015) 97−105. Available from: https://doi.org/10.1109/MRA.2014.2362863.

[16] M. Xiloyannis, L. Cappello, D.B. Khanh, S.C. Yen, L. Masia, Modelling and design of a synergy-based actuator for a tendon-driven soft robotic glove, in: Proc. IEEE RAS EMBS Int. Conf. Biomed. Robot. Biomechatronics, vol. 2016-July, IEEE, 2016, pp. 1213−1219, doi:10.1109/BIOROB.2016.7523796.

[17] K. Dinh, L. Cappello, M. Xiloyannis, L. Masia, Position Control using Adaptive Backlash Compensation for Bowden Cable Transmission in Soft Wearable Exoskeleton, in: Syst. (IROS), 2016 IEEE/RSJ, 2016, pp. 5670−5676.

[18] B.K. Dinh, M. Xiloyannis, C.W. Antuvan, L. Cappello, L. Masia, Hierarchical cascade controller for assistance modulation in a soft wearable arm exoskeleton, IEEE Robot. Autom. Lett. 2 (3) (2017) 1786−1793. Available from: https://doi.org/10.1109/LRA.2017.2668473.

[19] D. Park, K.J. Cho, Development and evaluation of a soft wearable weight support device for reducing muscle fatigue on shoulder, PLoS One 12 (3) (2017). Available from: https://doi.org/10.1371/journal.pone.0173730.

[20] P. Polygerinos, Z. Wang, K.C. Galloway, R.J. Wood, C.J. Walsh, Soft robotic glove for combined assistance and at-home rehabilitation, Rob. Auton. Syst. 73 (2015) 135−143. Available from: https://doi.org/10.1016/j.robot.2014.08.014.

[21] H.K. Yap, J.C.H. Goh, R.C.H. Yeow, Design and characterization of soft actuator for hand rehabilitation application, IFMBE Proc. 45 (2015) 367−370. Available from: https://doi.org/10.1007/978-3-319-11128-5-92.

[22] B.T. Quinlivan, L. Sangjun, P. Malcolm, D.M. Rossi, M. Grimmer, C. Siviy, et al., Assistance magnitude vs. metabolic cost reductions for a tethered multiarticular soft exosuit, Sci. Robot. 2 (2) (2016) 1−17. Available from: https://doi.org/10.1126/scirobotics.aah4416.

[23] S.M.M.D. Rossi, J. Bae, K.E.O. Donnell, K.L. Hendron, K.G. Holt, T. Ellis, et al., Gait improvements in stroke patients with a soft exosuit, Proc. Gait Clin. Mov. Anal. Soc. Meet, 2015, pp. 2−3.

[24] J. Bae, S.M.M. De Rossi, K. O'Donnell, K.L. Hendron, L.N. Awad, T.R. Teles Dos Santos, et al., A soft exosuit for patients with stroke: Feasibility study with a mobile off-board actuation unit, in: IEEE Int. Conf. Rehabil. Robot., vol. 2015-September, 2015, pp. 131−138, doi: 10.1109/ICORR.2015.7281188.

[25] D.J. Magermans, E.K.J. Chadwick, H.E.J. Veeger, F.C.T. Van Der Helm, Requirements for upper extremity motions during activities of daily living, Clin. Biomech. 20 (6) (2005) 591−599. Available from: https://doi.org/10.1016/j.clinbiomech.2005.02.006.

[26] I.A. Murray, G.R. Johnson, A study of the external forces and moments at the shoulder and elbow while performing every day tasks, Clin. Biomech. 19 (6) (2004) 586−594. Available from: https://doi.org/10.1016/j.clinbiomech.2004.03.004.

[27] M.A. Buckley, A. Yardley, G.R. Johnson, D.A. Cams, Dynamics of the upper limb during performance of the tasks of everyday living: a review of the current knowledge base, J. Eng. Med. 210 (4) (1996) 241−247. Available from: https://doi.org/10.1243/PIME_PROC_1996_210_420_02.

[28] C.E. Clauser, J.T. McConville, J.W. Young, Weight, volume, and center of mass of segments of the human body, Natl. Tech. Inf. Serv., 1969, pp. 1−112, doi: AMRL-TR-69-70(AD710622).

[29] H. In, H. Lee, U. Jeong, B.B. Kang, K.-j Cho, Feasibility study of a slack enabling actuator for actuating tendon-driven soft wearable robot without pretension, ICRA, Seattle, WA (2015) 1229−1234. Available from: https://doi.org/10.1109/ICRA.2015.7139348.

[30] F. Ikhouane, J. Rodellar, Systems with hysteresis analysis, identification and control using the bouc wen model, in: Control, Hoboken, New Jersey, United States, 2007, p. 199.

[31] H. Olsson, K. Åström, C. Canudas de Wit, M. Gäfvert, P. Lischinsky, Friction models and friction compensation, Eur. J. Control 4 (3) (1998) 176−195. Available from: https://doi.org/10.1016/S0947-3580(98)70113-X.

[32] M.C. Cirstea, Compensatory strategies for reaching in stroke, Brain 123 (5) (2000) 940−953. Available from: https://doi.org/10.1093/brain/123.5.940.

[33] A. Berardelli, M. Hallett, J.C. Rothwell, R. Agostino, M. Manfredi, P.D. Thompson, et al., Single-joint rapid arm movements in normal subjects and in patients with motor disorders, Brain 119 (1996) 661−674. Available from: https://doi.org/10.1093/brain/119.2.661.

[34] R.E. Challis, R.I. Kitney, Biomedical signal processing (in four parts). Part 1. Time-domain methods, Med. Biol. Eng. Comput. 28 (6) (1990) 509−524. Available from: https://doi.org/10.1007/BF02442601.

[35] L.E. Kahn, P.S. Lum, W.Z. Rymer, D.J. Reinkensmeyer, Robot-assisted movement training for the stroke-impaired arm: does it matter what the robot does?, J. Rehabil. Res. Dev. 43 (5) (2006) 619−630. Available from: https://doi.org/10.1682/JRRD.2005.03.0056.

[36] H.M. Lee, Y.Z. Huang, J.J.J. Chen, I.S. Hwang, Quantitative analysis of the velocity related pathophysiology of spasticity and rigidity in the elbow flexors, J. Neurol. Neurosurg. Psychiatry 72 (5) (2002) 621−629. Available from: https://doi.org/10.1136/jnnp.72.5.621.

[37] L. Alibiglou, W.Z. Rymer, R.L. Harvey, M.M. Mirbagheri, The relation between Ashworth scores and neuromechanical measurements of spasticity following stroke, J. Neuroeng. Rehabil 5 (2008). Available from: https://doi.org/10.1186/1743-0003-5-18.

Human Body Communication—Based Wearable Technology for Vital Signal Sensing

Jingjing Shi[1] and Jianqing Wang[2]

[1]*Northeastern University, Shenyang, China* [2]*Nagoya Institute of Technology, Nagoya, Japan*

11.1 Introduction

With the rapid aging of the global population, wireless wearable health-status monitoring system is becoming widespread in our daily life. It could be utilized in hospital, at home, or even in a vehicle. Human body communication (HBC) technology is one of the promising candidates for data communication in the wearable system. It utilizes the human body as a transmission medium to share information between the wearable sensor devices, which constitutes the wireless body area network (BAN) [1−5]. As shown in Fig. 11.1, such an HBC-based wireless wearable health-status monitoring system is usually composed of three parts, transceiver unit, receiver unit, and control unit. The transceiver unit may consist of several different wearable medical devices or vital sensors with communication functions such as blood pressure, blood glucose, electrooculogram, electrocardiogram (ECG), electroencephalogram, electromyography (EMG), and so on. They gather the vital signals regarding the current health state from the people being monitored and then transfer the data to the receiver unit by HBC technology. The receiver unit plays a role of data server, which may be a smartphone or a personal computer (PC). It also can continuously transmit the stored data to a client PC in a hospital or medical center for further processing and display via wireless local area network (WLAN). In the control unit, a restricted doctor or caregiver can access the server PC and see the monitoring data of patients or elderly people, and then provide more appropriate and professional diagnoses or health-care administration. Such an HBC-based health-status monitoring system can effectively reduce the inconvenience of wire links, and save unnecessary expenses on time and human resources.

On the other hand, as a typical emerging application with HBC technology, in-vehicle health-status monitoring has also been paid much attention to driver assistance systems [6].

Wearable Technology in Medicine and Health Care.
DOI: https://doi.org/10.1016/B978-0-12-811810-8.00011-7

215

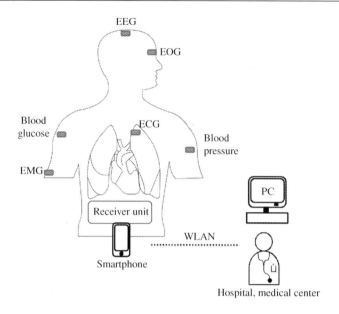

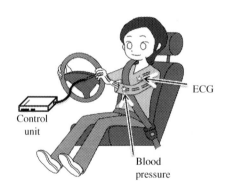

Figure 11.1
BAN applications of health-care monitoring system.

In recent years, in order to satisfy users' requirements, vehicle manufacturers have established mechanisms and devices able to transform this mechanical machine to a healthy and secured environment. More than a transport tool, vehicles must be comfort, convenience, and mainly safety and security from now on. By attaching the above-mentioned wearable sensor such as ECG sensor to the driver in a vehicle, the vital data can be automatically collected, and then forwarded to the steering wheel, which has a built-in receiver unit. According to the real-time monitoring vital data, the control unit in vehicle can grasp the driver's physical condition accurately and give an advance warning to the driver or start the automatic vehicle-control system to prevent accident, if necessary.

However, in real site environments such as homes, hospitals, and vehicles, wireless communication between the medical wearable devices may be degraded by radiated electromagnetic interference (EMI) and electromagnetic compatibility (EMC). So far, the EMI problems for medical devices are often investigated in hospital in mobile communication frequency bands [7]. If taking a case of wearable ECG, the EMI problem is mainly focused on the interference in commercial power frequency bands [8,9]. In fact, near the ECG signal frequencies, many frequency bands exist below 1 MHz for the usage of not only the commercial power frequency, but also broadcast, electronic security check. Besides that, wireless power transfer—as a promising technology for wireless battery charging and power supply utilized in home appliances such as cellular phones, PC, or electric vehicles—is also coming into the public horizon. Its operating frequencies are typically in the low MHz range [10,11]. Therefore, the wireless power transfer system will also become a potential EMI source with wearable devices. For example, the electromagnetic (EM) fields in these frequencies can couple into a wearable ECG through the sensing electrodes to cause an interference voltage in the detected ECG signal. On the other hand, 2014, hailed as a "Year of the Wearable," new wearable products and electronics were highly sought after and expected to become a mainstream product in the near future. Owing to this, the corresponding EMC issues will always rise in the area of body-centric wireless communications or BAN systems. Consequently, how to solve the mutual coupling and interference problems inside the wearable devices still will be a challenge to realize a highly reliable health-status monitoring system.

This chapter introduces an HBC-based wearable technology for sensing and transmitting real-time vital signals in health-care and medical applications. The technology can provide high reliability and security in a wearable system. It also makes the common use of vital signal sensing electrode and transmitting electrode possible so that wearable device structure can be effectively miniaturized. Moreover, this chapter also discusses possible EMI problems occurred in the wearable devices, with a focus on a two-step approach to quantify the interference voltage in a wearable ECG, particularly applied in a wireless power transfer system. The first step in the two-step approach is to derive the EMI voltages produced between the ECG sensing electrodes and the ground plane by EM field simulation/measurement. The second step is to evaluate the interference voltage in the detected ECG signal by an electric circuit analysis or simulation/measurement. The findings will provide an effective means to quantitatively evaluate the EMI for a wearable device, and furthermore give some guidelines for EMI suppression to help engineers in the design stage or practical usage. At last, an EMC test method for wearable devices with pseudo-biological signal generator and biological tissue-equivalent phantom is proposed to provide a solution to assist engineers in the EMC experiments without using realistic human body.

11.2 HBC-Based Wearable ECG System

With respect to the medical and health-care monitoring application, the wearable sensors which gather physical information from patient or elderly people are concentrated on some vital signals such as heart beat, pulse rate, brain waves, and so on. Thanks to this, a very high transmission speed up to several Mbps is not required in principle, so that a relative lower transmission speed below several hundreds of Kbps is sufficient to fulfill a respected performance. From this view of point, HBC technique has remarkable advantages in comparison with traditional wireless communication of low propagation loss, high reliability, as well as high confidentiality [12–15]. This section introduces our developed impulse radio (IR) type HBC transceiver incorporated with a wearable ECG detection system for transmitting the ECG signals, and then demonstrates the real-time vital data transmission in a vehicle in comparison with a commercially available radio frequency (RF) ECG to show its validity and usefulness. Compared to conventional 12-lead clinical ECG or commercial 24-hour portable Holter ECG system [16,17], the HBC-based wearable ECG system lacks any leads or wires and its low profile design allows the users to wear it with minimal interruption of daily activities.

Fig. 11.2 shows the structure of our developed HBC-based wearable ECG [18]. It consists of an ECG detector with two sensing electrodes, one ground electrode, an analog-to-digital (AD) converter that converts analog ECG signals to digital format, an HBC transmitter that transmits the ECG data by HBC technology. It can be realized as a patch-monitoring device which can be attached on the surface of left pectoral region during the measurement and monitoring.

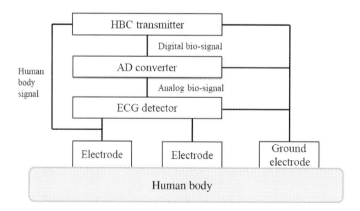

Figure 11.2
Block diagram of HBC-based wearable ECG.

11.3 ECG Detector

The ECG signals are acquired by the two sensing electrodes attached on the human body, then forwarded into the ECG detector. In the ECG detection unit, as can be seen in Fig. 11.3, the ECG detector comprises low pass filter (LPF), band elimination filter (BEF), and differential amplifiers. Since most of the important ECG signals are concentrated on the frequencies below 100 Hz, the two LPFs with a cutoff frequency of 100 Hz are used respectively to obtain the ECG signals below 100 Hz, moreover to suppress the high-frequency interference from HBC signals. While the high pass filters (HPFs) with a cutoff frequency of 10 Hz are used to effectively prevent the invasion of drift noise components, BEFs with 50/60 Hz are used for elimination of commercial power frequency noise which was considered as common-mode noise. In order to increase the input impedance, the voltage followers are employed to connect with a differential amplifier, in which the amplification factor is approximately 60 dB. The amplifier-adjusted ECG signals finally fall into AD converter, and then the digitized ECG signals are sent to the HBC transmitter for real-time data transmission. The AD converter samples the analog ECG signals with a sampling frequency of 500 Hz, quantization level of 10 bits, and amplitude of $0-3$ V.

11.4 Transmitter Structure

In the HBC transmitter, the AD-converted ECG signals are modulated with wide band pulses based on IR scheme. The transmitting pulses are generated by a 4.9 MHz oscillator with a pulse width of 10 ns. In conjunction with the generated pulses through an AND logic gate, the binary ECG signals are modulated by on—off keying scheme, and then forwarded to a band pass filter (BPF). Therefore, its main spectrum components range from 10 to 60 MHz. The clock is obtained by a division circuit with the oscillation frequency of 4.9 MHz, so the corresponding data rate is 1.2 Mbps. In addition, we measured the radiated

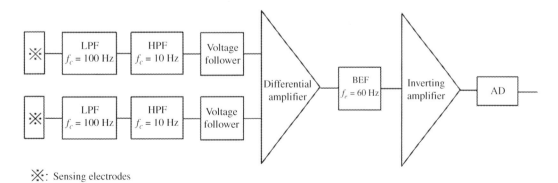

Figure 11.3
Structure of ECG detection circuit.

electric field intensity of the HBC transmitter output in an anechoic chamber and it was confirmed that the measured radiated electric field strength at a distance of 3 m from the transmitter antenna is below 28 dB μV/m. That means we do not need any license to use this HBC transmitter, because the radiated EM field satisfies the Japanese extremely weak radio laws [19].

Moreover, the common use of the sensing and transmitting electrodes based on time sharing and capacitive coupling was realized to simplify and miniaturize the HBC-based ECG structure. On one hand, the two signal electrodes attached on human body are used for ECG signal detection; and, on the other hand, one of the two signal electrodes also acts as signal transmitting electrode used for HBC transmitter. The ground electrode always acts as the ground for both ECG detector and HBC transmitter. So we connect one of the signal electrode to both the ECG detector and the HBC transmitter, and the other signal electrode to only the ECG detector. This structure realizes the common use of the electrodes for both ECG sensing and HBC transmission. The realized IR-type HBC transmitter has a square measure of 3 cm $\times$ 3 cm. In addition, by attaching the ground electrode to the human body, the circuit stabilization can also be improved largely.

11.5 Receiver Structure

In the HBC receiver, the received signal detected by receiving electrode is filtered and amplified, and then adjusted to an adequate level by an automatic gain controller. After the BPF, the shaped signal is demodulated by envelop detection, and then sent to a PC via USB interface incorporated in the HBC receiver. In the PC, we implemented a program to extract the received ECG waveform with a moving average filter and simple error correction technology by means of linear interpolation. When the error data are detected, the mean value of the error data before and after will be taken as interpolation points for correction. After the simple error correction, the received ECG data can be eventually displayed on the PC monitor.

11.6 ECG Transmission Experiment

Fig. 11.4 shows the ECG signal transmission experiment by HBC-based or a commercially available RF wearable ECG in a driving car. The experiment participant sat in the front seat of the car, wearing the HBC-based ECG on the surface of left chest. The HBC receiver was placed before the front glass and connected to a PC next to it by USB cable. By touching the receiving electrode of the HBC receiver, ECG data detected from the experiment participant were continuously sent to the PC in real time. The measured bit error rate (BER) was at a level of 10^{-3} before error correction, which can be expected to realize an error-free communication after employing a forward error correction technology.

Figure 11.4
View of ECG signal transmission in a driving car.

As an instance of health-care monitoring, ECG signals are generally used for heart rate variability (HRV) analysis. As an indicator of the regulation of autonomic nervous system functions, HRV is a physiological phenomenon of variation in the time interval between heartbeats and is measured by the variation in the beat-to-beat interval (RRI). Besides that, there are also other representative parameters for evaluating the autonomic nervous system functions such as the standard deviation of RRI (SDNN) and RR50, which indicates the ratio of number of adjacent intervals differing by over 50 ms in 1-minute period.

In order to verify the validity of HBC-based wearable ECG, we carried out transmission experiment simultaneously with a commercially available RF-ECG and further compared the representative parameters between the HBC-based ECG and the RF-ECG. It was found that, RRI, RR50, and SDNN of HBC-based ECG are in good agreement with RF-ECG. For example, the average difference is no more than 15% for RR50, and only 6% for SDNN. Moreover, as can be seen in Fig. 11.5, the comparison result of RRI is plotted between HBC-based ECG and commercially available EF-ECG. The derived RRI data are almost distributed around one straight line, and the correlation coefficient between them was found to be larger than 0.93. These comparison results indicate that our developed HBC-based ECG can provide the same performance as the commercially available RF-ECG, and can be considered as a good alternative in home and in-vehicle health-care monitoring system.

11.7 EMC Assessment for Wearable ECG

In a real environment using wearable ECG on human chest, disturbances generated by EM fields are unavoidable due to the wide use of commercial power frequency, low frequency, and medium frequency. These frequency sources may couple into ECG detection circuit, affect the performance of the circuit, cause unexpected error functions, or even stop it from

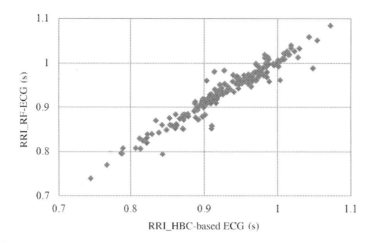

Figure 11.5
Comparison of RRI between HBC-based ECG and RF-ECG.

operations. To take a case of a person who wears an ECG sensor standing on a ground plane, the incident EM field will produce a common-mode voltage between the human body and ground. This is because the human body can be considered to be almost the same as a conductor within our considered frequency range. The common-mode voltage induced by EM field is actually an offset voltage to both of the two input sensing electrodes and cannot be rejected by the differential stage. Therefore, this common-mode voltage will cause EMI problems with the ECG detection circuit if it appears at the output of the detected ECG signal [20].

In order to investigate the influence of the EM field—induced common-mode interference voltage on ECG signal, the ECG detection circuit can be rewritten equivalently as Fig. 11.6. V_c is the common-mode voltage between the ground plane and two sensing electrodes. We here denote the ground plane as earth ground, to make it different from ECG circuit ground. Z_{ea} and Z_{eb} indicate the contact impedance between the two sensing electrodes and human body surface, respectively. When the sensing electrodes are completely contacted with the human body, the contact impedance can be represented as two contact resistance R_{ea} and R_{eb}, respectively. However, when the sensing electrodes are not completely contacted with the human body, the contact impedance can be represented as two coupling capacitance C_{ea} and C_{eb}, respectively. A contact impedance, which is denoted as Z_{eg}, also exists between the ground electrode and the human body surface, and C_s indicates a stray capacitance between the isolated ECG circuit ground and the earth ground. For simplicity, there is no consideration of the functions of various filters, and the amplification factor of differential amplifier is taken as 1. On the basis of differential amplifier circuit theory in this case, the differential output voltage V_{ab} for an ideal op-amp can be written as

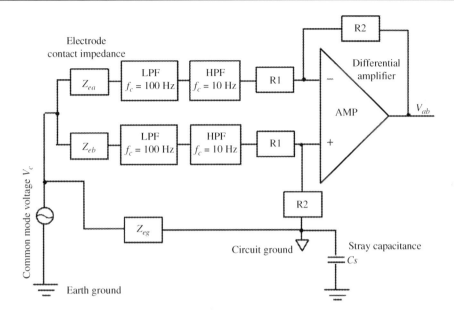

Figure 11.6

Equivalent circuit for investigating the EMI influence of the common-mode interference voltage.

$$V_{ab} = \frac{j\omega C_s R_2 Z_{eg}(Z_{ea} - Z_{eb})}{(Z_{ea} + Z_{eb} + 2R_1)Z_{eg} + (Z_{ea} + R_1)(Z_{eb} + R_1 + R_2) + j\omega C_s(Z_{ea} + R_1)(Z_{eb} + R_1 + R_2)Z_{eg}} V_c$$

(11.1)

In this equation, if Z_{ea} equals Z_{eb}, which means the two contact impedance are balanced, the common-mode input voltage V_c cannot be converted into a differential-mode output voltage V_{ab}. However, generally speaking, this assumption always fails, because the two contact impedance Z_{ea} and Z_{eb} are usually unbalanced or not equal according to different attachment conditions between themselves and the human body. That is the reason why the common-mode voltage V_c can be changed into a differential-mode voltage as an interference voltage at the output of the differential amplifier.

As described earlier, the imbalance of contact impedance Z_{ea} and Z_{eb} is the main causing factor which leads to a differential-mode interference voltage on the detected ECG signal. Therefore, to quantitatively make an EMI evaluation for the wearable ECG, a two-step approach was proposed as follows:

- Quantification of the EM field—induced common-mode voltage V_c between the ECG sensing electrodes and the ground plane
- Quantification of the common mode—converted differential-mode interference voltage V_{ab} appeared at the ECG detection circuit output.

For the first step, the common-mode voltage can be derived by using EM field analysis incorporated with an anatomical human body model [21]. The finite difference time domain (FDTD) method was adopted as the numerical EM field analysis tool in view of its flexibility in modeling anatomical human body model. For example, Fig. 11.7 shows the derived common-mode voltages versus different frequencies below 1 MHz. In this case, the human was assumed standing on the ground plane, irradiated by a uniform plane wave with electric field intensity of 1 V/m. Since the human body was almost considered as a perfect conductor during this frequency range, the induced common-mode voltage between the ground plane and human body was found to be nearly 0.7 V, and seems to be almost flat after all. On the other hand, for the second step, simulation program with integrated circuit emphasis (SPICE) known as a general-purpose circuit simulator was employed to obtain the interference voltage at the output of differential amplifier by using the calculated common-mode voltage in Fig. 11.7. On account of the influence of unbalanced contact impedance for the two sensing electrodes, we performed the circuit simulation with SPICE for different imbalance cases of both the contact resistance (R_{ea}, R_{eb}) and coupling capacitance (C_{ea}, C_{eb}).

As can be seen in Fig. 11.8, as a function of frequency, the differential-mode interference voltage V_{ab} for imbalance contact resistance of 10%, 20%, and 30% is plotted in (A), whereas the case for imbalance coupling capacitance of 10%, 20%, and 30% is plotted in (B), under Z_{eg} equals infinity. These results were derived under an assumption of incident electric field intensity of 10 V/m [22]. Taking the average value of contact resistance R_{ea} and R_{eb} as $R_e = 100$ kΩ, an imbalance of 10% represents, for example, $R_{ea} = 110$ kΩ and $R_{eb} = 90$ kΩ in this way, and so forth. It can be observed from Fig. 11.8 that the differential interference voltage V_{ab} greatly depends on the

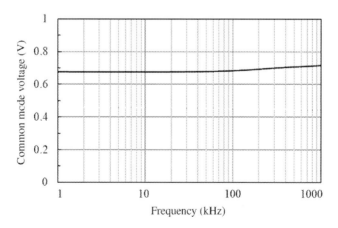

Figure 11.7
Common-mode voltages at the sensing electrodes versus frequency with plane-wave incident electric filed intensity of 1 V/m.

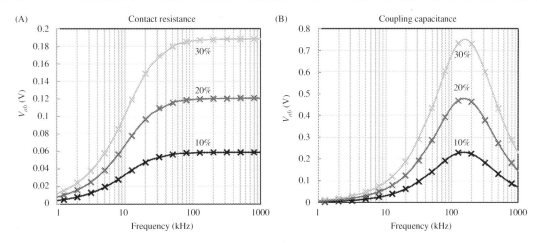

Figure 11.8

Calculated and SPICE-simulated differential-mode interference voltages at the output of differential amplifier versus frequency with plane-wave incident electric field intensity of 10 V/m. The imbalances of the contact impedance are 10%, 20%, and 30%, respectively. Solid line: calculated from Eq. (11.1); symbol: SPICE. (A) $R_{e=}(R_{ea}+R_{eb})/2 = 100$ kW; (B) $Ce_{=}(C_{ea}+C_{eb})/2 = 300$ pF.

imbalance of contact impedance, and increases gradually with the increasing imbalance percentage from 10% to 30% for both of contact resistance case and coupling capacitance case. In addition, in the case of contact resistance, common-mode voltage V_c increases with frequency from 1 to 100 kHz, and keeps constant after 100 kHz, which exhibits an HPF characteristic. In the case of coupling capacitance, V_c shows a maximum value around 150 kHz, yielding a BPF characteristic. These features can also be explained by the equivalent circuit in Fig. 11.6 or the theory expression of Eq (11.1). The solid lines in Fig. 11.8 indicate the results calculated by using Eq. (11.1), while the symbols indicate the results simulated by SPICE with circuit in Fig. 11.6. The calculated V_{ab} shows a good agreement with the simulated one, which ensures the validity and usefulness of both Eq. (11.1) and SPICE simulation in quantifying the EMI interference voltage.

Fig. 11.9 shows the influence of the average contact impedance R_e and C_e on the interference voltage V_{ab} with a plane-wave irradiation at a frequency of 145 kHz. From Fig. 11.9A, the interference voltage V_{ab} decreases with the increasing average contact resistance, and the resistance imbalance effect was found to be more sensitive for smaller contact resistance, i.e., when $R_e = 10$ kΩ, the interference voltage V_{ab} reaches 0.6 V with an imbalance of 30%. On the other hand, Fig. 11.9B shows that V_{ab} may further ascend to nearly 1 V with an imbalance of 30% at 150 pF when changing the coupling capacitance.

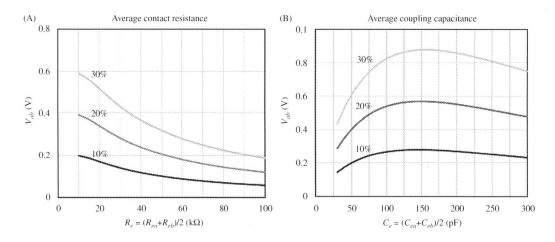

Figure 11.9
Differential-mode interference voltages V_{ab} versus average contact impedance R_e and C_e with plane-wave incident electric field intensity of 10 V/m. (A) Influence of R_e on V_{ab}; (B) Influence of C_e on V_{ab}.

11.8 Application Possibility in Wireless Power Transfer

As for the rising concern and good application prospects on wireless power transfer system, we thus here give an example to illustrate EMI evaluation for wearable devices under in-house wireless power transfer environment at a typical 6.8 MHz, which was considered to be a good candidate for home appliance applications [23,24]. The previously mentioned two-step approach was applied as the evaluation method in this example.

Fig. 11.10 shows a numerical model of 6.8 MHz wireless power transfer system with an anatomical human body model for EMI evaluation. The human body is irradiated by the single-sided wireless power transfer system operating at 6.8 MHz. Here we regard the single-sided wireless power transfer system as the transmitter, and it consists of a single-turn drive loop and a high-Q transmit coil. The single-turn drive loop has an outer diameter of 305 mm, while the transmit coil has a spiral shape with 6.1 turns, an outer diameter of 580 mm, and a pitch of 10 mm. The drive loop and transmit coil are set with either horizontal or vertical arrangement, and the nearest distance between the transmit coil and human body surface is denoted as d.

In the first step, with the numerical model in Fig. 11.10, the common-mode voltage V_c was calculated between the ground plane and human body by using full-wave FDTD simulation. The drive loop was fed with a voltage source with an inner resistance of 50 Ω and a series capacitance of 450 pF for matching. Based on our findings, it was found that the electric field distributes mainly outside the human body, whereas quite weak inside the human body. On the other hand, the vertical placement of the transmit coil exhibits a wider EM distribution range along the human body compared with that of horizontal placement. This also makes a larger common-mode voltage than that of horizontal placement. The corresponding results for 1 A_{RMS} transmit coil current in the 6.8 MHz wireless power transfer system are tabulated in Table 11.1. It can be seen that in either the case of $d = 1$ or 10 cm, the common-mode voltage V_c increases when changing the placement of drive loop and transmit coil from horizontal plane to vertical plane. In addition, if shortening the nearest distance between the transmit coil and human body d from 10 to 1 cm, the common-mode voltage V_c may increase by approximately 10%−30%.

By using the derived common-mode voltage given in Table 11.1, we then further calculated differential-mode interference V_{ab} using Eq. (11.1) and SPICE simulation with reference to Fig. 11.6 for the cases of contact resistance (R_{ea}, R_{eb}) and coupling capacitance (C_{ea}, C_{eb}), respectively. With regard to the influence of contact impedance of the sensing electrodes, an imbalance from 10% to 50% was taken into consideration to investigate the differential-mode interference voltage V_{ab}. Moreover, in order to verify the validity of the calculated interference voltage, we succeeded in fabricating a common-mode equivalent circuit and experimentally measured the differential-mode voltage V_{ab} for different imbalance cases of

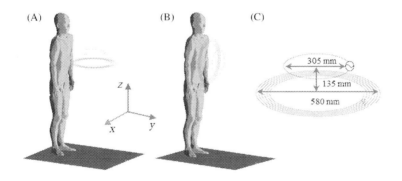

Figure 11.10
Numerical wireless power transfer system at 6.8 MHz near a human body.
(A) Horizontal placement of drive loop and transmit coil. (B) Vertical placement of drive loop and transmit coil. (C) Structure of drive loop and transmit coil.

Table 11.1: Common-mode voltage V_c (V) induced by 6.8 MHz wireless power transfer system with 1 A_{RMS} transmit coil current.

Placement	$d = 1$ cm	$d = 10$ cm
Horizontal	0.50	0.39
Vertical	1.36	1.24

contact resistance of R_{ea} and R_{eb}. The comparison result of V_{ab} with respect to V_c between the SPICE simulation and experimental measurement is shown in Fig. 11.11, in which solid lines represent SPICE-simulated results, and symbols represent measured ones. It can be seen that the simulated interference voltages along with the average contact resistance R_e are consistent with the measured ones with reasonable accuracy, suggesting the validity of the employed two-step approach and the simulated results. In addition, it was also found that the differential-mode interference voltage V_{ab} is mainly caused by contact imbalance of the sensing electrodes, because for each case of imbalance, V_{ab} seems flat and not sensitive to the contact resistance itself. This means that making the contact impedance as small as possible may be the most effective method to reduce the common-mode produced interference voltage for a wearable device. Meanwhile, Table 11.2 summarizes the derived differential-mode interference voltage V_{ab} for 1 A_{RMS} transmit coil current under different considered cases. Overall, the produced differential-mode voltage for the case of contact resistance is larger than that for the case of coupling capacitance, and the vertical placement of the drive and transmit coil also results in a larger interference voltage value than horizontal placement. Moreover, with the increasing distance of the transmit coil from human body, the voltage magnitudes become small obviously. For 1 A_{RMS} transmit coil current, even at a distance of 10 cm from human body, the differential-mode interference voltage may achieve more than 0.6 V for contact resistance case, and 0.1 V for coupling capacitance case, with an imbalance of 50%.

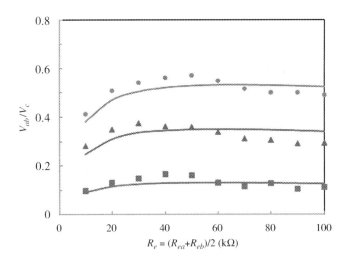

Figure 11.11
SPICE-simulated and experimentally measured differential-mode interference voltage V_{ab} due to the imbalances of contact resistance of R_{ea} and R_{eb} for common-mode input voltage V_c. Lines: SPICE simulated; symbols: experiment measured.

Table 11.2: Differential-mode interference voltage V_{ab} (V) with 1 A_{RMS} transmit coil current for cases of contact resistance and capacitance coupling.

(a) Contact Resistance, $R_e = (R_{ea} + R_{eb})/2 = 100$ kΩ and $R_{eg} = 100$ kΩ

	$d = 1$ cm		$d = 10$ cm	
Imbalance	Horizontal	Vertical	Horizontal	Vertical
10%	0.063	0.172	0.049	0.157
30%	0.171	0.465	0.133	0.424
50%	0.262	0.714	0.205	0.651

(b) Coupling Capacitance, $C_e = (C_{ea} + C_{eb})/2 = 30$ pF and $C_{eg} = 30$ pF

	$d = 1$ cm		$d = 10$ cm	
Imbalance	Horizontal	Vertical	Horizontal	Vertical
10%	0.007	0.018	0.005	0.016
30%	0.022	0.059	0.013	0.054
50%	0.044	0.119	0.034	0.109

11.9 Basic Design Guidelines for Wearable Devices

From the earlier findings, common-mode noise of the order of nearly 0.6 V may occur in a wearable device if the imbalance of two sensing electrodes exceeds more than 30%, either in plane-wave EM field irradiation with an operating frequency range below 1 MHz or a typical wireless power transfer system at 6.8 MHz. Such a level of noise can be able to mask over the detected ECG signals, because the power-amplified ECG signals after the differential amplifier are usually of the order of 1 V. To suppress such EMI interference, one should

- reduce the imbalance of contact impedance of sensing electrodes as much as possible;
- make the impedance of ground electrode as small as possible;
- adopt optimal filters such as LPF, HPF, and BEF to eliminate the common-mode voltage at a specific frequency.

Moreover, introducing a variable impedance after the sensing electrodes, which has a capability of impedance control function based on the detected amount of imbalance between sensing electrodes, may also be a useful means to make them in a balanced state.

11.10 EMC Test Method for a Wearable Device

Wearable devices operated in a body-centric network system require EMC testing to ensure that they can be used for vital signal collection and transmission without failing or causing other devices to fail. So in such a humanized environment, existence of human body cannot be disregarded for EMC test. However, basically in a practical EMC experiment, realistic human body is not permitted, so that the generation of a series of pseudo-signals which can simulate realistic vital signals of human body is necessary, particular in an EMC immunity testing.

Fig. 11.12 shows the composition of our developed pseudo-vital signal generator. The vital signal such as ECG or EMG signal, gathered as a series of digital data with a sampling frequency of 2 kHz and quantization level of 8 bits, is stored as preliminary preparation in a PC for its reproduction. After conversion with optimal amplitude, this digitized vital signal will be sent to the control circuit through serial port, and then the digital-to-analog (DA) converter to reconstruct the analog vital signal with a shaping filter. In this way, it is possible to simultaneously output the reproduced vital signals at a maximum of four channels. As a verification example, the EMG signal reproduced by the pseudo-vital signal generator is shown in Fig. 11.13, to compare with the realistic EMG signal. A perfect match can be seen obviously there, and the correlation coefficient between the reproduced pseudo-vital signal and realistic signal was found to be as high as 0.99. Such a well-reproduced pseudo-EMG signal can provide a good insight into the immunity testing for application of myoelectric-controlled arm prosthesis. In this case, the pseudo-EMG signal can be learned and translated into information by myoelectric prosthetic arm, and then the electric motors can use the translated information to control the artificial arm movements as one expected. Of course, the possibility of using a pseudo-EMG signal in an immunity testing for myoelectric-controlled arm prosthesis has been confirmed experimentally.

For the wearable devices attached on human body, we thus can provide a basic immunity test method as depicted in Fig. 11.14, an illustration of immunity testing system using pseudo-vital signal generator and biological tissue-equivalent phantom. A solid biological

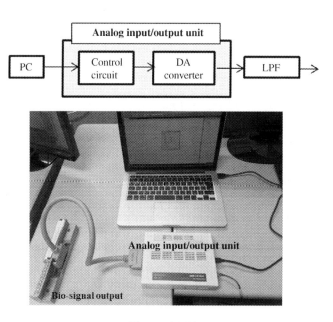

Figure 11.12
Composition of pseudo-biological signal generator for EMC testing.

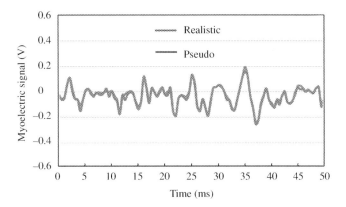

Figure 11.13

Comparison of the myoelectric signal reproduced by the pseudo-biological signal generator and the realistic one.

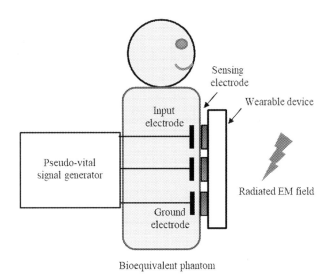

Figure 11.14

Mechanism of EMC testing for a wearable device based on pseudo-vital signal generator and biological tissue-equivalent phantom.

tissue-equivalent phantom is used as the substitute of realistic human body. In order to make the biological tissue-equivalent phantom act as human, the pseudo-vital signal generator is embedded into the phantom by means of connecting with two output signal electrodes inside the phantom. In this way, immunity experiments for the wearable devices under test can be conducted to verify whether or not they cause malfunction for the detection and transmission of pseudo-biological signal. Therefore, instead of realistic human body, our suggested testing method by using a biological tissue-equivalent phantom incorporated with a pseudo-vital signal generator can provide a good solution to deal with these kinds of wearable devices for engineers in the corresponding EMC tests. Since the international standards on EMI and EMC test are still under way, this work also contributes to providing suggestions and promoting international standardization process for wearable medical devices.

11.11 Conclusions

In this chapter, we have introduced an HBC-based wearable ECG for transmitting real-time vital signals on the human body in assisting home and vehicle monitoring systems. Its main feature lies in the common use of sensing and transmitting electrodes that has greatly simplified and miniaturized the structure of the wearable ECG. Such a wearable ECG has been clarified to possess a comparable performance with a commercial available RF-ECG, so that its feasibility in actual scenarios is worth looking forward to.

In accordance with wide-scale use of wearable products in recent years, a two-step approach has been proposed to quantitatively evaluate the EMI problem occurred in a wearable health-care device. The two-step approach involves a combination of EM field analysis for common-mode noise by using FDTD method and electric circuit analysis for converted differential-mode noise by SPICE simulation or fabricated circuit measurements. It has been demonstrated that the main reason of the EMI interference for a wearable device is due to an imbalance between the contact impedance of the two sensing electrodes. As a result, several basic design guidelines for wearable devices have also been suggested to suppress this kind of interference. This two-step approach should be an effective means for engineers to solve EMI problems in the wearable device circuit design stage.

Moreover, we have introduced an EMC test method with regard to a solution for investigating whether the wearable devices attached on the human body work or not in an immunity experiment. Instead of realistic human body, a pseudo-biological signal generator, which can simulate realistic bio-signals of human body, has been incorporated into a human tissue-equivalent phantom through two output electrodes. Such a suggested EMC test method without using realistic human body can be expected to provide a good solution to deal with these kinds of wearable devices for engineers in the corresponding testing experiments.

References

[1] J. Wang, Q. Wang, Body Area Communications, Wiley-IEEE, Singapore, 2012.
[2] IEEE Std 802.15.6-2012: IEEE Standard for local and metropolitan area networks—Part 15.6: Wireless Body Area Networks, February 2012.

[3]   P. Bonato, Wearable sensors and systems—from enabling technology to clinical applications, IEEE Eng. Med. Biol. Mag. 29 (3) (2010) 25−36.

[4]   E. Monton, J.F. Hernandez, J.M. Blasco, T. Herve, J. Micallef, I. Grech, et al., Body area network for wireless patient monitoring, IET Commun. 2 (2) (2008) 215−222.

[5]   E. Nemati, M.J. Deen, T. Mondal, A wireless wearable ECG sensor for long-term applications, IEEE Commun. Mag. 50 (1) (2013) 60−66.

[6]   I. Misumi, et al., A study on designs of wearable systems for car drivers' fatigue evaluation, Micromechatronics 47 (2) (2003) 1−10.

[7]   A. Baranchuk, J. Kang, C. Shaw, D. Campbell, S. Ribas, W.M. Hopman, et al., Electromagnetic influence of communication devices on ECG machines, Clin. Cardiol. 32 (10) (2009) 588−592.

[8]   E.M. Spinelli, N.H. Martinez, M.A. Mayosky, A transconductance driven-right-leg circuit, IEEE Trans. Biomed. Eng. 46 (12) (1999) 1466−1470.

[9]   A.C.M. van Rijn, A. Peper, C.A. Grimbergen, High-quality recording of bioelectric events, Med. Biol. Eng. Comput. 28 (5) (1990) 389−397.

[10]   F. Musavi, W. Eberle, Overview of wireless power transfer technologies for electric vehicle battery charging, IET Power Electrics 7 (1) (2014) 60−66.

[11]   L. Xie, Y. Shi, Y.T. Hou, A. Lou, Wireless power transfer and applications to sensor networks, IEEE Wireless Commun. 20 (4) (2013) 140−145.

[12]   J. Wang, et al., Analysis of on-body transmission mechanism and characteristic based on an electromagnetic field approach, IEEE Trans. Microwave Theory Tech. 57 (10) (2009) 2464−2470.

[13]   M. Shinagawa, et al., A near-field-sensing transceiver for intrabody communication based on the electro-optic effect, IEEE Trans. Instrum. Meas. 53 (12) (2004) 1533−1538.

[14]   H. Baldus, et al., Human-centric connectivity enabled by body-coupled communications, IEEE Commun. Mag. (2009) 172−178.

[15]   Y. Kado, M. Shinagawa, AC electric field communication for human-area networking, IEICE Trans. Electron E93-C (3) (2011) 234−243.

[16]   L. Su, S. Borov, B. Zrenner, 12-Lead Holter electrocardiography. Review of the literature and clinical application update, Herzschrittmacherther Elektrophysiol. 24 (2013) 92−96.

[17]   http://www.expressdiagnostics.co.uk/patients/heart-tests/24-hour-holter-ecg/

[18]   J. Wang, T. Fujiwara, T. Kato, D. Anzai, Wearable ECG based on impulse radio type human body communication, IEEE Trans. Biomed. Eng. (2015). Available from: https://doi.org/10.1109/TBME.2015.2504998.

[19]   http://www.tele.soumu.go.jp/j/ref/material/rule/ (in Japanese).

[20]   W. Liao, J. Shi, J. Wang, An approach to evaluate electromagnetic interference with wearable ECG at frequencies below 1 MHz, IEICE Trans. Commun E98-B (8) (2015) 1606−1613.

[21]   T. Nagaoka, S. Watanabe, K. Saurai, E. Kunieda, S. Watanabe, M. Taki, et al., Development of realistic high- resolution whole-body voxel models of Japanese adult males and females of average height and weight, and application of models to radio-frequency electromagnetic-field dosimetry, Phys. Med. Biol. 49 (1) (2004) 1−15.

[22]   International Standard IEC61000-4-3, Electromagnetic compatibility—Part 4-3: Testing and measurement techniques—radiated, radio-frequency, electromagnetic field immunity test, February 2006.

[23]   A. Sample, D. Meyer, J. smith, Analysis, experimental results, and range adaptation of magnetically coupled resonators for wireless power transfer, IEEE Trans. Ind. Electron. 58 (2) (2010) 544−554.

[24]   A. Christ, M.G. Douglas, J.M. Roman, E.B. Cooper, A.P. Sample, B.H. Waters, et al., Evaluation of wireless resonant power transfer systems with human electromagnetic exposure limits, IEEE Trans. Electromagn. Compat. 55 (2) (2013) 265−274.

CHAPTER 12

Wearable Technologies for Personalized Mobile Healthcare Monitoring and Management

Sandeep Kumar Vashist[1] and John H.T. Luong[2]

[1]*Immunodiagnostic Systems, Liege, Belgium* [2]*Sektion Experimentelle Anaesthesiologie, University Hospital Ulm, Ulm, Germany*

12.1 Introduction

Wearable technologies (WT) enable cost-effective and affordable personalized mobile healthcare monitoring and management (PMHMM) at any place and time, which obviates the need for skilled healthcare professionals and advanced healthcare facilities. Therefore, they are ideal for delivering mobile healthcare (mH) to remote, resource-deficient, decentralized and personalized settings. The smartphones (SPs) have become truly advanced as they are equipped with sophisticated features, data processing capabilities, and various sensors, such as light detectors, proximity sensors, fingerprinting, accelerometer, and high-resolution cameras. They have excellent connectivity to the personal Cloud and secure central server. Therefore, they can collect and store the real-time spatiotemporal tagged personalized data, which can be assessed by certified healthcare professionals from remote locations. This would be very useful for telemedicine applications, especially in tackling epidemics and emergency cases [1,2]. Cell phones have become ubiquitous with more than 7 billion cell phone subscribers worldwide out of which about 3 billion have the internet also on their cell phones [3]. Moreover, the continuously increasing capabilities of SP and their evolving technological features with additional functionalities provide a major impetus to the emerging trend towards WT-based PMHMM. This is further supported by the fact that about 70% of the SP users are from the developing countries, where there is an enormous need for cost-effective mH. Therefore, SP-based WT (SPWT) would play a critical role in PMHMM as they would establish a new trend in healthcare delivery, which would cut down the healthcare costs significantly and would

Wearable Technology in Medicine and Health Care.
DOI: https://doi.org/10.1016/B978-0-12-811810-8.00012-9

lead to better health outcomes [4,5]. Moreover, such WT would lead to the creation of new businesses and opportunities.

This chapter provides a comprehensive review of commercial WT for PMHMM including their technology features and prospective applications. Only the main commercial SPWT devices that capture a significant market share are mentioned. The product specifications and features are taken from the product manuals and literature while the prices are taken directly from the websites. WT enable the real-time and frequent monitoring of basic physiological parameters for critically improved personalized mH [6−8]. The personalized mH monitoring is critical in the management of chronic health conditions such as diabetes, obesity, hypertension, and psychological stress [1,9−27]. Apart from increasing the healthcare outreach by connecting isolated remote laboratories [8,28], the centralized PMHMM provided by WT would improve the patients' adherence to checkups, treatment and medication [29−34]. WT would enable the effective management of chronic diseases [35,36], lead to improved communication between healthcare professionals [37,38], and assist in the prevention of infectious diseases [39,40]. Other prospective applications are the delivery of education [41,42], large-scale screening of community for a particular disease [43], improved adherence of parents to the immunization schedules [44] of their children, and improved healthcare management [45]. Moreover, WT are highly useful in tacking emergencies as they facilitate the spatiotemporal mapping of disease incidence [46].

12.2 Wearable Technologies

A wide range of WT has been developed for the monitoring of basic physiological parameters. Most of the WT are developed by leading industrial giants, such as iHealth Labs, Runtastic, Apple, and Samsung, but several other companies with novel product portfolio, such as Cellmic and AliveCor, have also emerged during the recent years.

12.2.1 Blood Glucose Meters

iHealth has developed two blood glucose meters, i.e., Wireless Smart Gluco-Monitoring System and iHealth Align (Fig. 12.1), which are Food and Drug Administration (FDA) approved and Conformité Européenne (CE) compliant, and meet the ISO 15197:2003 in vitro blood glucose monitoring requirements. They are operated by iHealth Gluco, a dedicated smart application that is available free at iOS and Android stores, that requires users to log in using their specific details. The devices enable effective diabetes management by enabling more frequent blood glucose monitoring that enables diabetics to keep their blood glucose level within the normal physiological range by physical activity and/or diet-based interventions.

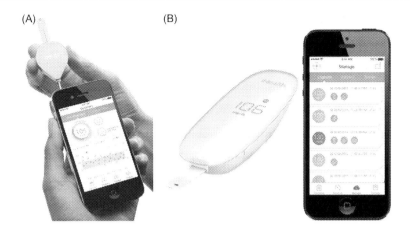

Figure 12.1
Smartphone-based blood glucose meters developed by iHealth Inc. (A) iHealth Align. (B) Wireless Smart Gluco-Monitoring System. *Reproduced with permission from iHealth Labs, Inc.*

The iHealth Align plugs into the headphone jack of the SP and is connected to the SP by Bluetooth technology. It is the smallest glucose meter available on the market that detects the entire diabetic blood glucose pathophysiological range (1.1−33.3 M) in 5 seconds using 0.7 μL of fresh capillary whole blood. It requires the users to perform the blood glucose measurement by pricking their fingers and uploading the sample onto the test strip plugged into the iHealth Align. The glucose measurement is performed by standard amperometric detection of blood glucose using glucose oxidase. The results are displayed on the SP's screen in a few seconds and stored securely inside the smart app, iHealth Gluco. The device can use only iHealth test strips (costs US$ 12.50 for a pack of 50 strips) and can be purchased online for US$ 16.95. The iHealth Gluco app enables users to view trends and statistics of their blood glucose readings for up to 90 days and automatically determines the remaining test strips in the vial and their expiry based on which it alerts the users to buy a new vial. The app acts as an automatic logbook where users can add notes to their stored readings and set up reminders for medication and insulin dosages, and recording of premeal and postmeal glucose. The users can share their results with family members, friends, caregivers, and doctors.

The Wireless Smart Gluco-Monitoring System is a miniaturized standalone blood glucose meter that can be purchased online for US$ 29.95. It is interfaced wirelessly with the SP via the Bluetooth technology and the smart app. It comes with a light emitting diode (LED) display and is powered by a built-in, rechargeable battery. The standalone device once synced with the SP can be used for glucose measurements without the SP for up to 500 readings that are stored in the device. The measurements are then transferred to the SP when they are connected by Bluetooth.

12.2.2 Blood Pressure Monitors

The frequent monitoring of blood pressure (BP) is desired for effective hypertension management and treatment as it enables the users and healthcare professionals to predict the cardiovascular risks. It is also an important physiological parameter to be monitored frequently for personalized healthcare management. Various FDA-cleared, and European Society of Hypertension (ESH) and CE certified BP monitors have been developed by iHealth (Fig. 12.2). It requires the users to turn on the Bluetooth on their SP, start the iHealth MyVitals app (interfaced it to the BP monitor), and perform the BP measurement. The SP's screen guides the user for an appropriate hand posture for the measurement that takes a few seconds. The BP results are presented in the form of the visual chart based on the classification provided by World Health Organization (WHO) and stored on the SP. The WHO classification provides an easy-to-read chart with an indication for normal, moderate or high BP ranges along with the trend in the BP readings. The BP monitors have integrated rechargeable batteries that ensure their use for many days, but they can be recharged via the provided USB cable. The Wireless BP Wrist Monitor (sold at US$ 79.95) is watch-shaped with an inflatable wristband, which monitors the BP on the wrist and displays the reading on the SP. The iHealth View Wireless BP Wrist Monitor (priced at US$ 99.99) is the newest model with the same function that shows the reading on the device itself. The other wearable device is the Wireless BP Monitor (available at US$ 99.95) that works like the conventional clinical BP monitor. It performs the BP measurement on the upper arm but is operated via the Bluetooth-interfaced SP using the

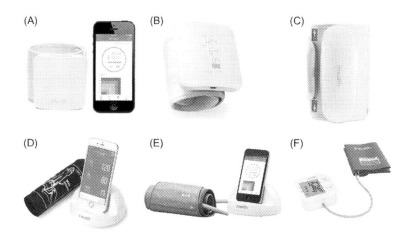

Figure 12.2
Smartphone-based blood pressure (BP) monitors developed by iHealth Inc. (A) Wireless BP Wrist Monitor. (B) iHealth View Wireless BP Wrist Monitor. (C) Wireless BP Monitor. (D) iHealth Wireless Ease BP Monitor. (E) iHealth BP dock. (F) iHealth Track. *Reproduced with permission from iHealth Labs, Inc.*

iHealth MyVitals app. The iHealth Wireless Ease BP Monitor (priced at US$ 39.99) and BP dock (sold at US$ 39.99) are the previous versions of BP monitors that are sold by iHealth. The iHealth Track is the most recent smart BP monitor that can be used even without any SP. It is a Class IIa medical device that is CE certified and can store up to 60 measurements before it requires synchronization with the SP via the iHealth MyVitals app. It is available in European Union (EU) at the price of 39.95€ and not yet available for sale in the United States.

12.2.3 Weighing and Body Analysis Scale

The wireless body composition scale, iHealth Core (Fig. 12.3A), is an FDA and CE approved PMHMM device that measures the body analysis parameters using advanced data processing algorithms and electronic sensors for measuring body compositions. The measured body analysis parameters are weight, basal metabolic index (BMI), body fat, lean mass, muscle mass, bone mass, body water, daily calorie intake (DCI), and visceral fat rating. The iHealth MyVitals app is opened after switching on the Bluetooth, which interfaces it to the scale. The measurements taken on the scale are then transmitted to the SP. The scale, sold in the United States at a price of US$ 129.95, requires four 1.5 V AAA batteries and weighs about 2.5 kg. It measures body weight, body fat, body water, and visceral fat in the ranges of 5−180 kg, 5%−65%, 20%−85% and 1−59, respectively. The wireless body analysis scale (priced at US$ 99.95) and iHealth Lite weighing scale (available for US$ 79.95) are the previous models. Runtastic Libra (available online for US$ 121.31) is a Bluetooth enabled digital body analysis scale developed by Runtastic GmbH, Austria (Fig. 12.3B). It measures the body weight (up to 180 kg), body fat, body water percentage, muscle mass, bone mass, BMI, basal metabolic rate (BMR), and active metabolic rate (AMR). The scale is made up of a highly resistant glass plate coated with

Figure 12.3

Smartphone-based body analysis/weighing scales. (A) (left to right) iHealth Core, Wireless body analysis scale, and iHealth Lite. (B) Runtastic Libra weighing scale. *(A) Reproduced with permission from iHealth Labs, Inc. (B) Reproduced with permission from Runtastic GmbH.*

indium tin oxide (ITO) electrodes, weighs 2.5 kg and runs on 3×1.5 V AAA alkaline batteries. The Bluetooth Smart technology transfers the measurements from Runtastic Libra to the SP/tablet located within 25 m via the Runtastic Libra app. The device employs bioelectrical impedance analysis (BIA), using an imperceptible, completely harmless and safe alternating current, for the measurements of various body metrics. The scale can be used by up to eight users, who can log on to their account at www.Runtastic.com to get detailed statistics and in-depth analysis. However, it should not be used by persons with medical implants (e.g. pacemakers), children below 10 years old, and pregnant women. It has potential interference from strong magnetic fields, which interferes with the signal transmission.

12.2.4 Activity and Sleep Monitors

12.2.4.1 iHealth Labs

The iHealth Edge (priced at US$ 69.95) (Fig. 12.4A) is the wireless daily activity and sleep tracker by iHealth Inc., which is worn on the wrist like a watch or is put in the back pocket of the pants. It has advanced power-saving algorithms, which enables its use for many days on a single recharge. A USB charging cable with the magnetic contactor, which contains contact pads that align and attach to the charge pins provided at the device base, charges the internal rechargeable 3.7 V lithium-polymer battery 100 mAh. The device has a three-axis accelerometer that detects the 3D-motion patterns and counts the number of steps taken, the distance traveled, and the burnt calories. The data is stored continuously in the device's internal memory and transferred to the iHealth MyVitals app on SP when the device is synced via Bluetooth. The data can be shared with family members, friends, doctors, or caregivers via the app. The device is sweat, rain, and splash proof, and made of hypoallergenic and skin-friendly thermoplastic polyurethane (TPU) rubber, i.e., latex- and polyvinyl chloride (PVC)-free. It also monitors continuously the sleep efficiency score and the sleep duration by automatically switching modes. The company has recently launched iHealth Wave (sold at

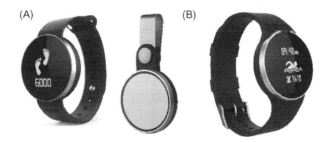

Figure 12.4
Physical activity and sleep monitors developed by iHealth Inc. (A) iHealth Edge. (B) iHealth Wave.
Reproduced with permission from iHealth Labs, Inc.

79.95 € in Europe) (Fig. 12.4B), the recent version of the device, which has the same functions but can also track swimming parameters such as swim strokes (breaststroke, crawl, and backstroke), number of movements, movements per minute, and burnt calories.

12.2.4.2 Runtastic GmbH

Runtastic GmbH [47] has developed various FDA-cleared, and CE-marked global positioning system (GPS) enabled WT devices for sports and fitness (Fig. 12.5). These include Runstastic Heart Rate Combo Monitor, Runtastic GPS Watch, and Heart Rate Monitor. The Runtastic activity tracker and the Heart Rate Monitor (priced at US$ 149.99) tracks and monitors the pace, speed, duration, distance, lap times, training zones, burnt calories and heart rate together with the elevation change, and target heart rate training zones. The device is charged via a USB cable containing six pins. The watch weighs 57.5 g while the monitor weighs 60 g. The Watch has a battery life of up to 14 h (while using GPS), a customizable display, night light function, a reliable compass with navigation functions, and other functions. The fitness data from the device is transferred to the user's personal account on www.Runtastic.com that can be assessed through a computer, SP or tablet connected to the internet. The website provides a detailed analysis of the results that is shared online with the Runtastic community. The device synchronizes automatically with the smart app on the SP by Bluetooth Smart technology. The company has launched a

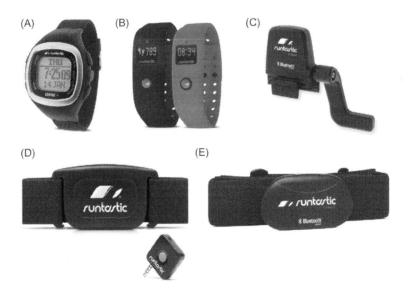

Figure 12.5
Wearable mH devices developed by Runtastic GmbH. (A) Runtastic GPS Watch. (B) Runtastic Orbit. (C) Runtastic Speed and Cadence Sensor. (D) Bluetooth Low Energy Chest Strap of Runstastic Heart Rate Combo Monitor. (E) Chest Strap and Dongle of Runtastic Heart Rate Monitor. *Reproduced with permission from Runtastic GmbH.*

series of GPS activity watches, the most recent version being the Runtastic Orbit (sold at US$ 99.99), which is a 24-hour activity, fitness and sleep tracker that is also equipped with time and alarm modes. This waterproof device has an integrated ambient light sensor for advanced sleep tracking. Runtastic Speed and Cadence Sensor (sold at US$ 58.62) enable the cyclist to convert their SP into a professional cycling computer in combination with the Runtastic Road Bike or Runtastic Mountain Bike app. It records the speed, cadence, GPS position, distance, altitude, and other performance parameters. The WT gadgets from other companies, such as Fitbit [48], Garmin [49], and Nuband [50,51], are very similar to the described WT and therefore, these are not discussed here.

The Runtastic Heart Rate Combo Monitor (sold at US$ 69.99), comprising a transmitter and a splash-proof chest strap powered by lithium batteries, measures the heart rate. The heart rate measurements from the chest strap (transmitter) are transmitted to the SP via a low power Bluetooth Smart technology. The chest strap sends the heart rate signal at a transmission frequency of 5.3 kHz, which is compatible with most training devices in the gym but requires a separate Runtastic receiver (charged by a lithium battery). The contacts of the transmitter are moistened by an ECG gel and connected to the chest strap with the help of two push buttons. The elastic strap is then adjusted so that the sensors contact the chest below the pectoral muscles. The users have to download the Runtastic app on their SP and set up their personalized account before they can use the devices that are interfaced to the SP by Bluetooth. The app displays the heart rate measurements in real-time during the exercises. The plugging in of the Runtastic receiver into the headphone jack of SP activates it to provide the real-time heart measurements from the transmitter immediately on to the SP after enabling the receiver ON in the SP settings. The receiver's battery lasts about 2.5 years if it is used daily for 1 hour. However, the chest strap might contain latex components that can trigger allergic reactions such as skin irritations and redness in some persons. Further, the device must be protected from the strong magnetic field that interferes with the signal and decreases its accuracy. Persons with pacemakers should consult their doctor before using the device.

12.2.4.3 Apple

Apple Watch [52], the latest WT for PMHMM, is a useful device that can be purchased available from Apple Store at a price ranging from US$ 549–1099 based on the customized accessories and specifications. It monitors the physical activity, heart rate, sleep, and calories burnt throughout the day using robust and highly sensitive sensors such as Heart Rate Monitor, accelerometer, gyroscope, barometer, ambient light sensor, and microphone. Moreover, it has solid data analysis and power-saving algorithms. It provides an easy-to-read dashboard of the health and fitness data of the user on the screen in addition to providing notifications and message alerts for better compliance. The Workout app provides a personalized list of exercises together with the last and best results for each activity. It facilitates navigation by tapping on the wrist whether the user has to take a right or left turn, and has Siri as the virtual assistant. The users can create their emergency card that provides information on their blood type, allergies, and disease conditions, which is available even on the locked screen of the Watch. They have complete control over their data and can share their individual health and fitness data with friends, family, and careers. The data can also be backed up in the iCloud account of the user. The company has stringent data security and privacy guidelines in compliance with the regulatory requirements.

The user can also make a phone call from the Watch after interfacing and connecting to the SP by Wi-Fi. Apple provides HealthKit to developers for making health and fitness applications. It also provides the ResearchKit and CareKit, which are open source software frameworks for medical researchers, doctors, and users. The developers can use ResearchKit to create applications for medical researchers to gather the desired data for their studies. Similarly, they can make applications for users willing to take more active role in their own well-being. The company is pursuing intensive research in human health and sports at highly specialized labs, which would enable the implementation of robust health monitoring technologies and sophisticated algorithms in the Apple Watch and iPhone after fine tuning the sensors, radios, and other components. Most users are looking for more healthcare capabilities, such as noninvasive technologies, a pulse oximeter, and a glucose sensor, in the Apple Watch 2. But this substantiates the need for taking the FDA approval. Apple recently announced the Breathe app for Apple Watch 2, which guides the users through various breathing exercises to calm them down in stressful situations resulting in a feeling of complete relaxation within a minute. The new operating system would also have the SOS app that would get activated just by holding the side button. It will send the user's location and a message to a registered contact and the emergency services.

12.2.4.4 Samsung

Gear S2 [53] (available online at a price range of US$ 299.99−449.99), a smart watch by Samsung, has very similar functions as Apple Watch but it is circular and has a rotating bezel for navigating easily through notifications, applications, and widgets. The user can receive calendar notifications, text alerts and new updates on Gear S2 after interfacing and connecting it to SP. Gear S2 has integrated accelerometer, gyroscope, heart rate sensor, ambient light sensor, and barometer. A dedicated smart app, S Health app, allows users to track their steps, heart rate and get alerts to breaking down their sedentary behavior. The Gear S2 is charged by putting it onto the wireless charging dock and has the IP68 rating for its resistance to dust, dirt, water, and sweat. It has 4 GB internal memory, 512 MB RAM, Dual-Core 1.0 GHz processor, 250 mAh Li-ion battery, Bluetooth v4.1, NFC. and Wi-Fi (802.11 b/g/n 2.4 GHz).

Samsung Gear Fit2 [54] (sold at US$ 179.99) is a slimmer built-in GPS based smart watch that enables the user to track his step count, floors climbed, sleep quality, and heart rate throughout the day. The built-in GPS provides real-time stats of the physical activity of the user on the map. Samsung Gear Fit2 automatically recognizes the types of activities of the user, such as cycling, running, strength training, yoga, pilates, working out on the elliptical, and other activities. The user receives all notifications from desired apps on the phone and can also respond to calls and text. The charging is done simply by placing it on the charging cradle and a single charge can last up to 4 days. It has an IP68 rating for resistance to dust, dirt, water, and sweat. It has 4 GB internal memory, 512 MB RAM, Dual-Core (1 GHz Exynos 3250) processor, 200 mAh Li-ion battery, lightweight (just 30 g), Bluetooth v4.1, and Wi-Fi. (802.11 b/g/n).

Samsung has also developed an open source reference design platform called Simband [55], which includes an open reference sensor module that integrates various advanced sensing technologies. The product has the open software, hardware, and mechanical architecture that facilitates the development of new digital health solutions. The potential sensors, integrated into Simband in the multi-sensor module called Simsense, are multichannel photoplethysmograph (PPG), ECG, bioimpedance (Bio-Z), galvanic skin response (GSR), accelerometer, and temperature. The multichannel PPG measures the changes in the blood flow at the microvascular level and determines the BP, heart rate and other physiological parameters. ECG complements the working of the PPG sensor by including the pulse arrival time into the estimation of BP while bioimpedance monitors a range of parameters such as blood flow and body fat. GSR sensors monitor the electrical conductivity of the skin that could be used for the determination of stress levels. Simband has an empty bucket to house customized sensor modules close to the arteries in the anterior part of the wrist. This would enable the digital healthcare professionals to place their desired sensor module inside the Simband utilizing the power and a secure communication provided by Simband. Moreover, it has Bluetooth and Wi-Fi communication to provide access to the Cloud and Samsung Architecture for Multimodal Interactions (S.A.M.I.) data platform. Additionally, Simband can be operated in various modes i.e. monitoring mode, collection mode, continuous collection mode, and fitness mode for different applications.

12.2.5 Pulse Oximeters

12.2.5.1 iHealth Labs

The Wireless Pulse Oximeter (Fig 12.6A), an FDA-cleared and CE-marked device priced at US$ 69.95, is a lightweight and miniaturized device for the noninvasive determination of blood oxygen saturation (SpO$_2$) and the pulse rate at the fingertip by shining two light beams into the finger capillaries. The device runs on an integrated rechargeable battery

(A)   (B)

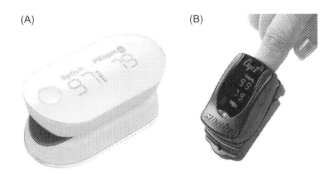

Figure 12.6

Pulse oximeters. (A) Wireless Pulse Oximeter developed by iHealth Lab Inc. *Reproduced with permission from iHealth Lab Inc. (B) Onyx II Model 9560 Finger Pulse Oximeter developed by Nonin Medical, Inc. Reproduced with permission from MDPI (http://www.alivecor.com/home).*

(3.7 V Li-ion, 300 mAh) that can be recharged using the provided USB cable. It is beneficial for patients with breathing difficulties (pulmonary dysfunction), COPD (chronic obstructive pulmonary disease), coronary heart diseases, and other vascular conditions. Moreover, the athletes or users can understand the working of their bodies during recreation activities and high-intensity sports or exercises. The device is interfaced and connected to the SP by the Bluetooth and iHealth MyVitals app. The current measurement is displayed on the oximeter's screen and recorded into the device's memory while the iHealth MyVitals app shows the trends in measurements and can share the information with other family members, friends, doctors or caregivers. The SpO$_2$ level indicating the amount of oxygen in the blood as a percentage of the maximum carrying amount is between 96% and 99% for a healthy individual but can be affected by many factors including high altitudes. The device measure SpO$_2$ in the range of 70%–99% with the accuracy of $\pm$ 2%. The normal resting pulse rate in humans is between 60 and 100 beats per minute (bpm), but it is also dependent on the fitness level, body weight, emotional state, medication, body position, and the involvement in physical activities. Therefore, the optimum reading guidelines provided in the product insert must be followed as several factors can affect the reading and lead to inaccuracy in results such as cold hands, fingernail polish, acrylic nails, hand movements, and weak pulse.

12.2.5.2 Nonin Medical, Inc.

Nonin Medical, Inc. [56] developed Onyx II Model 9560 Finger Pulse Oximeter (Fig. 12.6B), a compact and lightweight device sold at US$ 666, which measures SpO$_2$ and the pulse rate of well or poorly perfused patients, such as patients with congestive heart failure, asthma, COPD, and other cardiovascular disorders. The technology involves noninvasive pulse oximetry method [57,58], where the red (660 nm) and infrared (910 nm) lights are passed through the perfused tissue, and the fluctuating signal due to arterial pulses is detected. The SpO_2 levels are determined from the difference in color that is analyzed by the ratio of the absorbed red and infrared light. The well-oxygenated blood is characterized by bright red color while the poorly oxygenated blood is dark red. The device can perform the measurements on a large number of subjects with the thickness of the fingers varying from 0.8−2.5 cm and can be used in persons having light to dark skin tones and good to low perfusion. The device stores at least 20 single point measurements and transmits the results wirelessly up to 100 m. The advanced SmartPoint algorithm enables the automated determination of SpO_2 and the pulse rate and employs a sophisticated power-saving feature that automatically adjusts the transmitted power based on the distance between the device and the main unit. The PureSAT signal processing technology ensures high precision by effectively removing the noise, artifacts, and interferences. It measures SpO_2 and the pulse rate in the ranges of 0%−100% and 18−321 bpm, respectively, and works for over 1 year using two AAA batteries. The average root mean square (RMS) accuracies for the SpO_2 range of 70%−100% and the pulse rate of 20−250 bpm are ± 2 and ± 3, respectively. The device is certified to Microsoft HealthVault, a free online platform that communicates and receives data from the device for personalized healthcare management. It complies with the IEC 60601-1-2 for electromagnetic compatibility, part 15 of the Federal Communications Commission (FCC) Rules as a Class B digital device, which can be sold by or on the order of a licensed practitioner.

12.2.6 Electrocardiogram

12.2.6.1 iHealth Labs

iHealth Rhythm (sold at US$ 150) (Fig. 12.7) is smart one lead ECG that can be provided only by the doctor. It is a Class IIa regulated medical device that carries CE0197 mark while the FDA approval is still pending. The small (10 cm) and lightweight (20 g) device consists of a single ultra-flat recorder that clips onto a consumable three-electrode patch. Being light, wearable, and invisible under the clothes, the device monitors the cardiac activity of the patient for up to 72 hours, when placed on the patient sternum. It can detect four types of arrhythmia as well as atrial fibrillation and has an 'event recorder' button that marks the time whenever the symptoms of abnormal heart activity appear. The patients can press the 'event recorder' button on the device when they are experiencing abnormal chest pain, which will enable the device to precisely indicate the anomaly on the electrocardiogram (ECG) report. It synchronizes the ECGs taken by iHealth Rhythm via Bluetooth connection on the free iHealth PRO app for iPAD, which enables the healthcare professionals to visualize, store and organize ECG results. The reports can be edited in PDF format and shared with colleagues. The device would not interfere with the normal patient activities as the subject can wear it even when taking a shower.

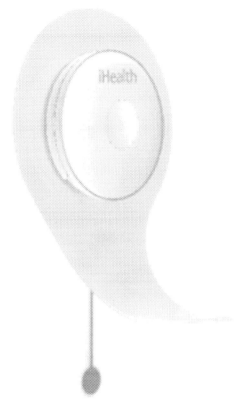

Figure 12.7
Electrocardiogram developed by iHealth Lab Inc. *Reproduced with permission from iHealth Lab Inc.*

12.2.6.2 AliveCor, Inc.

Kardia is the first FDA-cleared and CE-marked device for SP and smart watch based ECG, developed by Dr. David Albert and his team at AliveCor, Inc. [59], which has been cleared for sale in most countries including the US. It is the clinically validated mobile ECG that allows users to capture medical-grade ECG in just 30 seconds and interpret it immediately via instant ECG analysis (employing FDA-cleared machine learning algorithms) and consultation with board-certified cardiologists. The device can record, display, store, and evaluate single-channel ECG rhythms and heart rates. Kardia Mobile (sold for US$ 99) is a miniaturized credit card sized and lightweight (18 g) ECG device with two stainless steel electrodes (2.3 cm × 3 cm). Kardia accessories are the SP cases for iPhone 5/5s and iPhone 6/6s (sold for US$ 10 and US$ 15, respectively), which have dedicated slots for inserting Kardia mobile to have a SP cover equipped with ECG. The Kardia band is the wristband for Apple Watch that contains the mobile ECG but it is not yet approved for sale in the US as it is pending 510k clearance. The devices run via a dedicated smart app, i.e., Kardia App that can be downloaded freely from the App Store and Google play but requires an initial account set-up to make it personalized. The ECG data and information are stored via Cloud on a secure server that can be remotely accessed.

Kardia Mobile runs on a 3V CR2016 coin cell battery with 200 hour of operation that can last for one year. The single lead ECG is generated by resting the electrodes on the fingers from each hand. The device can take ECG of 30 seconds to 5 minutes duration with 300 samples per second sampling rate, 16-bit resolution, and 10 mV peak-to-peak input dynamic range. The measurement involves putting the fingers from the left and right hands on the left and right electrodes, respectively, of Kardia Mobile. The Kardia App initiates the ECG recording after establishing electrode–finger contact. The Kardia Mobile employs a proprietary technology, which converts the electrical impulses into the ultrasound signals. The ultrasound signals are transmitted to the SP's microphone using an enhanced filtering technology that minimizes artifacts and provides medical-grade Lead I ECG. Several clinical trials have already demonstrated the clinical accuracy of the device, which is highly cost-effective for monitoring the patient's heart rate and rhythm. Moreover, it does not add any extra cost to the users as the device is paid in the US by Medicare and other private insurance companies. The Kardiac Premium services, available at US$ 10 per month, provide detailed ECG reports and visualizations over time and personalized reports supplemented with analysis and actionable advice to maintain a healthy heart.

12.2.7 Cardiovascular Health Monitors

The recently launched iHealth CardioLab (Fig. 12.8), a CE-marked and FDA-approved Class IIa regulated medical device sold for €599 in Europe by iHealth Inc., enables the general practitioners (GP) to obtain a detailed cardiovascular assessment of their patients in less than two minutes. It results in better and sustainable cardiovascular healthcare for the patients by providing the physicians a quick and detailed cardiovascular evaluation, which allows them to anticipate many health problems such as the early detection of peripheral arterial disease (PAD). It provides a simple, accurate and efficient way of calculating the risk of PAD, strokes, and infarcts. The device will enable the GP to refer the patient to a cardiologist based on the detection of early stage abnormalities in the evaluation.

Figure 12.8
iHealth CardioLab, a cardiovascular health monitor developed by iHealth Lab Inc. *Reproduced with permission from iHealth Lab Inc.*

The device has two measurement cuffs for the upper arm and lower leg, each weighing 135 g and using a universal cuff of circumference 22−42 cm. It is powered by an integrated rechargeable battery (DC: 5 V, 1 A, 1 V × 3.7 V, Li-ion 400 mAh) that can be charged by the provided USB cable. The measurements are done using an oscillometric method involving automatic inflation and deflation, which enables the measurement of various cardiovascular health parameters such as BP, heart rate, mean arterial pressure (MAP), ankle brachial index (ABI), pulse pressure (PP), stroke volume (SV), and cardiac output (CO). These parameters are calculated from BP measurement using algorithms validated and used by cardiologists. The ABI, one of the most effective indicators for the diagnosis, screening, and prevention of PAD, is calculated by the ratio of systolic pressure at the ankle to the systolic BP of the arm. The device measures systolic pressure, diastolic pressure, and pulse rate in the ranges of 60−260 mm Hg, 40−199 mm Hg, and 40−180 bpm, respectively. The results are transmitted in real-time by Bluetooth to the iHealth PRO app on the iPad, which measures and calculates all these parameters.

12.2.8 Smart Biomedical Diagnostic Devices

12.2.8.1 Cellmic

Cellmic, formerly Holomic LLC, developed the CE-marked SP-based Holomic Rapid Diagnostic Test Reader (HRDR-200) [60] (Fig. 12.9A) for the readout of lateral flow assays (LFA) with high sensitivity and precision. HRDR-200 is equipped with a SP, an integrated reader housing, smart application, and access to secure Cloud Services and Test Developer. It is light-weight, handheld, economical, compliant with ISO 13485 and registered with FDA as a Class 1 medical device. The reader enables real-time diagnostic data integration via a secure Cloud service and integration with electronics health records (EHRs), laboratory information system (LIS), and hospital information system (HIS). HRDR-200 is a universal reader that can read chromatographic and fluorescent assays in lateral flow, flow-through and dipstick test formats with high precision. It can also read tests of various sizes or formats, including strips,

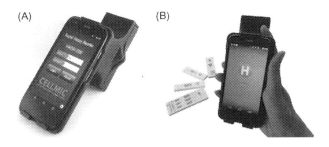

Figure 12.9
Smartphone-based diagnostic readers developed by Cellmic. (A) Holomic Rapid Diagnostic Test Reader (HRDR-200). (B) HRDR-300. *Reproduced with permission from Cellmic.*

cassettes, and multiplexed tests. It can be used either standalone or connected to a network. The device can identify the test, lot number, and patient data as it is equipped with QR code scanner for automatic entry of test and patient data, and analyzing the lot expiry. The data analysis is performed on a PC with Test Explorer Software while the test results can be printed, e-mailed or texted. It is connected to laptops, PCs, printers, routers, and EHRs via wireless, Bluetooth, and USB. The device prototype and its working have been demonstrated in a pioneering publication by Prof. Ozcan's group [61]. HRDR-300 (Fig. 12.9B) is the recently developed fluorescent reader that enables the multi-color imaging based readout of the control and test lines of LFA with high accuracy using an advanced image processing algorithm. The users can select the excitation and emission in the regions between 350−700 nm and 300−800 nm, respectively. The device can also be customized according to the needs of the multiplexed fluorescent assay, such as the requirement of multiple excitation sources. All other features are the same as those of chromatographic HRDR-200. The device is also ISO 13485 compliant and registered with the FDA as a Class I medical device. The company is working on the development of HRDR-400, a dual reader for chromatographic and fluorescent rapid tests in a single device. Holomic Substance Test Assistant (HSTA) is a handheld, economical and high-performance tool, which acts as a single chromatographic and fluorescent viewer for alcohol and substance abuse tests of different sizes or formats, including saliva tests. All the device features are similar to that of HRDR-200.

The development and working of the rapid diagnostic test (RDT) reader prototype [61] used for HRDR-200 is briefly illustrated here. The RDT reader has a plano-convex lens, a microcontroller, and three LED arrays (two located underneath the RDT tray for reflection imaging and one at the top for transmission imaging). It analyzes various types of lateral flow immunoassays and RDTs in reflection or transmission imaging modes under diffused LED illumination and provides a quantitative determination of analytes. The device can run on external batteries or the SP battery via USB connection. The highly sensitive RDT reader can detect even the minor color signal variation that cannot be observed by visualization. A smart application processes the raw images taken by the SP's camera within <0.2 seconds and shares the results with a central server also (in addition to storing the information on the SP), which can be assessed by a remote computer using web browsers. The software application provides a dynamic spatiotemporal map and real-time statistics for various diseases that can be diagnosed by RDTs, thereby enabling healthcare professionals and policy makers to monitor, track and analyze emerging diseases and outbreaks.

12.2.8.2 Mobile Assay, Inc.

Mobile Assay, Inc. developed a cost-effective mobile diagnostic reader (MReader) [62] (Fig. 12.10) based on Mobile Image Ratiometry (MIR) and Instantaneous Analysis software. The SP-based MReader provides rapid quantification of lateral flow test strips [63,64] as it can read multiple tests simultaneously down to 1 ppb with higher accuracy and significantly higher sensitivity than the human eye. It requires no additional attachments

Figure 12.10

Mobile diagnostic reader (MReader) developed by Mobile Assay, Inc. Image provided by Michael Williams, Mobile Assay, Inc. *Reproduced with permission from MDPI (http://www.alivecor.com/home).*

and has an integrated advanced light level compensation and camera linearity to provide a highly reproducible reading that is unaffected by ambient light fluctuation. For processing image analysis of the dye signal on the test strips, the MIR subtracts the background noise, selects the signal bands and provides the pixel density ratio. The device is operated by the smart application that can be run on Apple, Android and Windows-equipped SP/gadget. The geotagged and time-stamped results are uploaded to the mobile diagnostic Cloud via Wi-Fi or a cellular network with the push of a button. The data is stored securely on the portal provided by the company using advanced encryption and data storage techniques. The stored data can be assessed by a secure "sign in" to the portal. The results stored in the Cloud are analyzed for trend analysis of the particular test. The MReader-based rapid diagnostic assays have been developed for the detection of drugs (e.g., cocaine and benzoylecgonine), food pathogens, and aflatoxin. It detects $0.1-300$ ng mL^{-1} of cocaine, $0.003-0.1$ ng mL^{-1} of benzoylecgonine, and food pathogens [65], which facilitates the rapid and efficient tracking of the origin and severity of outbreaks.

12.2.9 CellScope

CellScope [66], San Francisco, USA, founded by Prof. Daniel Fletcher's research group at the University of California, Berkeley, developed two optical attachments that modify the

SP into a diagnostic-quality imaging system for healthcare and skincare applications. "CellScope Oto" is a clip-on digital Otoscope for taking high-resolution images of the middle ear to probe if the subjects have an ear infection (Fig. 12.11). The device comprises a custom designed iPhone case, an Otoscope attachment, an Otoscope case, five-tip cases, and an iPhone app. The smart app transfers the images captured by the SP to an HIPAA-compliant website for reviewing, comparing, and transmitting the result of ear examination. The automated analysis of the results helps in minimizing the doctor visits for the parents as the smart app enables the uploading of high-resolution images of the ear canal and eardrums on the CellScope's web platform, which can be accessed remotely by a doctor for diagnosis, treatment, and monitoring. This device significantly reduces the healthcare costs considering about 30 million doctor visits by the parents every year in the United States. The device can screen sick children in school or daycare facilities for ear- infection. The "CellScope Derm" is another clip-on Dermascope that facilitates the remote diagnosis of patient's skin conditions by capturing and transmitting highly magnified diagnostic quality images of the skin. It has an illumination system and lower-magnification optics for capturing the images at a wider field.

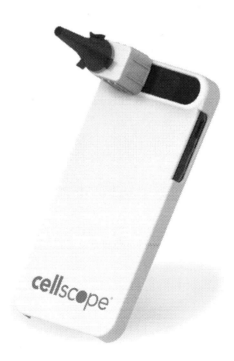

Figure 12.11
CellScope Oto developed by CellScope. Image provided by Cori Allen, CellScope, Inc. *Reproduced with permission from MDPI (http://www.alivecor.com/home).*

12.2.9.1 GENTAG, Inc.

GENTAG [67] offers a broad range of WT for PMHMM such as near-field communication (NFC)-based low-cost disposable wireless sensors, the spatiotemporal tagged data from which can be read by NFC-capable SP/tablet/PC (Fig. 12.12A). It facilitates the reading of NFC-based sensors or other customized sensors within the range of 15 miles. The company has developed wireless NFC SensorLinkers, which are portable, lightweight (74 g), and battery-powered devices for home medical monitoring applications, smart homes and machine-to-machine applications (Fig. 12.12B). The NFC SensorLinker has the NFC NXP PN544 Reader integrated circuit (IC) with a proprietary long-range antenna, 2.4 GHz Wi-Fi, Bluetooth 4 Dual Mode, USB port, and a lithium-ion rechargeable battery. It comes with a rechargeable cradle and a USB charger. The SensorLinker runs on any SIM-based GPRS or 3G WCDMA cellular network. It can be paired with existing Bluetooth devices or bundled with custom made NFC sensors. Its primary applications are the wireless monitoring of elderly people and children, increasing the compliance with treatment, monitoring of medication, and use in wireless hospital discharge kits and hybrid Wi-Fi/Bluetooth/NFC sensor applications. The technology also offers the direct reading and monitoring of implantable NFC sensors and devices such as pacemakers, but with an ultra-low power

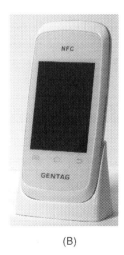

(A)   (B)

Figure 12.12

Smartphone-based technologies developed by GENTAG, Inc. (A) NFC Smart Cell Phone and (B) NFC Sensorlinker. Images provided by Dr. John P. Peeters, GENTAG, Inc. *Reproduced with permission from MDPI (http://www.alivecor.com/home).*

requirement. The NFC SensorLinker technology can also be integrated with FDA-approved general healthcare devices such as those for measuring BP, weight, glucose, oxygen, activity, etc., which employ the interface with the Bluetooth technology at the moment.

The company has used NFC to develop disposable wireless skin patches, personal drug delivery systems, and SP-based sensors. The wireless skin patches are waterproof, showerproof and nonallergic, and are based on low-cost passive technology (battery-less). The FDA-cleared adhesive lasts up to 2 weeks and employs the frequency of 13.56 MHz that is the global standard in healthcare. Further, the patches have a unique ID and a Radar Responsive (RR) tag that enables a precise wide area nonGPS wireless geolocation and geofencing over a radius of several miles. The technology can be used for several prospective applications for the monitoring and the location of patients with Alzheimer's dementia. GENTAG has further developed several products and applications such as (1) a NFC-radio frequency identification device (RFID) sensor, (2) NFC immunoassay for prostate cancer, (3) a NFC tag to prevent counterfeit drugs or products, (4) NFC sensors to monitor temperature, radiation, chemicals, and pressure, (5) SP-based devices for home monitoring, such as NFC thermometer; NFC-Bluetooth weight management kit; NFC custom blister packs and medication sensors; and, remote wireless monitoring of elderly or at-risk patients directly in their homes by taking BP, fever, and medication data, (6) RR tags, and (7) NFC diagnostic skin patches for fever monitoring, drug delivery, glucose monitoring, post orthopedics surgery/post-hospital discharge monitoring, and the prevention of hospital errors (due to mismatches from patient−surgery, patient−drug delivery/medication, mother−baby, etc.).

GENTAG has developed low-cost, disposable and painless skin patches with wireless sensors for the SP-based transdermal glucose sensing and monitoring [68], which are more cost-effective than the existing glucose monitoring devices [69−71] that still require finger-pricking. They offer annual savings of US$ 3000 and US$ 300 for Type 1 and Type 2 diabetics per patient, respectively. The SP can be programmed for the delivery of insulin and the geolocation of patients in case of emergency.

The company developed an SP-based disposable noninvasive immunoassay for the early stage diagnosis of prostate cancer based on the detection of a proprietary PCADM-1 biomarker in urine. Considering the cost-effective and more frequent analysis of prostate cancer by this assay, it is better than the existing prostate specific antigen (PSA)-based blood test being used in hospitals, which is highly affected by various physiological factors [72]. The company also developed a customizable spectroscopic radiation detection cellphone that discriminates hazardous γ-rays from normal γ-rays.

12.2.10 Next-generation Smart Devices

Cicret Bracelet [73] is the smart bracelet by Cicret, which can perform multifarious tasks on the arm of the users, similar to what they can do with a tablet but without any need to pick up the SP. The users can answer their phone calls, read e-mails, play games, check the weather, navigate, and perform many other tasks. The water-resistant device employs a removable battery, functions with an Android operating system, and comes with an internal memory of 32 or 64 GB. The product is currently priced at US$ 250 and would be commercialized in the first quarter of 2017. It has a pico projector for projecting the SP's screen onto the arm, and eight long-range proximity sensors that make the user's skin as a touchscreen. When the user puts a finger on the screen that appears on the arm, one of the sensors is stopped, which sends the information back to the processor in the bracelet. The Circlet Bracelet has an accelerometer, Bluetooth, Wi-Fi component, LED, processor, vibrator, micro USB port, removable battery, memory card and ROM, and a snap button. The technology will be used in many forthcoming healthcare applications.

12.3 Conclusions and Prospects

WT are indispensable for PMHMM as they obviate the need for skilled medical professionals and deliver cost-effective mH at any place at any time. They empower the users to effectively monitor and manage their health and lead to better and sustainable healthcare benefits. Most of the WT are interfaced to SP that have become ubiquitous and are now considered as an ideal point-of-care technology platform for delivering mH to the developing nations, where it is critically required due to the immense shortage of medical professionals and infrastructure. The WT-based devices have contributed significantly to increasing the awareness worldwide for improved PMHMM. They empower the users to better understand and analyze their general healthcare data and take informed decisions. They would also enable them to take physical activity, nutritional, relaxation and other lifestyle interventions that would improve their general health and lead to sustainable better health outcomes. They would significantly cut down the healthcare costs and motivate the users to participate in effective prevention and management of chronic health diseases, such as diabetes, obesity, cardiovascular disorders, stress, and cancer, by implementing such simple lifestyle interventions. Most of the WT-based devices are CE-marked while FDA approval has also been granted to some. They have widely penetrated the consumer market and are used by millions of people worldwide. However, several challenges that still need to be counteracted are the high clinical accuracy, high reproducibility, robustness, and adaptability to various SP models.

The recent years have witnessed tremendous advances in Cloud computing, which is a prerequisite for WT as it leads to tremendous cost savings by minimizing the infrastructure costs [74]. However, there are potential concerns about data security and privacy [75,76]. Most nations want to store the data physically within their national boundaries [77], a limitation that has been solved by Amazon Web Services allowing companies to store their data within their national boundaries. The Google's Government Cloud also permits governments to store their data in their countries as per their data security guidelines.

But there is a need for creating international Cloud computing standards, which has led to several initiatives such as EuroCloud and Google's Data Liberation Front. The user's personal data in personal health records (PHRs) could be exposed to third party servers and unauthorized parties, which is a potential threat according to the ethical guidelines that grant patients full control over their PHRs. This has been counteracted by attribute-based encryption [78] and using a trusted third party [77]. Several Cloud computing models have also been developed to deal with the regulatory requirements of security and privacy [79]. The benefits of electronic health records (EHRs) in improving caregivers' decisions and patients' outcomes are clearly evident [80]. The Health Information Technology for Economic and Clinical Health Act (HITECH) in the US has authorized incentive payments via insurance agencies to the healthcare practitioners, who would use EHRs to achieve specified improvements in care delivery [81]. The SP-based mobile Cloud computing [82] being employed in WT would facilitate critically improved PMHMM [83]. The coming years will witness growing number of WT coming to the market that would lead to next-generation healthcare boom.

12.4 Conflicts of Interests

The authors declare no conflict of interest.

Acknowledgments

The authors would like to thank Mr. Louie Lyu from iHealth Labs, France, and Dr. Onur Mudanyali from Cellmic Inc., USA for providing the high-resolution images of their products.

References

[1]   A. Ozcan, Mobile phones democratize and cultivate next-generation imaging, diagnostics and measurement tools, Lab Chip (2014).
[2]   S.K. Vashist, et al., Cellphone-based devices for bioanalytical sciences, Anal. Bioanal. Chem. 406 (14) (2014) 3263–3277.
[3]   D. Thilakanathan, et al., A platform for secure monitoring and sharing of generic health data in the Cloud, Fut. Gener. Comp. Syst. 35 (2014) 102–113.
[4]   S.K. Vashist, et al., Emerging technologies for next-generation point-of-care testing, Trends Biotechnol. 33 (11) (2015) 692–705.
[5]   S.K. Vashist, J.H. Luong, Trends in in vitro diagnostics and mobile healthcare, Biotechnol. Adv. 34 (3) (2016) 137–138.
[6]   B. Martínez-Pérez, I. de la Torre-Díez, M. López-Coronado, Mobile health applications for the most prevalent conditions by the World Health Organization: review and analysis, J. Med. Internet Res. 15 (6) (2013).
[7]   G. Phillips, et al., The effectiveness of M-health technologies for improving health and health services: a systematic review protocol, BMC Res. Notes 3 (1) (2010) 250.
[8]   A.S. Mosa, I. Yoo, L. Sheets, A systematic review of healthcare applications for smartphones, BMC Med. Inform. Decis. Mak. 12 (1) (2012) 67.

[9] P.Y. Benhamou, et al., One-year efficacy and safety of Web-based follow-up using cellular phone in type 1 diabetic patients under insulin pump therapy: the PumpNet study, Diabetes Metab. 33 (3) (2007) 220–226.

[10] T. Botsis, G. Hartvigsen, Current status and future perspectives in telecare for elderly people suffering from chronic diseases, J. Telemed. Telecare 14 (4) (2008) 195–203.

[11] P.M. Carrera, A.R. Dalton, Do-it-yourself healthcare: the current landscape, prospects and consequences, Maturitas 77 (1) (2014) 37–40.

[12] M.C. Carter, et al., Adherence to a smartphone application for weight loss compared to website and paper diary: pilot randomized controlled trial, J. Med. Internet Res. 15 (4) (2013) e32.

[13] A. Coulter, Engaging Patients in Healthcare, McGraw-Hill International, United Kingdom, 2011.

[14] T. Donker, et al., Smartphones for smarter delivery of mental health programs: a systematic review, J. Med. Internet Res. 15 (11) (2013) e247.

[15] M.B. Duffy, Humanizing the healthcare experience: the key to improved outcomes, Gastrointest. Endosc. 79 (3) (2014) 499–502.

[16] M. Fiordelli, N. Diviani, P.J. Schulz, Mapping mHealth research: a decade of evolution, J. Med. Internet Res. 15 (5) (2013) e95.

[17] S. Franc, et al., Telemedicine and type 1 diabetes: is technology per se sufficient to improve glycaemic control? Diabetes Metab. 40 (1) (2014) 61–66.

[18] C. Free, et al., The effectiveness of mobile-health technology-based health behaviour change or disease management interventions for health care consumers: a systematic review, PLoS Med. 10 (1) (2013) e1001362.

[19] C. Free, et al., The effectiveness of mobile-health technologies to improve health care service delivery processes: a systematic review and meta-analysis, PLoS Med. 10 (1) (2013) e1001363.

[20] A. Honka, et al., Rethinking health: ICT-enabled services to empower people to manage their health, IEEE Rev. Biomed. Eng. 4 (2011) 119–139.

[21] J. Joe, G. Demiris, Older adults and mobile phones for health: a review, J. Biomed. Inform. 46 (5) (2013) 947–954.

[22] R.M. Kaplan, A.A. Stone, Bringing the laboratory and clinic to the community: mobile technologies for health promotion and disease prevention, Annu. Rev. Psychol. 64 (2013) 471–498.

[23] D.D. Luxton, et al., mHealth for mental health: integrating smartphone technology in behavioral healthcare, Prof. Psychol.-Res. Pract. 42 (6) (2011) 505–512.

[24] G.A. O'Reilly, D. Spruijt-Metz, Current mHealth technologies for physical activity assessment and promotion, Am. J. Prev. Med. 45 (4) (2013) 501–507.

[25] S. Pagoto, The current state of lifestyle intervention implementation research: where do we go next? Transl. Behav. Med. 1 (3) (2011) 401–405.

[26] M. Price, et al., mHealth: a mechanism to deliver more accessible, more effective mental health care, Clin. Psychol. Psychother. 21 (5) (2014) 427–436.

[27] J. Stephens, J. Allen, Mobile phone interventions to increase physical activity and reduce weight: a systematic review, J. Cardiovasc. Nurs. 28 (4) (2013) 320–329.

[28] L. Bellina, E. Missoni, Mobile cell-phones (M-phones) in telemicroscopy: increasing connectivity of isolated laboratories, Diagn. Pathol. 4 (2009) 19.

[29] L. Dayer, et al., Smartphone medication adherence apps: potential benefits to patients and providers, J. Am. Pharm. Assoc. 53 (2) (2013) 172.

[30] P.E. Hasvold, R. Wootton, Use of telephone and SMS reminders to improve attendance at hospital appointments: a systematic review, J. Telemed. Telecare 17 (7) (2011) 358–364.

[31] R.T. Lester, et al., Effects of a mobile phone short message service on antiretroviral treatment adherence in Kenya (WelTel Kenya1): a randomised trial, Lancet 376 (9755) (2010) 1838–1845.

[32] J.M. Montes, et al., A short message service (SMS)-based strategy for enhancing adherence to antipsychotic medication in schizophrenia, Psychiatry Res. 200 (2–3) (2012) 89–95.

[33] M.A. Thomas, P.R. Narayan, C. Christian, Mitigating gaps in reproductive health reporting in outlier communities of Kerala, India—A mobile phone-based health information system, Health Policy Technol. 1 (2) (2012) 69—76.

[34] N. Tripp, et al., An emerging model of maternity care: smartphone, midwife, doctor? Women Birth 27 (1) (2014) 64—67.

[35] A.P. Demidowich, et al., An evaluation of diabetes self-management applications for Android smartphones, J. Telemed. Telecare 18 (4) (2012) 235—238.

[36] A. Rao, et al., Evolution of data management tools for managing self-monitoring of blood glucose results: a survey of iPhone applications, J. Diabetes Sci. Technol. 4 (4) (2010) 949—957.

[37] M. Migliore, Smartphones or tablets for a better communication and education between residents and consultant in a teaching hospital, J. Surg. Educ. 70 (4) (2013) 437—438.

[38] K.F.B. Payne, H. Wharrad, K. Watts, Smartphone and medical related app use among medical students and junior doctors in the United Kingdom (UK): a regional survey, BMC Med. Inform. Decis. Mak. 12 (1) (2012) 121.

[39] C. Lunny, et al., Short message service (SMS) interventions for the prevention and treatment of sexually transmitted infections: a systematic review protocol, Syst. Rev. 3 (1) (2014) 7.

[40] K.E. Muessig, et al., Mobile phone applications for the care and prevention of HIV and other sexually transmitted diseases: a review, J. Med. Internet Res. 15 (1) (2013) e1.

[41] A.W.M. Lee, et al., Lecture rule no. 1: Cell Phones ON, Please! a low-cost personal response system for learning and teaching, J. Chem. Educ. 90 (3) (2013) 388—389.

[42] S. Wallace, M. Clark, J. White, 'It's on my iPhone': attitudes to the use of mobile computing devices in medical education, a mixed-methods study, BMJ Open 2 (2012) 4.

[43] J.K. Lau, et al., iPhone ECG application for community screening to detect silent atrial fibrillation: a novel technology to prevent stroke, Int. J. Cardiol. 165 (1) (2013) 193—194.

[44] J.L. Peck, M. Stanton, G.E. Reynolds, Smartphone preventive health care: parental use of an immunization reminder system, J. Pediatr. Health Care 28 (1) (2014) 35—42.

[45] G.J. Doyle, B. Garrett, L.M. Currie, Integrating mobile devices into nursing curricula: opportunities for implementation using Rogers' diffusion of innovation model, Nurse Educ. Today 34 (5) (2014) 775—782.

[46] M.O. Lwin, et al., A 21st century approach to tackling dengue: crowdsourced surveillance, predictive mapping and tailored communication, Acta Trop. 130C (2013) 100—107.

[47] Y. Yang, et al., Silicon-based hybrid energy cell for self-powered electrodegradation and personal electronics, ACS Nano 7 (3) (2013) 2808—2813.

[48] https://www.fitbit.com/us.

[49] http://www.garmin.com/en-GB/.

[50] http://www.nubandsports.com/nuband01/.

[51] http://www.nubandsports.com/nuband-active/.

[52] http://www.apple.com/watch/.

[53] http://www.samsung.com/global/galaxy/gear-s2/.

[54] http://www.samsung.com/global/galaxy/gear-fit2/.

[55] http://www.samsung.com/us/ssic/pdf/Samsung_Simband_Backgrounder.pdf.

[56] http://www.nonin.com/PulseOximetry/Finger/Onyx9560.

[57] A. Jubran, Pulse oximetry, Crit. Care 3 (2) (1999) R11—R17.

[58] J.E. Sinex, Pulse oximetry: principles and limitations, Am. J. Emerg. Med. 17 (1) (1999) 59—67.

[59] http://www.alivecor.com/home.

[60] http://www.cellmic.com/.

[61] O. Mudanyali, et al., Integrated rapid-diagnostic-test reader platform on a cellphone, Lab Chip 12 (15) (2012) 2678—2686.

[62] C.F. Woolley, M.A. Hayes, Emerging technologies for biomedical analysis, Analyst 139 (10) (2014) 2277—2288.

[63] D. Tseng, et al., Lensfree microscopy on a cellphone, Lab Chip 10 (14) (2010) 1787—1792.

[64] D.C. Cooper, et al., Mobile image ratiometry: a new method for instantaneous analysis of rapid test strips, Nat. Preced. (2012). Available from: https://doi.org/10.1038/npre.2012.6827.1.

[65] D.C. Cooper, Mobile image ratiometry for the detection of *Botrytis cinerea* (gray mold), Nat. Preced (2012). Available from: https://doi.org/10.1038/npre.2012.6989.1.

[66] W.H. Wu, et al., MEDIC: medical embedded device for individualized care, Artif. Intell. Med. 42 (2) (2008) 137−152.

[67] http://gentag.com/.

[68] J.P. Peeters,, Diagnostic Radio Frequency Identification Sensors And Applications Thereof, 2011, U.S. Patent No. 8,077,042 (13 December 2011).

[69] S.K. Vashist, et al., Technology behind commercial devices for blood glucose monitoring in diabetes management: a review, Anal. Chim. Acta 703 (2) (2011) 124−136.

[70] S.K. Vashist, Non-invasive glucose monitoring technology in diabetes management: a review, Anal. Chim. Acta 750 (2012) 16−27.

[71] S.K. Vashist, Continuous glucose monitoring systems: a review, Diagnostics 3 (4) (2013) 385−412.

[72] J. Gudmundsson, et al., Genetic correction of PSA values using sequence variants associated with PSA levels, Sci Transl Med 2 (62) (2010) 62ra92. p.

[73] https://cicret.com/wordpress/.

[74] S. Marston, et al., Cloud computing—the business perspective, Decis. Support Syst. 51 (1) (2011) 176−189.

[75] S. Subashini, V. Kavitha, A survey on security issues in service delivery models of cloud computing, J. Network Comp. Appl. 34 (1) (2011) 1−11.

[76] D.W. Sun, et al., Surveying and analyzing security, privacy and trust issues in cloud computing environments, Proc. Eng. 15 (2011) 2852−2856.

[77] D. Zissis, D. Lekkas, Addressing cloud computing security issues, Fut. Gener. Comp. Syst.—Int. J. Grid Comput. Esci. 28 (3) (2012) 583−592.

[78] M. Li, et al., Scalable and secure sharing of personal health records in cloud computing using attribute-based encryption, IEEE Trans. Parallel Distrib.Syst. 24 (1) (2013) 131−143.

[79] E.J. Schweitzer, Reconciliation of the cloud computing model with US federal electronic health record regulations, J. Am. Med. Inform. Assoc. 19 (2) (2012) 161−165.

[80] D. Blumenthal, M. Tavenner, The "meaningful use" regulation for electronic health records, N. Engl. J. Med. 363 (6) (2010) 501−504.

[81] D. Blumenthal, Launching HITECH, N. Engl. J. Med. 362 (5) (2010) 382−385.

[82] H.T. Dinh, et al., A survey of mobile cloud computing: architecture, applications, and approaches, Wireless Commun. Mobile Comput. 13 (18) (2013) 1587−1611.

[83] M.N. Boulos, et al., How smartphones are changing the face of mobile and participatory healthcare: an overview, with example from eCAALYX, Biomed. Eng. Online 10 (1) (2011) 24.

CHAPTER 13

Patient-Generated Health Data: Looking Toward Future Health Care

Patrick Slevin and Brian Caulfield

Insight Centre for Data Analytics, University College Dublin, Dublin, Ireland

13.1 Introduction

Patients generating data about their health is not a new phenomenon. Since the development of early self-monitoring devices such as the weighing scale and thermometer, and later the home-pregnancy test, glucometer and blood pressure (BP) monitor, patients have long been capturing data about their health. But the thing that is new, very new in fact, is the profound transformation of these traditional data due to rapid digitization. The digitization of health-related data is occurring at the nexus of several interrelated socio-technical factors, namely the development of broadband infrastructure, advancements in smart technology allowing for affordable smartphones, the proliferation of wearable sensing devices, and the growing public desire to track, measure, and understand various aspects of our being. At an individual level, the digital data footprint may be considered as "small data." However, when we consider it in terms of the potential contained in longitudinal data sets from large sections of the population, we think in terms of unlocking the value from "big data." The promises of these "big data" are receiving heightened attention as healthcare systems and industry pursue data-driven models of care that can provide health management stakeholders with actionable information that drives a more proactive healthcare model to ease the financial and resource burdens associated with aging populations and the prevalence of chronic disease. The term patient-generated health data (PGHD) has emerged from the discourses regarding the use of data in health care. Shapiro et al. defined PGHD as *"health-related data—including health history, symptoms, biometric data, treatment history, lifestyle choices, and other information—created, recorded, gathered, or inferred by or from patients or their designees to help address a health concern"* [1]. It is clear that this refers to a very wide spectrum of data types and sources and a detailed discussion of all would be beyond the scope of this chapter. The focus of this chapter, therefore, will be in the benefits of digitally generated PGHD from wearable sensors and mHealth applications from the perspective of both patient and health-care professionals.

Wearable Technology in Medicine and Health Care.
DOI: https://doi.org/10.1016/B978-0-12-811810-8.00013-0

The focus on digitally generated PGHD corresponds with the "digital patient culture" that is developing at a staggering pace. In 2018, it is estimated that 135 million wearable units will be shipped worldwide—up from 9.7 million units in 2013 [2]. In terms of ownership, approximately 29% of Americans have a wearable device, while in the United Kingdom research has shown that 3 million wrist-worn wearable units were sold in 2015, up 118% in 2014 [3,4]. As wearable devices are frequently accompanied by a dedicated application (app), of equal importance to the wearable industry is the growing number of smartphone users, which is forecasted to reach 2.08 billion worldwide by 2019 [4]. These upward trends are also seen in relation to mHealth app market with between 165,000 and 259,000 apps now available for download [5,6]. Furthermore, the remote patient monitoring market is also expanding at an enormous rate. A recent report shows a 44% rise in the number of patients been remotely monitored to 7.1 million in 2016, which is set to grow to 50.2 million by 2021 [7]. The report also found that the number of mHealth monitoring devices with integrated wireless connectivity increased from 3 million in 2015 to 4.9 million in 2016 allowing for greater interoperability with wearable and smartphone technologies [7]. As the digital health ecosystem concretizes, the optimal circumstances for the rapid digitization of PGHD is fast establishing. The knock on social effect is that a new type of health consumerism is also emerging—one where individuals are keener than ever to adopt a more technologically driven personalized approach to health which is creating opportunities to collect health-related data that is unparalleled to any period in history.

Society is on the cusp of fully realizing the digital patient and the nature and potential of the PGHD they produce is demanding our attention, especially when solutions to ease the climate of uncertainty in health care remain elusive. In this context, new questions arise in relation to the future of health care; how will PGHD disrupt our conceptions of patienthood or the health-care professional (HCP)? Will traditional health-care practices suffice as patients demand more emphasis being placed on the data they generate as they go about their daily lives? If PGHD is to become part of the health-care fabric, what are the key barriers facing their integration? And, what, if any, are the potential benefits of leveraging digitally produced PGHD for health care, patients, and HCPs? The following sections will attempt to open up a space for new discussions regarding the potential benefits of PGHD in a time when society is faced with a pressing need to rethink the future of health care.

13.2 PGHD—What Are the Benefits?

For so long, patients were only expected to share one type of information with their HCP—how they were feeling. However, as digitally produced PGHD gain traction, patients will be armed with a plethora of self-data to utilize in their daily health-care routines. The following sections will try to unpack some of the potential benefits that could arise from leveraging PGHD for patients and HCPs.

13.2.1 Patient Engagement

Patient engagement has gained significant attention in health-care policy and research in the past decade. Evidence suggests that increased patient engagement enhances the quality of patient safety, reduces health-care costs, and improves health outcomes [8,9]. The increased emphasis on engaging patients comes at a time when there is a growing acceptance that HCPs are only one of the integral cogs required to tackle illness, health, and well-being. Patients are now expected to play an essential role in the management of their health and their engagement is central to achieve it. Recent approaches to reforming the provision of health care such as patient-centered and personalized care are examples of the turning tide toward trying to effectively engage patients in their care. The philosophy underpinning these approaches centers on transforming the patient's role from a passive recipient of care to that of an engaged and active participant. An engaged patient is one who feels empowered to partake in collaborative discussions about the trajectory of their care and is one who is empowered to take control in the management of their health with the support of their HCP [10].

But patient engagement is a complex matter; it requires the intertwined responsibility of the patient, the HCP and the health-care system to all function in harmony. The problem is however, HCPs are often left as the sole drivers to engage patients—consultations frequently involve information flows that are one way, from the HCP to the patient, without collaborative dialogue which only further deepens the culture of patients as passive recipients of care [10]. Achieving satisfactory levels of engagement has proven problematic globally [11]. The breakdown in patient engagement is seen more so in chronic disease populations—studies have shown that 50% of patients do not take medications as prescribed, while low patient adherence to pulmonary and cardiac rehabilitation is also a growing issue [12–16]. So, how can PGHD help with patient engagement?

Proponents of PGHD believe that as patients increasingly participate in the collection of PGHD, they will be enabled to achieve a more active status in their health. For example, when a patient decides to collect self-data, they impart on themselves an element of mastery and control that was not there before as it is they (or their caregiver) and not the HCP who is generating their health data. By taking on this responsibility they are no longer a passive recipient waiting to be assessed and told what to do. Instead they are self-assessing and self-directing. Interestingly, within the very practice of collecting PGHD, another form of patient engagement emerges as a by-product of the patients' motivation to capture these data. In other words, a patient who is motivated to capture self-data is, in a sense, a patient who is already engaging with their health care. Thus, as a generation of patients starts to capture PGHD, they are simultaneously self-activating their engagement with their health [17], and, in a sense, initiating a virtuous cycle.

Patient self-activation relates to the idea of the engaged patient explained earlier—but instead of the HCP leading the engagement of patients, self-activated patients are driven by their *own* motivation, knowledge, skills, and confidence to make decisions in the management of their health [18]. For HCPs, the self-activated patient will provide them with an opportunity to venture down previously unexplored avenues to discuss engagement in a more person-centered manner. For instance, patients who are motivated to collect PGHD through digital devices might be more willing to explore these technologies as a medium for delivering person-centered treatment plans and engagement strategies. Moreover, as self-activated patients (or their caregiver) turn their motivation to acting on these data, such as sharing it with their HCP or demanding their integration into an electronic health record (EHR), levels of self-efficacy will improve as patients demonstrate an increasing confidence to partake in discussions and decisions regarding the trajectory of their health based on their PGHD.

Patients increasingly want to be more involved in the health-care decision-making process [19]. Health care has responded to this need by placing a stronger emphasis on shared decision making (SDM) as mechanism to further engage patients in their health care. SDM refers to a dialogue centered on sharing information between the HCP and the patient—the HCP provides options, describes the risks and benefits, and the patient is encouraged to express their preferences and values; a consensus is then collaboratively reached on how to proceed [20,21]. SDM is important as it can increase patient knowledge, lower anxiety regarding the care process, and creates greater harmony between the patients' values and preferences and the course of their care [22].

13.2.2 PGHD and the Interappointment Period—Bridging the Gap

Patients manage their health primarily in the home. As Riggare pointed out in her landmark paper on her personal experience of dealing with illness, she spends 1 hour a year receiving her neurological care and the other 8765 hours self-managing her Parkinson's disease [23]. Clinical visits are thus episodic, and HCPs are increasingly recognizing that the intermittent data captured by clinical assessments only offers a snapshot of the patient's health. The main issue is the interappointment period, as it represents a time when understanding the patient's health status is beyond the resources of many HCPs. Finding solutions to help fill this knowledge gap has been receiving significant consideration in recent years particularly as health-care systems have begun implementing chronic disease management strategies that aim to shift care provision from being purely an acute-based endeavor to a more integrated model of care that focuses on leveraging primary care resources in an effective manner [24,25].

As the shift is happening, the need to understand the health status of patients outside the clinical environment is high on the agenda. The use of self-monitoring devices such as the BP monitor and glucometer were early attempts to gather data about patients outside of the clinic. The logic of these devices is that patients would be engaged by the information to help positively influence and support their self-management strategies. The data captured from these devices, however, only provides the patient and the HCP with disease-specific information. These data answer the "What?" but not the "How?" and the "Why?" attempts to answer the "Why?" usually happen through further tests and assessments or conducting a more detailed patient interview placing further financial and time burdens on patients as well as on the resources of HCPs and health-care systems.

However, when we consider the ubiquitous nature of smartphones and wearable technologies, the spatial and temporal boundaries for the collection of patient data are widening. Answering the "How and Why?" can be supported by PGHD as they can reduce the interappointment knowledge gap particularly by providing HCPs with access to previously unattainable lifestyle and psychosocial data. These data include activity, sleep, mood, smoking or alcohol habits, exercise, diet, calorie intake to name just a few. Each discrete PGHD element on its own has value, but it is the combination of data points where the greatest promise is envisioned. An example of this would be potential benefits to be gained from combining PGHD with already existing medical record data. Carolinas HealthCare System in the United States has recently launched an "app" that enables patients to integrate their fitness and well-being data from approximately 70 discrete devices directly into their care management plan via their personal health record [26]. The platform also facilitates the aggregation of longitudinal PGHD, which is an important aspect as it provides a broader context for data generated by disease-specific monitoring devices.

For example, a patient visits their general practitioner (GP) after noticing their BP has been higher than normal for the past 2 weeks. Upon speaking to the patient, the GP suspects that the patient has not been as active as usual so they access the patient's activity data and discover that has indeed been reducing steadily for the past month. Furthermore, the GP might notice that the patient's disposition is not their usual so they access the patient's mood data and observe that they have been logging depressive mood ratings for the past 2 months. The GP might be able to deduce here that as the patient's mood lessened so too did their activity levels, which may have affected their BP levels. The point is, by having access to more nuanced PGHD such as lifestyle and psychosocial data, HCPs can leverage otherwise unattainable contextual information to fill in the knowledge gaps relating to the "How and Why?" questions. Looking forward, leveraging PGHD to fill in the interappointment gap will extend to ensuring patient safety; PGHD can help HCPs to holistically understand patient responses to new medications—patient reported outcomes combined with physiological PGHD will have an important role to play if, and when, adverse reactions occur. Finally, the more PGHD points HCPs can access the less clinical visits they will require, especially with the establishment of eHealth ecosystems that include video-consultations, EHRs, patient portals, and ePrescribing all of which will facilitate the remote alteration of treatment plans.

13.2.3 Can PGHD Help Lower Health Insurance?

Vehicle telematics is not a term often cited in health-care literature but it is a well-known concept in the auto-insurance world. Vehicle telematics refers to the collection, transmission, and analysis of data generated from an on-board unit (OBU) fitted to a motor vehicle [27]. For decades, auto-insurance premiums were based on demographic measures such as driver's age, occupation, place of residence, car type, engine size, or expected mileage. With the use of OBUs however, auto-insurance companies are now basing premiums on real-time dynamic measures of driver behavior; e.g., actual miles traveled, acceleration, deceleration (including braking habits), time spent on the road, trip duration, location and driving style—this is referred to as pay-how-you-drive (PHUD) insurance [27,28].

PHUD insurance relies on monitoring aspects of the driver's performance to create safer driving practices through the provision of constructive feedback from their insurance company regarding safer driving techniques. The ultimate aim is to identify the low-risk drivers from the high-risk ones by mining driver data to determine if a customer's driving behavior justifies a lower or higher insurance price [29]. The benefits of PHUD insurance go beyond purely the financial; environmentally, reduced fuel consumption and optimized traffic patterns are possible outcomes while socially, increased driver education through constructive feedback has the potential to lower driving-related fatalities and the associated health-care costs of motor accidents [30].

Insurance models such as PHUD should be of interest to governments, health-care systems, consumers, and patients particularly in a time when out-of-pocket health-care expenditure and private health insurance premiums are on the rise [31]. When you consider the role of vehicle telematics data in the auto-insurance industry, it is not hard to imagine the role PGHD can have for health insurance customers and providers. As auto-insurers adopt OBUs to monitor their driver's behavior, it should come as no surprise that other industries have begun investigating the potential of wearables and self-monitoring devices to measure their customer's health behaviors and lifestyles.

On the back of the Affordable Care Act passed by the Obama administration in 2010, companies were granted permission to spend up to 30% of their annual insurance premiums on rewards for healthier behavior [32]. Thusly, similar trends to those in the auto-insurance industry have started to emerge elsewhere in the private sector primarily through the establishment of workplace employee wellness programs that leverage health data from wearable technology as part of their employee health insurance models. Such programs are implemented to incentivize healthier behaviors in their employees to improve overall staff productivity while lowering health insurance premiums. In 2013 for example, US company Cigna launched a pilot program with four of its US-based employer health plan clients [33].

Employees were provided with a wearable that tracked activity and calories burned among other physiological data. The data was then shared with health coaches who worked with the employees to adjust and motivate them to create healthier habits. Employees could earn points based on reaching healthy behavior targets, the more points they gained the further discounts they received on their premiums [33].

More recently, UnitedHealthcare are the latest insurers to offer wearable devices to their customers through their wellness program Motion [34]. Customers will use a Fitbit Charge 2 to track their progress across three domains of Motion's program: Frequency, Intensity, and Tenacity (FIT). Customer's daily progress is tracked and if daily goals are reached, they can earn financial credits which can be applied to their policy. Customers can earn up to $4 per day in credits as they achieve their FIT targets with a maximum of $1500 of credits achievable in a year. For example, customers are rewarded if they can reach 300 steps in 5 minutes an hour a part six times per day (Frequency); achieve 3000 steps in 30 minutes once a day (Intensity); or complete 10,000 steps per day (Tenacity) [34].

The evolution of behavior-based health insurance can be driven by PGHD. As increasing numbers of customers track their health, health insurers can plug into already existing data as means to attract a new customer base. Providers could incentivize customers, who already track or a willing to track, by offering the prospect of lower premiums if they have, or can adopt, healthier behaviors. Mining customer's PGHD will allow insurers to create fairer customer profiles, where traditional factors such as age, gender, and area of residence are not the main criteria for determining risk. This would pave the way for personalized health insurance policies and premiums which would help create a more competitive health insurance market.

Much like the PHUD insurance policies, health insurance driven by PGHD can have a positive impact beyond cheaper premiums; if customers are willing to share data regarding their health-care resource utilization, insurers can work with governments and health-care systems to offer insights into what types of patients are utilizing what services and when they are utilizing them (not only when, in terms of everyday use but also when, in relation to their life span or disease journey)—these insights could inform more effective health-care budgeting as well as enhancing the allocation of resources. Finally, if insurers create collaborative programs with the input of health-care systems and allied health groups, consumers will have access to expert medical advice and support regarding the creation of healthier behaviors based on their PGHD. This type of health insurance ecosystem would close the loop so to speak, with respect to scaling the management of individual health to population health. As increasing numbers of customers are empowered to manage their health through person-centered health insurance, healthier behavior change becomes possible at the populations level which could help tackle the social costs associated with lifestyle-related diseases.

13.2.4 The Wider Research Potential of PGHD

Another major benefit that can be derived from PGHD is its potential as a research resource that can be used to expand our knowledge base regarding human behavior and performance. If large populations of people are generating longitudinal data over time, and are willing to contribute this data to a shared database that can be mined, we can greatly expand our understanding of human behavior and its relationship with health. Our current understanding of human behavior and its relationship with health is derived from periodic questionnaire methodologies, a strategy that only provides a snapshot of behavior. PGHD offers the potential to provide a very rich picture of how people feel and behave on a longitudinal level. If this data is taken alone, we can learn a lot about patterns of behavior and perceived health or mood of the population, which gives us greater understanding of the health and social care needs of a population. If we can link this PGHD with other clinical, or even genetic, data, then we can derive greater understanding of the relationships between different domains of measurement of health. In particular, we can understand the interplay between lifestyle/behavior and biomedical/genetic markers of health.

13.3 Emerging Issues

The integration of PGHD into the daily health-care routines is of course not without its barriers, however. This section will explore some of the key emerging issues facing the use of PGHD.

13.3.1 Changing Roles and New Conversations

As more and more patients introduce PGHD into their day-to-day health-care regimes, traditional forms of information exchange in the clinic and hospital will undoubtedly change as well. With their PGHD in hand, patients will expect conversations about their data and how they can be used to inform decisions. Correspondingly, HCPs will find themselves entering unfamiliar territory as patient preferences turn to data-centric consultations. PGHD can have a positive impact on aspects, such as SDM, by engaging the patient to have an authoritative voice in the consultation. However, patients who collect PGHD will be informed about the capabilities of these data and in many cases, much more informed than their HCP. Thus, no longer will the HCP be the de facto leader of consultations—the patient, by virtue of being an authority on their data, will naturally begin to steer consultations with the expectation that their HCP will collaborate in these discussions. Moreover, as patients tap into the "Doctor Google" phenomenon, they will begin to couple their PGHD with health information further empowering them to offer options regarding their course of care or at minimum, to challenge those provided by their HCP.

The shifting cultures of health responsibility discussed earlier are enabling patients to be more assertive in deciding the appropriateness of care provision they receive which has been discussed elsewhere regarding the coproduction of health care [35,36]. With the introduction of PGHD into everyday clinical practices, patients and HCPs will organically begin to coproduce more empathetic care pathways brought about by the digitization of health data. However, to unlock the true potential of PGHD, HCPs need to improve their literacy regarding PGHD in order to evaluate how best to use these data in their patient's treatment and to offer patients an informed decision concerning how they can effectively use PGHD in their care.

13.3.2 Creating Actionable PGHD

As the research regarding health-related data has evolved, two interchangeable terms have emerged in the literature, "actionable data" and "actionable information." These terms are used to describe user-friendly data that can be acted upon by the end user in a manner that has a practical value while meeting their particular needs [37–40]. An issue facing the integration of PGHD into clinical use will be determining what constitutes practical value for the key stakeholders. But determining practical value will rely on other factors, such as identifying the needs of patients and HCPs across the various contexts and scenarios of use. With the increasing potential for people to bring PGHD into the clinical setting, of central importance, will be understanding how PGHD can be integrated into the workflows of HCPs in a manner that is empathetic to the resources and culture of their clinic. For instance, 6 months of activity data would hold little practical value to a GP if the data cannot be easily visualized and interpreted within the constraints of a 10-minute consultation. Other questions arise when we think about the relevant aspect of the actionable data argument. Each disease or illness require information from discrete types of data to help HCPs to make decisions; more work is required in terms of PGHD to understand what are the information needs of HCPs per disease. Gaining this insight could also be used to educate patients about the types of data relevant to their health profile—if patients are only bringing relevant PGHD to their clinical visits, this may help to reduce the occurrence of data overload for HCPs.

Another aspect that will affect the actionability of PGHD for HCPs will be their integration with EHRs. In countries where EHRs are operational, questions relating to the interoperability of PGHD remain mostly unanswered as do questions concerning the ways in which HCPs will make clinical decisions based on the syntheses of medical information and PGHD. Progress is being made however, with PGHD programs such as the one ongoing in the Carolinas HealthCare System mentioned earlier. Elsewhere, the University of California Davis Health System has recently begun a PGHD initiative called "Integrating Patient-Generated Health Data to Improve Health" [41]. The program involves integrating

1.4 million PGHD points captured from diabetes and better BP initiatives from patient-connected devices and integrates the PGHD points into their EHR system. Going forward, learning from ongoing PGHD programs like Carolinas will help shed light on issues relating to interoperability and workflow. We should not forget about those countries without operational EHRs for they far outweigh their counterparts. Creating actionable information for HCPs in this context will rely heavily on finding solutions that allow PGHD to be utilized when it cannot be seamlessly integrated with already existing medical records. This will be a complex issue to overcome given the embeddedness of the paper-based culture in many health-care practices.

From the patient perspective, health literacy will have a massive bearing on the actionable nature of PGHD particularly when we consider that lower levels of health literacy have been shown to negatively impact patients' ability to self-manage [42]. A core component of a person's health literacy is their health numeracy skills. Health numeracy refers to "the degree to which individuals have the capacity to access, process, interpret, communicate, and act on numerical, quantitative, graphical, biostatistical, and probabilistic health information needed to make effective health decisions" [43]. Given that many wearable and self-monitoring devices are accompanied by an app, and furthermore that the greater number of app-based data visualizations are numerical and graphical, health numeracy emerges as an important attribute for people to have if they are using PGHD as part of their health routines. For instance, if an individual is capturing weekly BP data, seeing these measurements will be of little use if they do not understand the significance of the numbers which will leave them without the vital knowledge that would otherwise empower them to act upon this data safely and appropriately. To avoid the risk of endangering patients, ensuring that users are (1) made aware of the necessity for health numeracy skills and (2) supported to develop adequate levels of health numeracy should be a priority for health-care systems and HCPs as the use of PGHD evolves.

13.4 Conclusion

The ongoing digital health phenomena have witnessed mobile and smart devices, including sensor technology, become commonplace technologies in people's daily routines. As their adoption grows, the numbers of people self-tracking are helping to create a significant pool of PGHD that is progressively catching the attention of health care and HCPs alike. This chapter has discussed several benefits associated with leveraging PGHD in health care while also exploring some of the emerging issues that should be considered as research in the area evolves. Understanding how small data sets such as PGHD will impact social institutions has immediate and future advantages especially in a time when data-driven solutions are showing great promise to propel service innovation. Indeed, for many, data is the fuel source that will empower future societies, but before then many questions still need to be answered particularly concerning the human factor elements of data use. Ultimately, unlocking the potential of small data, such as PGHD, is more of a social problem than it is a technological one and unless data-driven solutions can meet the needs of end users across the multiple contexts they occupy, it may, unfortunately, be a long time before we can deliver on the promises of big or small data.

References

[1]   M. Shapiro, D. Johnston, J. Wald, D. Mon, Patient-Generated Health Data. White Paper (Prepared for Office of Policy and Planning, Office of the National Coordinator for Health Information Technology), RTI International, Research Triangle Park, NC, 2012.

[2]   Statista, Number of smartphone users worldwide from 2014 to 2020. <http://www.statista.com/statistics/330695/number-of-smartphone-users-worldwide/>, 2016 (accessed 05.10.17.).

[3]   ZDNet, Forrester: Nearly 1 in 3 Americans will use a wearable device by 2021. <http://www.zdnet.com/article/forrester-nearly-1-in-3-americans-will-use-a-wearable-device-by-2021/> 2016 (accessed 05.10.17.).

[4]   Mintel, Brits step up to wearable technology: sales of fitness bands and smartwatches up 118% in 2015. <http://www.mintel.com/press-centre/technology-press-centre/brits-step-up-to-wearable-technology-sales-of-fitness-bands-and-smartwatches-up-118-in-2015>, 2016 (accessed 02.13.17.).

[5]   IMS Institute for Healthcare Informatics, Patient adoption of mHealth: use, evidence, and remaining barriers to mainstream acceptance. <http://www.imshealth.com/files/web/IMSHInstitute/Reports/Patient Adoption of mHealth/IIHI_Patient_Adoption_of_mHealth.pdf>, 2015 (accessed 05.11.17.).

[6]   Research2Guidance, mHealth app developer report 2016. <https://research2guidance.com>, 2016 (accessed 05.09.17.).

[7]   Research & Markets, mHealth and home monitoring—8th edition. <https://www.researchandmarkets.com/research/r5t6d4/mhealth_and_home>, 2017 (accessed 05.11.17.).

[8]   J.H. Hibbard, J. Greene, What the evidence shows about patient activation: better health outcomes and care experiences; fewer data on costs, Health Aff. 32 (2013) 207−214. Available from: https://doi.org/10.1377/hlthaff.2012.1061.

[9]   A. Coulter, J. Ellins, Effectiveness of strategies for informing, educating, and involving patients, BMJ. 335 (2007) 24−27. Available from: https://doi.org/10.1136/bmj.39246.581169.80.

[10]   K.L. Carman, P. Dardess, M. Maurer, S. Sofaer, K. Adams, C. Bechtel, et al., Patient and family engagement: a framework for understanding the elements and developing interventions and policies, Health Aff. (Millwood). 32 (2013) 223−231. Available from: https://doi.org/10.1377/hlthaff.2012.1133.

[11]   R. Osborn, D. Squires, International perspectives on patient engagement: results from the 2011 Commonwealth Fund Survey, J. Ambul. Care Manag. 35 (2012) 118−128. <https://doi.org/10.1097/JAC.0b013e31824a579b>.

[12]   G.G. Blackburn, J.M. Foody, D.L. Sprecher, E. Park, C. Apperson-Hansen, F.J. Pashkow, Cardiac rehabilitation participation patterns in a large, tertiary care center: evidence for selection bias, J. Cardiopulm. Rehabil. 20 (2000) 189−195. Available from: https://doi.org/10.1097/00008483-200005000-00007.

[13]   H.J. Bethell, S.C. Turner, J.A. Evans, L. Rose, Cardiac rehabilitation in the United Kingdom. How complete is the provision? J. Cardiopulm. Rehabil. 21 (2001) 111−115.

[14]   J.K. Lee, K.A. Grace, A.J. Taylor, Effect of a pharmacy care program on medication adherence and persistence, blood pressure, and low-density lipoprotein cholesterol: a randomized controlled trial, JAMA 296 (2006) 2563−2571. Available from: https://doi.org/10.1001/jama.296.21.joc60162.

[15]   T. Eaton, P. Young, W. Fergusson, L. Moodie, I. Zeng, F. O'Kane, et al., Does early pulmonary rehabilitation reduce acute health-care utilization in COPD patients admitted with an exacerbation?

A randomized controlled study, Respirology. 14 (2009) 230−238. Available from: https://doi.org/10.1111/j.1440-1843.2008.01418.x.

[16] N.J. Greening, J.E.A. Williams, S.F. Hussain, T.C. Harvey-Dunstan, M.J. Bankart, E.J. Chaplin, et al., An early rehabilitation intervention to enhance recovery during hospital admission for an exacerbation of chronic respiratory disease: randomised controlled trial, BMJ 349 (2014) g4315. Available from: https://doi.org/10.1136/bmj.g4315.

[17] G. Appelboom, M. LoPresti, J.-Y. Reginster, E. Sander Connolly, E.P.L. Dumont, The quantified patient: a patient participatory culture, Curr. Med. Res. Opin. 30 (2014) 2585−2587. Available from: https://doi.org/10.1185/03007995.2014.954032.

[18] J.H. Hibbard, J. Stockard, E.R. Mahoney, M. Tusler, Development of the patient activation measure (PAM): conceptualizing and measuring activation in patients and consumers, Health Serv. Res. 39 (2004) 1005−1026. Available from: https://doi.org/10.1111/j.1475-6773.2004.00269.x.

[19] S.M. Auerbach, Do patients want control over their own health care? A review of measures, findings, and research issues, J. Health Psychol. 6 (2001) 191−203. Available from: https://doi.org/10.1177/135910530100600208.

[20] C. Charles, A. Gafni, T. Whelan, Shared decision-making in the medical encounter: What does it mean? (Or it takes, at least two to tango), Soc. Sci. Med. 44 (1997) 681−692. Available from: https://doi.org/10.1016/S0277-9536(96)00221-3.

[21] M.J. Barry, S. Edgman-Levitan, Shared decision making—the pinnacle of patient-centered care, N. Engl. J. Med. 366 (2012) 780−781. Available from: https://doi.org/10.1056/NEJMp1109283.

[22] E.O. Lee, E.J. Emanuel, Shared decision making to improve care and reduce costs, N. Engl. J. Med. 368 (2013) 6−8. Available from: https://doi.org/10.1056/NEJMp1209500.

[23] S. Riggare, A patient perspective on self-care,, Nuff. Trust Blog (2015).

[24] T. Bodenheimer, E.H. Wagner, K. Grumbach, Improving primary care for patients with chronic illness, JAMA. 288 (2002) 1909. Available from: https://doi.org/10.1001/jama.288.15.1909.

[25] T. Bodenheimer, K. Lorig, H. Holman, K. Grumbach, Patient self-management of chronic disease in primary care, JAMA. 288 (2002) 2469−2475. Available from: https://doi.org/10.1001/jama.288.19.2469.

[26] mHealth Intelligence, Giving wearables a place in the patient record. <http://mhealthintelligence.com/news/giving-wearables-a-place-in-the-patient-record>, 2016 (accessed 06.09.17.).

[27] G. Vaia, E. Carmel, W. DeLone, H. Trautsch, F. Menichetti, Vehicle telematics at an Italian insurer: new auto insurance products and a new industry ecosystem, MIS Quart. 10 (2011) 115−117. Available from: https://doi.org/10.1108/02635570910926564.

[28] S. Husnjak, D. Peraković, I. Forenbacher, M. Mumdziev, Telematics system in usage based motor insurance, Procedia Eng. 100 (2015) 816−825. Available from: https://doi.org/10.1016/j.proeng.2015.01.436.

[29] D.I. Tselentis, G. Yannis, E.I. Vlahogianni, Innovative insurance schemes: pay as/how you drive, Transp. Res. Procedia 14 (2016) 362−371. Available from: https://doi.org/10.1016/j.trpro.2016.05.088.

[30] J. Bordoff, P. Noel, Pay-as-you-drive auto insurance: a simple way to reduce driving-related harms and increase equity, J. Risk Insur. 37 (2008) 25. Available from: https://doi.org/10.2307/251178.

[31] N.V. Motaze, C. Chi, P. Ongolo-Zogo, J. Ndongo, C. Wiysonge, Government regulation of private health insurance, Cochrane Database Syst. Rev. (2015). Available from: https://doi.org/10.1002/14651858.CD011512.

[32] K. Madison, H. Schmidt, K.G. Volpp, Smoking, obesity, health insurance, and health incentives in the Affordable Care Act, JAMA. 310 (2013) 143. Available from: https://doi.org/10.1001/jama.2013.7617.

[33] P. Olson, Wearable tech is plugging into health insurance. <https://www.forbes.com/sites/parmyolson/2014/06/19/wearable-tech-health-insurance/#50b0abe618bd>, 2014 (accessed 06.06.17.).

[34] H. Mack, UnitedHealthcare to offer customized Fitbit Charge 2 for incentive-based employee wellness program. <http://www.mobihealthnews.com/content/unitedhealthcare-offer-customized-fitbit-charge-2-incentive-based-employee-wellness-program>, 2017 (accessed 06.06.17.).

[35] P. Hyde, H.T.O. Davies, Service design, culture and performance: collusion and co-production in health care,, Hum. Relations. 57 (2004) 1407–1426. Available from: https://doi.org/10.1177/0018726704049415.

[36] N. Crisp, L. Chen, Global supply of health professionals, N. Engl. J. Med. 370 (2014) 950–957. Available from: https://doi.org/10.1056/NEJMra1111610.

[37] S. Foldy, S. Grannis, D. Ross, T. Smith, A ride in the time machine: information management capabilities health departments will need, Am. J. Public Health. 104 (2014) 1592–1600. Available from: https://doi.org/10.2105/AJPH.2014.301956.

[38] S.S. Woods, N.C. Evans, K.L. Frisbee, Integrating patient voices into health information for self-care and patient-clinician partnerships: Veterans Affairs design recommendations for patient-generated data applications, J. Am. Med. Informatics Assoc. 23 (2016) 491–495. Available from: https://doi.org/10.1093/jamia/ocv199.

[39] S. Kumar, W. Nilsen, M. Pavel, M. Srivastava, Mobile health: revolutionizing healthcare through transdisciplinary research, Computer (Long. Beach. Calif) 46 (2013) 28–35. Available from: https://doi.org/10.1109/MC.2012.392.

[40] L. Hood, N.D. Price, Demystifying disease, democratizing health care, Sci. Transl. Med. 6 (2014) 225ed5. Available from: https://doi.org/10.1126/scitranslmed.3008665.

[41] ICT & Health, UC Davis Health is integrating patient-generated health data to improve health. <https://www.ictandhealth.com/news/newsitem/article/uc-davis-health-is-integrating-patient-generated-health-data-to-improve-health.html>, 2017 (accessed 06.11.17).

[42] N.D. Berkman, S.L. Sheridan, K.E. Donahue, D.J. Halpern, K. Crotty, Low health literacy and health outcomes: an updated systematic review, Ann. Intern. Med. 155 (2011) 97–107. Available from: https://doi.org/10.7326/0003-4819-155-2-201107190-00005.

[43] A.L. Golbeck, C.R. Ahlers-Schmidt, A.M. Paschal, S.E. Dismuke, A definition and operational framework for health numeracy, Am. J. Prev. Med. 29 (2005) 375–376. Available from: https://doi.org/10.1016/j.amepre.2005.06.012.

CHAPTER 14

Evolution Map of Wearable Technology Patents for Healthcare Field

Serhat Burmaoglu[1], Vladimir Trajkovik[2], Tatjana Loncar Tutukalo[3], Haydar Yalcin[4] and Brian Caulfield[5]

[1]*Department of Health Management, Izmir Katip Celebi University, Izmir, Turkey* [2]*Faculty of Computer Science and Engineering, Ss Cyril and Methodius University, Skopje, FYR Macedonia* [3]*Faculty of Technical Sciences, University of Novi Sad, Novi Sad, Serbia* [4]*Department of Information Management, Izmir Katip Celebi University, Izmir, Turkey* [5]*Insight Centre for Data Analytics, University College Dublin, Dublin, Ireland*

14.1 Introduction

Wearable technologies have ignited a new type of human–computer interaction with the rapid development of information and communication technologies. This technology facilitates mobility and connectivity for users that they can access online information conveniently and communicate with others immediately while moving. Based on IDTechEx analysis report the market of wearable technology will see threefold growth from over $24 bn in 2017 to $70 bn by 2025 (https://www.idtechex.com/research/topics/wearable-technology.asp (accessed: 02.01.2018)). In addition to market statistics the applications of wearable technologies spread to different areas based on the map of Beecham Research's report. Dispersion of wearable technologies is demonstrated in Fig. 14.1.

It can be seen in Fig. 14.1 that eight sectors were determined by Beecham Research and all of them require appropriate human–machine interaction. The focus of this chapter is medical sector and it is demonstrated that there are many different products that function for different healthcare operations and tracking issues. It is thought that healthcare sector is more suitable application area for wearable technologies because of the human-centered approach of medical industry and growing trend on personalized healthcare. From the patents side it can be seen that medical technology was the field where the most patent applications were filed based on the European Patent Office's report. In this report digital communication, computer technology, electrical machinery, apparatus, and energy fields were coming after medical field [1].

Wearable Technology in Medicine and Health Care.
DOI: https://doi.org/10.1016/B978-0-12-811810-8.00014-2

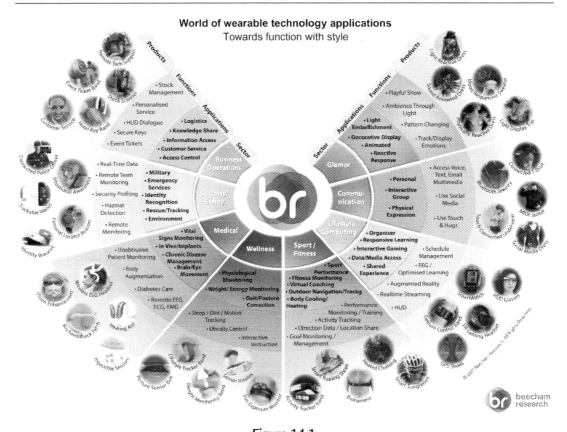

Figure 14.1

Dispersion of wearable technologies to different areas. *Source: From http://www.beechamresearch.com/ article.aspx?id = 20 (Accessed 30.03.17).*

In this chapter, the literature on wearable technologies and their usage in healthcare are reviewed first. Then, the patent analysis methodology and the applied model are explained. Analysis is performed in fourth section and findings and visuals are demonstrated. Finally, based on findings state of wearable technologies in healthcare is determined and the possible development fields are discussed.

14.2 Literature on Wearable Technologies and Usage in Healthcare

The methodology used for the selection and processing of the research papers is similar to the one described by Moher et al. [2]. Few publisher databases from Springer publishing digital libraries have been searched for publications containing the phrases: "wearable" and "connected health" and their synonyms obtained using semantic word net structures. A total of 6007 papers were found from the period 2006−16. From the found papers, 1074 papers with the highest quotation score were selected. These papers were subject to our further analyses.

Fig. 14.2 presents the trend of the papers with highest quotation marks per year. The absolute values of the papers per year are not shown due to the fact that we need to analyze the trend. The increasing trend on research of wearable in connected health is obvious. The decline of papers in the last year is due to the fact that papers need time to obtain quotations.

Using the identified papers we performed similar search within those papers seeking for different keywords (and their synonyms). We analyzed the domain of application of wearable devices (lifestyle, fitness, medical). The distribution of the paper that states the domain of application in one of those three categories per year is demonstrated in Fig. 14.3.

As can be seen in Fig. 14.3 the data indicates that research of the wearable technologies in the medical domain is gaining on popularity, especially in the last 5 years. The same stands for the other two domains, although they have begun to be the subject of the scientific

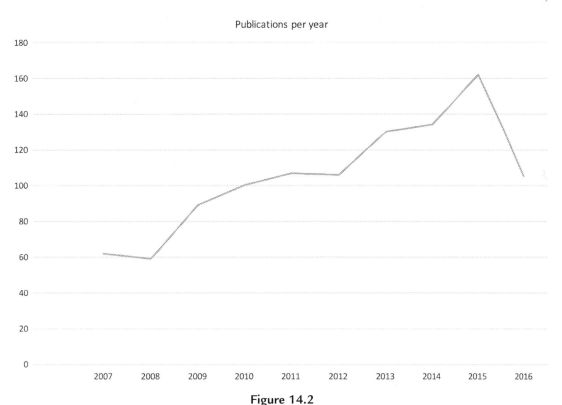

Figure 14.2
Trend in publishing papers on wearable in connected health.

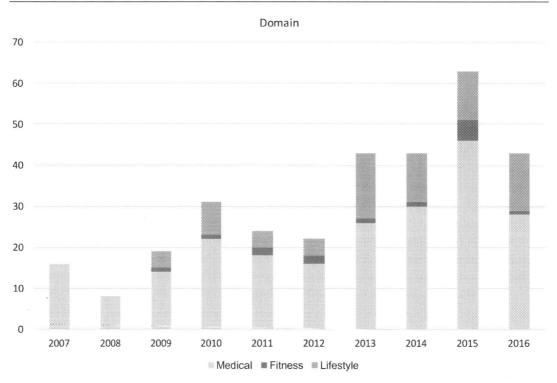

Figure 14.3
Distribution of papers that state domain: medical, fitness, and lifestyle per year.

research from 2009. Nevertheless the differentiation in the research related to the selected domains in the last years can be noted.

Technology related topics were investigated using the following keywords: sensor, protocol, and cloud-based systems. The results are presented on Fig. 14.4.

While protocol-oriented research has steady interest within the research community, the research interest for sensors is increasing, while cloud-based systems are becoming very popular in the last few years. This is in accordance with identified current research topics. The research in sensor-related topics has made most of the progress, yet still being very active in persuading the goals of transparency, self-healing, and self-powering. The research on communication protocols is expected to flourish again towards the expected adoption of 5G standard, offering the solution for the architecture, smart device communications, and antenna redesign. Cloud as a research topic is another key concept efficiently enabling future personalized pervasive health information management and mining.

Publications on different implemented services using wearable technologies in connected health: monitoring, detection, and recommendation have been considered. Fig. 14.5 presents the findings.

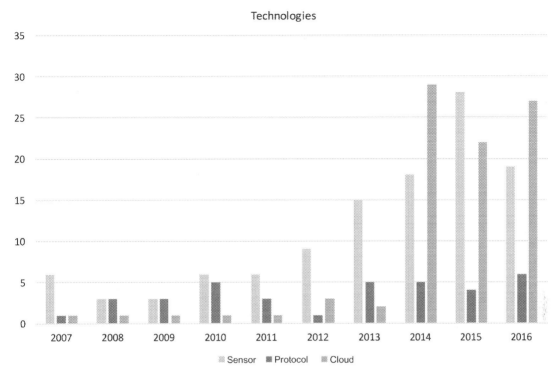

Figure 14.4

Distribution of technology research topics in papers per year.

It can be noted that monitoring services have an increasing trend, while the detection related topics vary in research interest over time. The recommendation service raises as slowly fluctuating, yet relatively constant interest among researchers.

Different user-related concerns were also investigated: quality of experience, security, privacy, and technology acceptance. Fig. 14.6 gives an overview of the obtained results.

The privacy and security issues are becoming more interesting as wearable technologies and services mature and penetrate the market. However, very little research is dedicated to the quality of experience and technology acceptance. As the wearable solutions increase in number, those issues will become more relevant, promoting the human-centered design and the ease of use as additional qualities.

The identified current and future research trends, the progress that has been made and the future challenges, indicate the revolution of wearables is yet to come. For the wearable future the research will have to tackle the current weaknesses and obstacles towards wider consumer acceptation, human-centered design, and resolving the users concern. The ultimate benefit of wearable technologies would be a reshaped, sustainable, pervasive, and

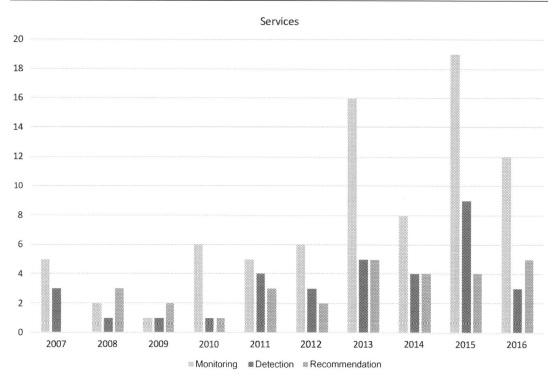

Figure 14.5
Distribution on the research on wearable technologies services per year.

personalized healthcare. The future, sustainable connected health paradigm aims to maximize healthcare resources and provide increased, flexible opportunities for consumers to engage with clinicians in better self-management of their health.

14.3 Patent Analysis and Research Methodology

Patent documents are an ample source of technical and commercial knowledge in terms of technical progress, business trends, and proprietary ownership and, thus, patent analysis has long been thought as a useful tool for R&D management in corporate setting and techno-economic analysis in macro context. Patent analysis became more popular with increasing complexity of high-technology innovation, short cycle of technological developments, and rapidly changing market demand [3].

A patent represents an invention in a particular field of technology [4] and provide not only legal protection for intellectual property rights but also include detailed information about the developed technology [5]. Organizations are interested in analyzing patents for determining novelty in patents; evaluating competitiveness of firms [6]; monitoring trends [7]; forecasting technological developments in special domains; identifying the promising

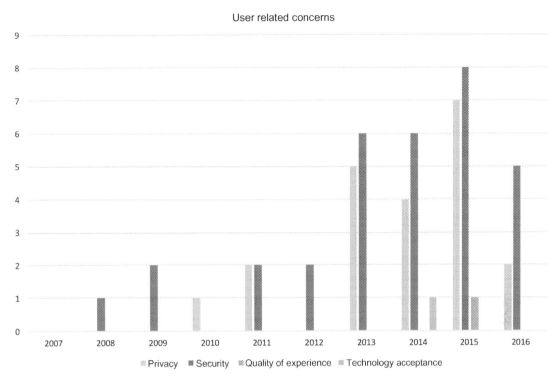

Figure 14.6
Distribution on research papers dealing with user concerns per year.

technologies; preparing technological roadmapping; identifying emerging technology streams and finding the competitors [4]. From the macro perspective, it can be said that, patent analysis has often employed to understand the linkage between technology development and economic growth [8].

Patent databases are an important source of information for innovators, R&D engineers, corporate executives, and policymakers in technology latecomer countries [9]. Innovators use patent information for understanding the prior art and assess if their inventions are commercially viable or not. R&D engineers use patent data to solve problems with existing technology [10,11]. Patent analysis is applied for visualizing and monitoring the trends in selected field and repetitive actions may be prevented. According to EU Commission report up to 30% of all R&D expenditures is wasted on redeveloping existing inventions [12].

There were two accepted perspectives in patent technology analysis; citation- and content-based approaches. Citation-based studies consider the citations between two patents as knowledge flows [13]. By using these knowledge flows main technological trends may be discovered. However, these visualizations neglected the patent contents and the quality of citation relationship cannot be recognized. On the other hand content-based studies used text mining techniques to measure the content similarity between pairs. Another alternative approach is network-based patent analysis, which prepared for overcoming drawbacks of citation analysis. Although network analysis shares some commonality with citation analysis, its relative advantage is substantial. First, network analysis shows the relationship among patents as a visual network and therefore assists the analyzer in intuitively comprehending the overall structure of a patent database. Second, network analysis enriches the potential utility of patent analysis because it takes more diverse keywords into account and produces more meaningful indicators [3]. Network analysis based on text mining, which decreases search time and cost. Therefore it can be said that by applying network analysis approach content- and citation-based studies are combined. A general patent analysis scenario may be demonstrated as in Fig. 14.7.

As can be seen in Fig. 14.7 a typical patent analysis steps begins with defining the scope, concepts, and purposes of the analysis. Second issue is deciding the search query strategy. After searching and downloading the related patents data should be segmented, cleaned, and normalized. In the fourth step, patent content should be analyzed to summarize their claims, topics, functions, or technologies. By clustering in the fifth step analyzed patents grouped or classified based on some used metrics. These groups are visualized in sixth step and then technology or business trends and relations predicted at last step. As can be seen this scenario requires the analyst to have a certain degree of expertise in information retrieval, domain-specific technologies, and business intelligence. This multidiscipline requirement makes such analysts hard to find or costly to train. Therefore automated technologies for assisting analysts in patent processing and analysis are thus in great demand [14].

In this study we seek to understand the evolution of wearable technologies by using patent data. The research methodology is demonstrated in Fig. 14.8.

As can be seen in Fig. 14.8 at first search query is decided and the data is retrieved from Derwent Innovation Database. In the second step patent trend is analyzed by using the analogy of growth curves. Growth curves represent the growths in performance over time.

- Task identification
- Searching
- Segmentation
- Abstracting
- Clustering
- Visualization
- Interpretation

Figure 14.7
A typical patent analysis steps.

Figure 14.8
Research methodology.

By using these curves, technology trends can be evaluated. In this study Gompertz model [15,16], exponential model and Fisher-Pry ([17]) model were used. The Gompertz model is often referred to as the "mortality model" in technology forecasting. Gompertz model produces an S-curve, which rises more sharply but begins to taper off earlier than the Fisher-Pry model. The Fisher-Pry model predicts characteristics very similar to those of biological system growth. This is the reason that it is commonly referred to as the "substitution model" based on its application in forecasting in which the rate of new technology will replace existent technology. The Fisher-Pry presents a slow beginning, a rapid slope and a tapering off at the finish [17]. After finishing growth trend analysis, cluster analysis is performed with country-, inventor-, and keyword-based approaches. Finally, findings are visualized, interpreted, and discussed. Cooccurrence networks are used to provide a graphic visualization of potential relationships between patents within last decade of wearable technologies for healthcare patents world. Network diagrams were created based on the codes used in the classification of patents and the mutual relations of the mainstream areas classified together with patents taken in the related area were visualized via PAJEK (http://vlado.fmf.uni-lj.si/pub/networks/pajek/).

14.4 Patent Data and Analysis

In Derwent Innovations Index, patents are divided into 20 broad subject areas or sections. These sections are grouped into three areas as; Chemical Sections (A−M), Engineering Sections (P−Q), Electrical and Electronic Sections (S−X). These sections are then further subdivided into classes. Each class consists of the section letter, followed by two digits. For example, X22 is the class designation for Automotive Electrics and C04 is the class for all Chemical Fertilizers. When used in combination with other search fields such as a Topic search, these classes allow you to precisely and effectively restrict your search to the relevant subject area (http://images.webofknowledge.com/WOKRS517B4/help/DII/ hs_derwent_class_code.html). An online search was conducted on the Derwent Innovation Index to obtain patent data on Wearable Health Technologies. In order to fetch the patent data of the registered trademarks in the field of wearable health technologies, an online search was carried out through IPC codes. Although there are very specific codes that contain "medical" or "health" in the codes, those who start with "A61" are preferred in order to increase coverage. The query used to access the data that has been analyzed in the light of this information is formed as follows: a total of 4085 patents have been obtained which satisfy the criteria for (TI = (wearable*) AND TS = (medical* OR health*)) OR (TI = (wearable*) AND IP = (A61*)). At first growth characteristics of these patents are analyzed and the growth trends are demonstrated in Fig. 14.9.

It can be seen in Fig. 14.9 that growth trend of publications is more similar to exponential curve. Based on the determinant coefficients the highest value is coming from the exponential growth curve ($R^2 = 0.98$) also. It can be interpreted that the growth of wearable technology patents will have great potential in future. When Fisher-Pry curve analyzed it can be seen that second highest determinant coefficient score ($R^2 = 0.89$) belongs to this curve. This can be interpreted as new technologies have been already substituting older ones and exponential growth is coming with an evolution. Finally, from the Gompertz curve the determinant coefficient score is the lowest and it can be interpreted that trend has not yet ready to mature.

Various levels of information about the patents on wearable technologies for healthcare examined have been reached. This information; in addition to Patent No., Document Title Inventor, Assignee, Derwent Primary Accession Number, Abstract, Equivalent Abstract, Derwent Class Code, Derwent Manual Code, International Patent Classification, Patent Details, Application Details and Date DCR Number, Markush Number, Ring Index Number, and Derwent Registry Number information have been compiled. A list of applicants of the patents registered in the relevant area for the purpose of the study has been prepared and the files with at least 10 times or more patent applications are given in Table 14.1.

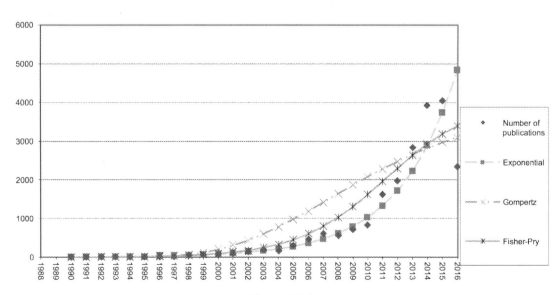

Figure 14.9
Growth trends.

Table 14.1: Most frequent assignees.

Assignee	N
Samsung Electronics Co Ltd (SMSU-C)	57
Konink Philips NV (PHIG-C)	42
Magic Leap Inc (MAGI-Nonstandard)	31
Univ Zhejiang (UYZH-C)	25
Procter & Gamble Co (PROC-C)	25
Zoll Medical Corp (ASAH-C)	24
Fitbit Inc (FITB-Nonstandard)	21
Apple Inc (APPY-C)	21
West Affum Holdings Corp (WAFF-Nonstandard)	18
Zhou C (ZHOU-Individual)	17
Hello Inc (HELL-Nonstandard) Hello Inc (HELL-Nonstandard)	13
Valencell Inc (VALE-Nonstandard)	13
LG Electronics Inc (GLDS-C)	12
Nippon Telegraph & Telephone Corp (NITE-C)	12
Univ Shanghai Jiaotong (USJT-C)	12
Chengdu Wise Sci & Technology Co Ltd (CHEN-Nonstandard)	11
Beijing Technology Inst (BEIT-C)	11
Physio-Control Inc (PHYS-Nonstandard)	10
Boe Technology Group Co Ltd (BOEG-C)	10
Equos Res KK (EQUS-C)	10
Seiko Epson Corp (SHIH-C)	10

When patent applicants are examined in Table 14.1 it is observed that companies such as Samsung, Konink Philips NV, Magic Leap Inc, Univ Zhejiang, Procter & Gamble Co, have taken the first rows. Fig. 14.10 presents the countries where the patents have been submitted. Gephi software was used to create visualization. Gephi is a data visualization program that makes ready-made network data visualized with different algorithms. In the visualization of the network maps, undefined networks type is used to visualize the force Atlas2 algorithm [18,19], which provides placement according to attraction and repulsion powers of the nodes. The nodes within the network are approximated by applying the Barnes-Hut approximate repulsion [19].

The colors of the nodes show the sets that are revealed depending on the modularity value, which indicates cluster separation. The sizes of the nodes are proportional to the values of betweenness centrality. When the data examined the ideas regarding wearable technologies are mostly patented by China, United States, and EU. These countries are technology leaders and the firms are mostly concentrated in these regions. In other words, data shows that commercial products patented in wearable health technologies are primarily registered in the United States, Europe, and the World Patents Offices. In terms of volume it can be said that the rate of registration in European countries and United States is the highest. Frequency creates insight about the patent map generally for medical wearables but understanding deeply network structure should be examined also.

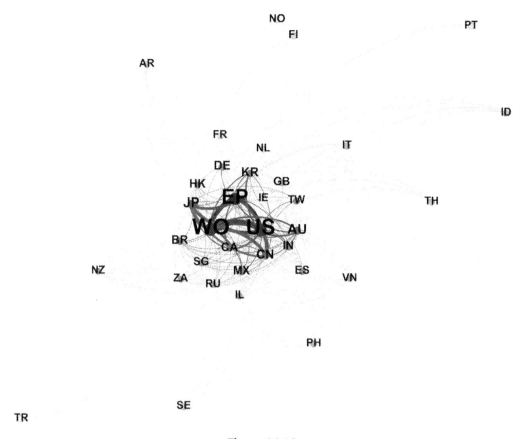

Figure 14.10
Cooccurrence analysis of wearable technology patents for countries. *AR*, Argentina; *AU*, Australia; *BR*, Brazil; *CA*, Canada; *CN*, China; *DE*, Germany; *EP*, European Patents; *ES*, Spain; *FI*, Finland; *FR*, France; *GB*, United Kingdom; *HK*, Hong Kong; *ID*, Indonesia; *IE*, Ireland; *IL*, Israel; *IN*, India; *IT*, Italy; *JP*, Japan; *KR*, Korea; *MX*, Mexico; *NL*, Netherlands; *NO*, Norway; *NZ*, New Zealand; *PH*, Philippines; *PT*, Portugal; *RU*, Russian Federation; *SE*, Sweden; *SG*, Singapore; *TH*, Thailand; *TR*, Turkey; *TW*, Taiwan; *US*, United States of America; *VN*, Vietnam; *WO*, Patent Cooperation Treaty; *ZA*, South Africa.

The same keyword in patent applications may have different meanings for different sectors. In other words, companies often use different names for the same invention, and additional variations may be introduced when the patent application is translated into different languages. International Patent Codes (IPC) have been used to make a more accurate analysis on the patents registered in the field for this reason. Some of the keywords can appear in many different contexts within patent titles. Therefore subject classification system is used for an effective patent searching and analyzing. In our study patent IPCs are visualized using the cooccurrence networks method in social network theory. Cooccurrence networks are the collective interconnection terms of their paired presence within a specified

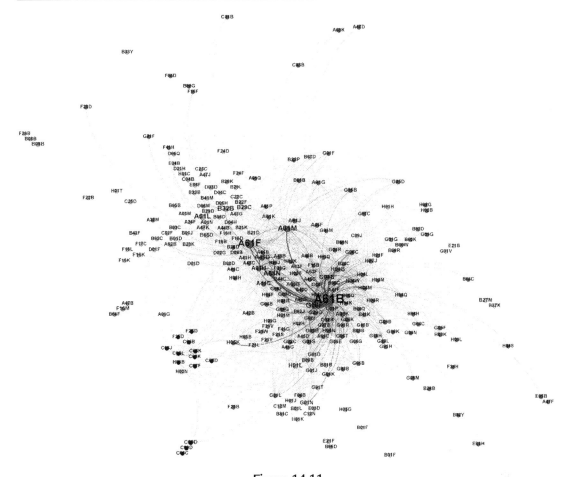

Figure 14.11

Cooccurrence analysis of wearable technology patents for patent classification codes.

unit of text. Networks are generated by connecting pairs of terms using a set of criteria defining cooccurrence [20]. It is now possible to see which patents on wearable health technologies concentrate on the subareas with which mainstream areas in Fig. 14.11 and network parameters of this figure is demonstrated in Table 14.2.

The nodes within the network are approximated by applying the Barnes-Hut approximate repulsion algorithm [19]. The colors of the nodes in Fig. 14.11 show us the different sets that are revealed depending on the modularity value, which indicates at the cluster separation. The sizes of the nodes are proportional to the values of betweenness centrality demonstrated in Table 14.2.

The clustering over the IPC based coexistence map reveals two main nodes: A61B (which stand for diagnosis; surgery; identification) and A61L (methods or apparatus for sterilizing

Table 14.2: Network parameters of wearable technology patents for patent classification codes.

Id	Label	Degree	Weighted Degree	Modularity Class	Component Number	Eccentricity	Closeness Centrality	Harmonic Closeness Centrality	Betweeness Centrality
2	A61B	155	7101.0	0	0	3.0	0.705	0.800	7797.8
4	A61M	93	1251.0	0	0	3.0	0.592	0.673	2625.8
6	A61F	119	1635.0	0	0	3.0	0.636	0.727	4985.2
7	G06F	106	3787.0	0	0	4.0	0.601	0.694	2217.5
8	A61N	92	1553.0	0	0	3.0	0.592	0.671	2077.7
10	B32B	49	531.0	2	0	4.0	0.522	0.578	1621.9
11	G08B	76	1252.0	0	0	4.0	0.561	0.634	933.0
13	A61K	34	283.0	0	0	4.0	0.514	0.553	281.1
14	G02B	45	901.0	0	0	4.0	0.513	0.566	208.0
15	A61H	62	1239.0	0	0	4.0	0.546	0.608	884.4
16	A63B	58	896.0	0	0	4.0	0.532	0.594	336.3
17	H04W	50	905.0	0	0	4.0	0.519	0.576	167.3
18	G06Q	47	583.0	0	0	4.0	0.515	0.570	118.0
19	H04B	63	695.0	0	0	4.0	0.540	0.606	678.0
21	H01L	50	469.0	3	0	4.0	0.513	0.574	842.9
22	B25J	27	525.0	0	0	4.0	0.484	0.523	55.6
23	A41D	72	768.0	0	0	4.0	0.561	0.629	1272.6
24	G08C	61	446.0	0	0	4.0	0.535	0.600	377.9
25	G06K	56	530.0	0	0	4.0	0.526	0.588	261.1

materials or objects in general; disinfection, sterilization, or deodorization of air; chemical aspects of bandages, dressings, absorbent pads, or surgical articles; materials for bandages, dressings, absorbent pads, or surgical articles). In other words patents in the field of wearable health technologies are combined with the more general class of the methods and techniques for sterilizing objects and materials.

14.5 Discussions and Conclusion

The analysis clearly shows that the ideas registered in the field of wearable health technologies are concentrated in the framework of methods and techniques (A61B) and methods of sterilization (A61L). This can be interpreted as an increase in the need to diagnose human health with real-time healthcare follow-up. It is noteworthy that among the applicants, some universities (Univ Zhejiang, Univ Shanghai Jiaotong, Beijing Technology Inst) are involved in the registration of a significant portion of the registration documents obtained in the field of wearable health technologies. Real-time patient identification practices also bring with it the requirements for personalized health services. It is necessary for the wearable health technology to be constructed in an infrastructure that allows for the collection of personal information on the person and storage in a ready-to-use state. It is considered that the wearable health technologies that have reached the critical intensity necessary for technology will enter into a close relationship with information and communication technologies in order to obtain the structure necessary for providing real-time information to decision makers and service provider circles. It is also obvious that there will be a struggle for the integration of Internet of Things applications and conceptual framework in establishing the necessary infrastructure in this context. From this point of view it is clear that the collection and compilation of personalized data has a significant influence on wearable health technologies. It is thought that wearable health technologies will have an important place in the studies of this potential and the health integration activities to be carried out on the micro-, mezo-, and macro-scale projects and it is considered that there will be a concentration towards the ideas in this context.

References

[1] EPO, Annual Report 2016, 2016.
[2] D. Moher, et al., Preferred reporting items for systematic reviews and meta-analyses: the PRISMA statement, PLoS Med. 6 (7) (2009).
[3] B. Yoon, Y. Park, A text-mining-based patent network: analytical tool for high-technology trend, J. High Technol. Manage. Res. 15 (1) (2004) 37−50.
[4] A. Abbas, L. Zhang, S.U. Khan, A literature review on the state-of-the-art in patent analysis. World Pat. Inf., 37 (2014) 3−13.
[5] I. Park, et al., Exploring potential R&D collaboration partners through patent analysis based on bibliographic coupling and latent semantic analysis, Technol. Anal. Strategic Manage. 27 (7) (2015) 759−781.

[6] F. Narin, E. Noma, R. Perry, Patents as indicators of corporate technological strength, Res. Policy 16 (2-4) (1987) 143—155.

[7] D. Archibugi, M. Planta, Measuring technological change through patents and innovation surveys, Technovation 16 (9) (1996). p. 451519-468.

[8] Z. Griliches, Patent Statistics as Economic Indicators: a Survey., National Bureau of Economic Research, 1990.

[9] F. Madani, C. Weber, The evolution of patent mining: applying bibliometrics analysis and keyword network analysis. World Pat. Inf. 46 (2016) 32—48.

[10] A.J.C. Trappey, S.C.I. Lin, A.C.L. Wang, Using neural network categorization method to develop an innovative knowledge management technology for patent document classification. Proceedings of the Ninth International Conference on Computer Supported Cooperative Work in Design, vols. 1 and 2, (2005) pp. 830—835.

[11] C.V. Trappey, H.Y. Wu, An evaluation of the time-varying extended logistic, simple logistic, and Gompertz models for forecasting short product lifecycles, Adv. Eng. Inf. 22 (4) (2008) 421—430.

[12] M. Kim, Y. Park, J. Yoon, Generating patent development maps for technology monitoring using semantic patent-topic analysis, Comp. Ind. Eng. 98 (2016) 289—299.

[13] B. Gress, Properties of the USPTO patent citation network: 1963—2002. World Pat. Inf. 32(1) (2010) 3—21.

[14] Y.H. Tseng, C.J. Lin, Y.I. Lin, Text mining techniques for patent analysis, Inf. Process Manage. 43 (5) (2007) 1216—1247.

[15] T. Daim, S. Jordan, A foresight based on scientific indicators: a framework drawn from the case of laptop battery alternatives, Foresight 10 (3) (2008) 43—54.

[16] B.M. Gupta, C.R. Karisiddappa, Modelling the growth of literature in the area of theoretical population genetics, Scientometrics 49 (2) (2000) 321—355.

[17] T.U. Daim, et al., Forecasting emerging technologies: use of bibliometrics and patent analysis, Technol. Forecasting Soc Change 73 (8) (2006) 981—1012.

[18] R. Alhajj, J. Rokne, Encyclopedia of social network analysis and mining, Springer Publishing Company, 2014.

[19] M. Jacomy, et al., ForceAtlas2, a continuous graph layout algorithm for handy network visualization designed for the Gephi software, PLoS One 9 (6) (2014).

[20] S. Wang, M. Skubic, Y.N. Zhu, Activity density map visualization and dissimilarity comparison for eldercare monitoring, IEEE Trans. Inf. Technol. Biomed. 16 (4) (2012) 607—614.

CHAPTER 15

The Interplay Between Regulation and Design in Medical Wearable Technology

Jayson L. Parker[1], Qasim Muhammad[1], John Kedzierski[1] and Sana Maqbool[2]

[1]University of Toronto Mississauga, Department of Biology, Mississauga, Canada [2]University of Toronto Mississauga, Communication, Culture, Information & Technology, Mississauga, Canada

15.1 Consumer Wearable Technology is not a Medical Device so What's the Point of this Discussion?

The explosion in wearable technology on the market does not fall into the category of a medical device. In the view of this author, there are two forces at work here that draw upon medical device regulation. The first, as made clear by manufacturers [1] is that these products were introduced based on a market analysis that assumed they would not be regulated as medical devices. To avoid medical device regulation, you need some understanding of that framework to know how far your design and marketing team can go before you fall under the jurisprudence of a health authority. This requires some knowledge of medical device regulation. The second consideration is motivated by fear and lack of expertise: the perception that regulation of a medical device is an all or nothing proposition with burdensome requirements that would diminish the business case for the proposed product in question. If this second issue can be addressed, this author believes that many firms recognize the business case for wearable technology for chronic diseases and aging at home, and would pursue these comparatively price insensitive markets if they had more confidence in what it meant for a product introduction to be regulated as a medical device.

15.2 Organization of this Chapter

In this chapter, we will spend some time covering medical device regulation independent of specific considerations for wearable technology. While there are plenty of sources on the basic elements of this topic, the nuances that are critical and have not been reviewed to our knowledge. Once we have done this, we will then move onto digital health regulation. It is important to realize that product regulation, especially of digital health technology, by a

Wearable Technology in Medicine and Health Care.
DOI: https://doi.org/10.1016/B978-0-12-811810-8.00015-4

health authority is constantly changing. By the time this work is read, there will already be changes. Part of the goal of this chapter is to convey the fundamental logic of product regulation, so the reader is empowered to find and readily understand upcoming changes to this area. If you feel you understand the basics of regulation—it is advised not to skip the basics covered here. Discussion will focus on the decision making by the US Food and Drug Administration (US FDA; the author here is Canadian so there is no patriotic bias). The US FDA long ago established itself as a regulatory leader globally through its prescient decision to decline the marketing of the notorious drug thalidomide [2]. In addition, the importance of the size of the US market makes it an essential as part of any product proposal. Fortunately, there are a lot of similarities across regulatory agencies as they appear to follow the US FDA as a best in class example (based on personal communications by the author with other agencies).

The author will refer to product designers in this chapter. Engineers are essential to this process, but other backgrounds are also needed to achieve insightful design. In addition, "patient" and "user" will also be used, reflecting the shifting divide between health and wellness products and traditional medical products. This divide between health and wellness products, versus medical products, is shifting. Wearable health products available to consumers today collect enough data on a range of body functions that it can become confusing whether they are intended for medical purposes. As a result of the increasing power, wearable technology available to consumers, a more nuanced approach to classifying medical devices is needed and that thinking is discussed in this chapter.

15.3 What is the Relevance of Device Regulation by a Health Authority?

Understanding how a product is to be regulated has far-reaching implications. Once we understand how a device is regulated by the health authority for the country in question, we can immediately get a sense of how long it will take the product to reach the market and the complexity of product testing required. While health authorities, such as the US FDA are concerned with ensuring products are safe, to varying degrees they are also concerned with the effectiveness of the product. The longer it takes a product to reach the market the more expensive it becomes, making its business case less attractive. A strong business case is relevant when seeking financial support as a start-up or championing a new product idea within a company. The complexity of the tests required to validate the product (safety and efficacy) negatively impact the costs to take a product idea to market. For medical devices and health and wellness products, we can iterate between the implications of a design choice and what it means for how it will be regulated, to determine the optimal design strategy. Wearable technologies have challenged regulatory agencies with how to manage these products where in the past the division between medical devices and consumer products was straightforward.

15.4 Medical Device Regulation—Not a Primer

For the reader, new to medical device regulation, other sources will be more helpful to introduce the subject matter [3]. This section is concerned with the underlying logic and critical nuances. Our quick journey will start with what some consider the gold standard of how medical regulation should be performed—drugs. Once we understand what rigorous regulation looks like, we can then consider the device world.

Drug regulation has Draconic requirements, but it is very straightforward strategically when considered against the process of medical device regulation. Drugs undergo safety testing, a pilot study, and then a formal study in respective phases I–III clinical trials. Phase II studies are small clinical studies that usually have a control group, that give the investigators some idea of whether the product is working to inform their decision on whether to proceed to a very expensive phase III trial. Phase III trials have large clinical populations (this will vary based on the indication) that are statistically powered to 80%.

The first rule of product regulation is that the key rules are often not specified—there is no document ("guidance document" is the specific term) that lays this out explicitly. We find such answers by looking at the history of approval decisions by the FDA on their website. We search for the product specifically (for medical devices there is a corresponding 3-letter code for each product category—once you have this is it is easier to search). There is no explicit statement from the FDA that phase III trial has to be statistically powered or necessarily what an appropriate endpoint would be for approval. Answers to these questions, just as in case law in legal practice, are inferred by looking at recent cases of product approval. Reviewing approval history for a specific product line is an essential tool to understand what is acceptable practice for product approval. The decision history of the FDA for a specific product line (e.g., reviewing approval decisions over the past several years) informs us on acceptable endpoints, whether a clinical study is necessary, sample sizes and so on.

If a drug is changed in anyway (unless a biosimilar compound; [4]) it must repeat the entire clinical trial testing process from phase I to phase III, which can easily take 8 years to complete [5]. The real issue to wrestle within drug development is how to design your clinical trial. Your path to market is laid out and largely invariant from one drug to another, as each follows each clinical testing phase (Fig. 15.1). The rule with drug regulation—there is a healthy respect for the unknown and thus any change in the structure of a drug, even if by a single atom, is greeted with extensive testing.

Medical device regulation is a very different thought process compared to drug regulation. This shift in thinking can make many uncomfortable. A new medical device navigating its path to market could conduct a pilot study and a pivotal study (a larger more formal study of product performance). However, a medical device may reach the market without undergoing testing in a pivotal trial. It may also reach the market with no clinical data

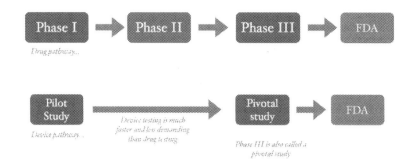

Figure 15.1

An overview of the basic differences in the approval pathways by the US Food and Drug Administration for drugs and medical devices.

Table 15.1: An overview of device risk class in the US. The key to risk class determination is to ask what the potential for injury is to the patient should the device fail to operate properly or be misused. This precept will be elaborated in more detail later, and will be a constant driver behind any changes to how medical devices are regulated.

Risk Class	Device Example	Potential of Serious Injury
Class I	Cotton swab for ear	low
Class II	X ray for dental imaging	medium
Class III	Cardiac stent	High

whatsoever—not even a pilot study (Fig. 15.1). This was most dramatically illustrated by hip-on-hip metal replacements [6] that reached the market without any clinical testing. The surgical revision rate reached four times the average in a registry for other hip replacements [6], or in some cases, the failure rate was as high as 50% [7]. Mounting evidence likely lead to the voluntary withdrawal of metal on metal hip implants. This was not a "mistake," there are numerous examples of medical devices reaching the market without clinical testing.

Unlike drugs, the medical devices regulatory paradigm can be summed up by saying regulators think can predict the consequences for users and patient with device design changes and thus formal testing requirements are often relaxed. This reflects a balance between keeping the costs of innovation down to meet the needs of society and public safety. This paradigm of our confidence in our ability to predict the consequences of device design changes for users and patients is expressed in the application of risk classes (Table 15.1). Risk is defined as the potential for harm to the patient (or even user of the device) should it fail to operate properly. In this paradigm, class I is the lowest risk, where a malfunctioning cotton swab to clean one's ear has very little risk (both in severity and probability—to be reviewed in hazard analysis later). On the other hand, a malfunctioning device implant, such as a cardiac stent can result in the death of the patient, and is thus class III (European and Canadian risk classes are slightly different but the concept is the same).

Heart—lung machines are class II [8] while external defibrillators are class III [9]. Does this make sense? Not in our opinion, which brings us to a central practical concept of medical device regulation: how a product is regulated is not just about the science and clinical reality. How a device is regulated also reflects externalities, such as special interest groups, political pressure, lack of resources to update or change questionable practices and cost control. If you simply try to deduce how a device should be regulated from a purely logical perspective you will be right most of the time—but there will still be many cases where you will be wrong due to these externalities. Again, the best plan of action is to check the recent approval history (decision history) for similar products to get your answer. Unlike most regulatory agencies, the US FDA has an extensive database with guidance documents and a good paper trail—so even if your regulatory environment is outside the US, the US FDA can be a great resource for material to help you think through how your product will be regulated (the paper trail for drugs is far superior to that for medical devices).

When is clinical data required? That depends on the decision history for the product line in question and fundamental analysis you can perform using hazard analysis. There is no consistent relationship, for the most part, between the device risk class and the requirement for clinical data as part of the approval process (Fig. 15.2). In this author's experience, with the exception of cardiovascular products, clinical data for pivotal studies with medical devices are rarely statistically powered. This is unlike the situation with drugs, and reflects again our different paradigm with medical devices. Most medical devices come to market without any clinical data [11].

The risk class is an important aspect of a medical device. But a medical device with class II may go to market through either a 510k pathway or, more rarely, a premarket application (PMA). The 510k pathway has no equivalent in the European or Canadian health agency regulatory frameworks. This pathway says if your proposed product is sufficiently similar to a product the US FDA has already approved (called a "predicate"), you can skip many of

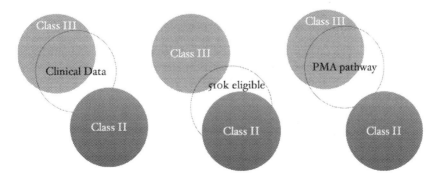

Figure 15.2

The risk class does not automatically tell us whether any clinical evidence has been required to support the approval of the device. Similarly, the risk class of the device does not automatically preclude or include the 510K or PMA pathway, at least for class II and class II products.

the testing requirements since a facsimile of your product is already on the market ("substantially equivalent"), and by inference, your product is safe and effective as well. This makes the FDA an attractive first market for device if the proposed product can be found to be substantially equivalent. If the FDA does not deem the product substantially equivalent, then the product may have to do a PMA (where clinical studies are almost unavoidable) or opt for a new pathway called the *de Novo* 510k (not to be reviewed here). Clinical data is not always required for 510k filings—it depends—you have to review the decision history for your product category (Fig. 15.2).

The 510k pathway lends itself to some design strategies that intentionally or unintentionally have been made use of across product lines. Originally called the "serial 510k problem," the author prefers to see this as an example of "design creep." If you are not eligible for the 510k pathway because your product is too dissimilar from approved products, including a prior generation of your own product already on the market, you need to do a PMA (e.g., a formal study involving clinical data beyond 510K requirements). Rather than introduce one big change to your product which could trigger the requirement of extensive testing with clinical studies, you could introduce that design change in smaller increments over successive product generations, avoiding this requirement (Fig. 15.3). In this strategy, the ancestor (first product) of the product line compared to the product today may not be substantially equivalent, but each minor increment of design change is considered substantially equivalent. In this manner, major design changes could be introduced over

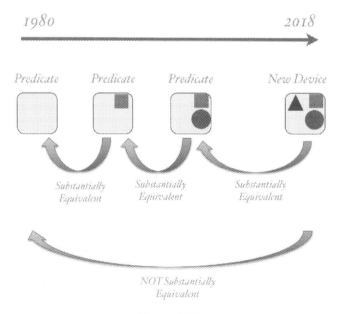

Figure 15.3
The serial 510k problem describes how minor design changes that require very little evidence, can accumulate over time to create a new product with large design changes that escaped proper evidence to establish safety.

time that avoid major testing requirements if they were introduced in a single generation of the device (this is an observation, not a practice the author supports). Surprisingly, there is good evidence this as taken place for a product as serious as heart−lung machines [10].

The 510k pathway possesses some interesting marketing challenges. How can a product claim to be superior to its competitors in a marketing context, if it was approved through the 510k pathway? If the FDA has reason to believe your product is superior to your predicate, you may not be found substantially equivalent. Even if you have a superior product, you want to show it has similar performance to your predicate. This can be achieved in part by relying on noninferiority tests in product testing, which makes it more difficult to detect superior performance by your product versus a comparator [11]. If a manufacturer can pull this off with a superior product, they can then rely on word of mouth and small published studies by customers that later demonstrate your product's superiority without jeopardizing your use of the 510k pathway (anonymous industry sources; observation not endorsed by the author).

The 510k pathway deems your product substantially equivalent to a predicate if you appear sufficiently "similar" with respect to safety, performance, design, and product labeling. If you are not sufficiently similar with respect to all of these criteria you are not, officially, substantially equivalent, and therefore do not qualify for the 510k pathway. However, how strict the FDA is in demanding similarity for each of the consideration looks like it varies with the product category in question (3-letter code in the US FDA website for searching). For example, a product for reviewing medical imaging on a mobile device was compared against a predicate of a desktop radiology viewer station (author is avoiding specific product /company references here to avoid friction with the community). The question has been asked how similar with respect to design a mobile viewer is to a desktop viewing station. While the FDA has accepted this as a predicate (eventually), many would hold the view that there are substantial design differences here. There are also examples of products deemed substantially equivalent, despite differences in their product label.

How do we come to terms with all of this? Again, review the decision history for your product category. Once you know how much tolerance the FDA allows for "similarity" for each of these criteria, this gives you the design margin of how different your new product concept can be and still be considered substantially equivalent to existing predicates (design creep logic). The *de Novo* 510k pathway in theory makes things easier as it requires no predicate, but this is a new pathway and with little guidance and decision history behind it, it is ripe for abuse and this author believes its application will be curtailed or short lived. Be mindful your predicate has not relied on the now no longer accepted split predicates (unlikely for wearable products) or a product that was forcibly removed from the market. Products voluntarily withdrawn from the market by the manufacturer are fair game for use as a predicate [11].

To recap the logic above combined with the strategy described for design creep, the goal of a designer is to create a product with the biggest design differences they can come up with that are still sufficiently similar to existing predicates to qualify for the 510k pathway. Class III devices are not eligible for the 510k pathway, but such a risk class is unlikely to come up for a wearable technology product. This design maneuverability room for the product engineer (in my view, not all skills for good design come from an engineering background) does not mean one has to go that far in creating a product. Some products approved by the 510k pathway, while similar to a predicate, are sufficiently different to the user that they are deemed superior once on the market. This may be due to ease of use, compatibility issues or undetected performance superiority during product testing through the use of noninferiority analysis of test data that the user is able to recognize. There is a close relationship between product design and regulatory requirements, and the designer needs to move back and forth between these two lenses to find the optimal balance. Predicates will not always be a competitor product. Depending on how flexible the FDA is, a predicate may come from another market area altogether (those new to this thinking are often quick to conclude that there are no predicates for their product). It must be emphasized that how similar the new product must be for each of the criteria for substantial equivalence will be different across product categories (Fig. 15.4). *The goal of product design is one that is superior to competitors in the eyes of the user, but similar in the eyes of the regulator.*

The risk class of a device is an important determinant of how it is regulated. However, the risk class does not just reflect design features. Product claims of user benefit and the type of user targeted are also factors, as well as how the data is shared from the manufacturer for products with data transmission. To appreciate these elements, let us turn to hazard analysis, the conceptual underpinning that allows us to adapt to future changing regulatory policies.

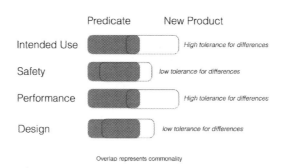

Figure 15.4

The rules of the 510k pathway appear to be applied unevenly to different product lines. For some product lines, there may be little tolerance for any deviations in the intended use before establishing a product is not substantially equivalent. Conversely, for other product lines there may be a lot of tolerance for such differences before concluding a product is not substantially equivalent. For each criteria of the 510K pathway, the historically permissible changes that are 510L eligible have to be established for a specific product category to determine what is possible in future design iterations such that they can be 510K eligible.

15.5 Combination Medical Device Products

For digital health, and perhaps wearable products in future, this has emerged as a reality. Hybrid or combination products involve both biological and medical device components. Or combination products can involve medical device and drug components. Classic examples of this would include a drug-eluting cardiac stent. The cardiac stent would be implanted to hold a narrowed blood vessel open. The drug is added so it elutes from the stent over time to help control local inflammation [12]. While the device may be composed of a combination of elements, there has to be a dominant classification. In this case, a cardiac stent is regulated as a medical device despite the fact it involves a drug. The dominant classification is based on the primary mode of action of the product [13]. The primary mode of a cardiac stent is mechanical as its purpose is to hold a narrowed artery open, while the drug merely enhances this action.

Some companies in the digital health space have combined a drug with a bioabsorbable transmitter, which sends a signal when the drug is ingested [14]. There has been some confusion here, and while the situation may change subsequent to this writing such a product, in our view, should be regulated as a drug. The primary mode of action of a drug is not changed with the addition of the transmitter. However, if the manufacturer sells the transmitter separately, and the user combines the transmitter with the drug at home for ingestion, then in our view the bioabsorbable transmitter may be regulated as a medical device since it is not sold as a combination product. However, the firm had developed the sensor and drug as a physically combined product. The firm submitted their combination product as a new drug application to the FDA and later rejected [15].

Wearable manufacturers are increasingly eyeing the diabetes market and blood glucose testing. They can avoid the draconic realm of drug regulation and retain their status as a medical device, so as long as they sell their hardware separately from any medication that diabetics would use as part of their treatment. There is bound to be controversy and changes here, but the essential principles above are not expected to change: a product's classification is defined by its primary mode of action.

A product is a medical device so long as its primary mode of action does not rely on metabolism [16]. For those with advanced training in biocompatibility, this knowledge can make lead to confusion when considering the primary mode of action of an implant. It is true that any object in contact with the interior of the body will trigger some chemical reactions, and may affect some biochemistry in the patient (or user). *However, this does not make it the primary mode of action unless such reactions are necessary for the device implant to function.*

15.6 When is a Wearable Technology a Medical Device? Hazard Analysis is the Key

The best way to understand regulation barring externalities, and the best way to anticipate future changes by the FDA, is through hazard analysis (sometimes called risk analysis). If the regulation of a product category of interest seems to be inconsistent with a hazard analysis, that is a red flag thing may change in the future. It is not enough to just apply current guidance, since it will inevitably change to some degree with time. When an area of technology opens up exposing users to new risks, or increases the likelihood of hazards previously known, that is a red flag regulation is about to change. *It's up to design teams to anticipate the thinking of the FDA.*

There are many tools for hazard analysis. We prefer a quick assessment which involves listing the possible adverse health outcomes, severity, probability, and cause (Tables 15.2–15.4 derived from Kam [17]). More involved tools, such as failure mode and evaluation analysis and human factor engineering, are used but require much more time to conduct [18].

Table 15.2: An overview of one example of hazard analysis. This is the first of the hazard analysis illustrated in this chapter. Here the severity of an adverse outcome as well as the probability of an event are given unitless measures. In this example the distances between categories are linear but this does not have to be the case. This establishes the numerical scale for probability and severity that is used later in this analysis.

Probability		Severity	
Rating	Meaning	Rating	Meaning
1	Extremely remote	1	Negligible
2	Remote	2	Minor
3	Occasional	3	Moderate
4	Reasonably probable	4	Critical
5	Frequent	5	Catastrophic

Table 15.3: An overview of one example of hazard analysis. Using the numerical scales established in Table 15.2, this is now applied to actual clinical adverse outcomes from device failure or misuse. For each adverse event, a clinical outcome is established that is quantified representing the hazard index.

Hazard	Health Outcome	Severity	Probability	Hazard Risk Index
Incorrect heart rate reading	Chest pain	Moderate (3)	Occasional (2)	6
Incorrect heart rate reading	Heart attack	Catastrophic (5)	Occasional (3)	15
Incorrect number of daily steps	Fatigue	Negligible (1)	Frequent (5)	5
Incorrect measure of daily standing time	Thrombosis	Critical (4)	Remote (2)	8
Aggressive exercise	Heart attack	Catastrophic (5)	Occasional (2)	10

Table 15.4: A numerical interpretation of the hazard index in terms of responses. Note that is possible for adverse events of low severity to be unacceptable if the probability is very high. In contrast, it is also possible that severe adverse events can be of low risk if the probability is very low. This table translates the hazard index into actionable responses for designers and clinicians.

15−25	Unacceptable risk	Corrective action has to be performed
10−14	High risk	Risk reduction should be performed
5−9	Medium risk	Mitigation is recommended
1−4	Acceptable low risk	Acceptable for implementation

Hazard analysis allows us to see, from fundamental analysis whether our wearable product may be considered a medical device. The official ruling from the FDA is that any product that treats or diagnoses a condition is a medical device [19]. This is also true for any app or software [20]. That is easy to identify. What complicates the picture is that products that mitigate the likelihood of the disease may also be regulated as a medical device. This creates a murky boundary in determining whether a product should be considered a medical device.

If we consider a wearable device marketed to patients who are recovering from myocardial infarction within the past 6 months, this may be considered a high-risk population. The consequences of faulty readings from an activity tracker may be irrelevant for an average person but crucial for this specific user population. The health outcomes (Table 15.3) reflect this specific user population, which are very different from what we would enter for healthy user. We can see a range of serious consequences here that are at least somewhat likely.

Health and wellness products are not regulated by the FDA but medical products are regulated [21]. In regulatory parlance, "health" and "medical" are two entirely different categories of products. While our wearable activity tracker is not marketed as a medical device, it may be considered as such for several different reasons. The first is our target user population as described above. The elevated risks associated with device error in this population alone could spur the FDA to regulated this as a medical device. However, if we marketed this to the general population, while some patients with recent myocardial infarction might use this device, this is not the same as targeting such users in a marketing strategy. Based on the hazard analysis presented here, we would have real concerns about the safety of the targeted user population.

Claims of user benefit can also elevate the risk of a product, which may cause it to be regulated as a medical device. In this example, if the wearable device makes a bold claim to prevent heart attack from exertion, this would give users a false sense of confidence and as such may engage in aggressive exercise well beyond their capacity (Table 15.3). Such a claim of user benefit, as we can see from the table, has unacceptable outcomes. In this context, such a bold claim, in the absence of any change in the design of the device, could cause it to be regulated as a medical device.

Potential abuse by the user also factors into the risk assessment of a product. Independent of bold claims, if users get a false sense of confidence with a device that can lead to unacceptable outcomes, this can elevate the risk class of the device. There is a parallel here with devices intended to inform users of peanut antigen for those with lethal allergies. While not marketed as a medical device, any false sense of confidence that the user gains could have catastrophic outcomes. In this discussion, we are forward looking at the wearable technology landscape. As yet, there are no cases yet that have transpired to illustrate this issue, but we can anticipate this could be an issue.

A more recent issue to emerge is the risk associated with biometric data sharing [22]. Such sharing would occur between the manufacturer and the user of the wearable device. Depending on the type of user and the type of data, the consequences of confusing data display could lead to harmful decisions by the user. In our example, any number of errors could instead just be the result of a confusing display to the patient, resulting in a similar hazard index (Table 15.3), such as incorrect daily number of steps or incorrect standing time.

In summary, the kind of user targeted, the claims of user benefit and the potential for user overconfidence all can cause a wearable device to become a regulated health product. Designers can use this to their advantage, by making sure their product concept risk class is not elevated due to user product misuse or confusion.

15.7 Ghost in the Machine: Software Considerations in Wearable Technology

Software is a medical device if it is involved in treating or diagnosing a disease [20]. If it is used in broad disease mitigation, it is less clear and we would use hazard analysis and above and review FDA decision history to determine if it is a medical product.

Software is either part of the product or interacts with the product from a distance. These two -design choices have two very different regulatory outcomes. The FDA has regulated software as medical device if it is connected to a medical device. Such software would have the same risk class as the device it is a party of (e.g., software is class III in implantable cardioverter defibrillators [23]). In contrast, an app or remote software may not be considered a medical device merely because it interacts with a medical device [20]. This can impact design where a high-risk (even if class II) wearable product interacted with software via the cloud, which would allow such software to have a lower risk class, or not be regulated as a medical device at all.

Displays of biometric data and the analysis of the data is also a relevant design consideration in a regulatory context. The FDA recognizes that accessories to a medical device may have a lower risk class or not be regulated products at all [24]. A display of data on a secondary viewer screen which is not involved in arriving at decisions in regard to patient care will have a lower risk class than the viewer of the same data used to arrive

Figure 15.5
The advent of digital health has created many new design permutations to existing products that
can be regulated differently. The software landscape continues to change and will be different
again by the time this chapter is published.

at decisions regarding activity and care. The mobile device or wearable technology could be
the secondary viewer screen, with the primary viewer in the physician's office (Fig. 15.5).
We can see hazard analysis play out here again, where, in principle, the secondary screen
may be irrelevant or at lower risk than the parent device. However, it can also be a source
of risk if confusing readings give the patient a false sense of confidence (user confusion in
viewing data has been reviewed for biometric data sharing [22]). Confusing readings giving
a false sense of confidence may lead the user to take vacations in remote areas that may
delay urgent access to care if they are unknowingly at risk of requiring such support. A big
issue with wearable technology is user adherence. Industry experience for some
manufacturers is that 3 months is the precipice, after which sustained use of digital
technology by users drops off substantially (personal communication). Adherence refers to
the continued use of the technology by the user. If a secondary viewer screen, perhaps
though a confusing display, gives the user a false sense of confidence in their progress this
could exacerbate adherence problems in the use of the wearable technology.

Software raises the intriguing design consideration of what to physically connect to the
device and what functionality should be performed at a distance. The separation of these
activities allows for lower risk classes of remote software. At the same time, this author
speculates that while secondary viewer screens are lightly regulated, a more cautious
approach based on hazard analysis is necessary. Secondary viewing screens could be a
problem if the displays are confusing that could impact a patient's behavior (who we can

also call a user if you will since they see the secondary screen) in unforeseen ways depending on their condition and lifestyle.

15.8 Coda

There is at times a Quixotic interplay between product regulation and choices to be made in wearable technology design. This is in part due to the questionable elements of regulation that reflect externalities that go beyond science and clinical considerations. While product regulation provides helpful guidelines for design teams, a deeper understanding is required to anticipate where regulation is heading so teams do not find their product regulated in a very different manner a few years from now. Software is the most rapidly changing area of health regulation, and wearable technology is likely to track that trend. Design teams have to take seriously a hazard analysis approach to their product concept and look at how they can leverage the regulatory landscape to avoid what some consider burdensome testing requirements. It is important that design teams are emboldened to enter specific markets, such as chronic illness or aging at home, where wearable technology may command a stronger value proposition. Having one's product regulated as a medical device is not necessarily burdensome as there is a spectrum of oversight. With smart design choices informed by the regulatory environment a product can be kept on the low end of the risk continuum.

References

[1]   S. Gottlieb. Why Apple dumbs down your smartphone. Forbes. Dec 4 (2015): https://www.forbes.com/sites/scottgottlieb/2015/12/04/why-apple-dumbs-down-your-smartphone/#579532883393

[2]   M. Hamburg. 50 years after Thalidomide: why regulation matters. US Food and Drug Administration. Feb 7th (2012): https://blogs.fda.gov/fdavoice/index.php/2012/02/50-years-after-thalidomide-why-regulation-matters/

[3]   US Food and Drug Administration. Overview of device regulation. https://www.fda.gov/MedicalDevices/DeviceRegulationandGuidance/Overview/ucm2005300.htm. (accessed 17.05.17).

[4]   US Food and Drug Administration. Information for consumers (biosimilars). https://www.fda.gov/drugs/developmentapprovalprocess/howdrugsaredevelopedandapproved/approvalapplications/therapeuticbiologicapplications/biosimilars/ucm241718.htm. (accessed 17.05.17).

[5]   K.I. Kaitin, J.A. DiMasi, Pharmaceutical innovation in the 21st century: new drug approvals in the first decade, 2000−2009, Clin Pharmacol Ther 89 (2) (2010) 183−188. Available from: https://doi.org/10.1038/clpt.2010.286.

[6]   B.M. Ardaugh, S.E. Graves, R.F. Redberg, et al., The 510k ancestry of metal on metal hip implant, NEJM 368;2 (2013) 97−100.

[7]   British Orthopaedic Association. Large diameter metal on metal bearing total hip replacements, 2011. Doi 10.2106/JBJS.K.01220

[8]   U.S. Food and Drug Administration, Centre for devices and radiologic and health. product classification. https://www.accessdata.fda.gov/scripts/cdrh/cfdocs/cfpcd/classification.cfm?ID = 764. (accessed 31.05.17).

[9]   US Food and Drug Administration, Centre for devices and radiologic and health. product classification. https://www.accessdata.fda.gov/scripts/cdrh/cfdocs/cfpcd/classification.cfm?id = 849. (accessed 31.05.17).

[10] J.L. Parker, R. Abdelmegid. Have heart lung machines undergone proper safety testing? (in preparation).

[11] US Institute of Medicine of the National Academies. Medical devices and the public's health: the FDA 510(k) clearance process at 35 years. The National Academies of Sciences, Engineering, Medicine (2011). https://www.nap.edu/catalog/13150/medical-devices-and-the-publics-health-the-fda-510k-clearance. (accessed 17.05.17).

[12] T. Htay, M.W. Liu, Drug eluting stent: a review and update, Vasc Health Risk Manag 1 (4) (2005) 263–276.

[13] US Food and Drug Administration. Office of combination products. https://www.fda.gov/combinationproducts/aboutcombinationproducts/ucm101496.htm. (accessed 31.05.17).

[14] http://www.proteus.com. (accessed 31.05.17).

[15] FDA issues complete response letter for digital medicine new drug application. Business Wire (2016): http://www.businesswire.com/news/home/20160426006993/en/FDA-Issues-Complete-Response-Letter-Digital-Medicine.

[16] US Food and Drug Administration. Guidance for industry and staff: the interpretation of the term chemical action in the definition of device under Section 201(h) of the federal food, drug and cosmetic act. (2011): https://www.fda.gov/downloads/RegulatoryInformation/Guidances/UCM259068.pdf. (accessed 31.05.17).

[17] D. Kamm. An introduction to risk/hazard analysis for medical devices. http://www.fda-consultant.com/risk1.pdf. 2005. (accessed 01.06.17).

[18] US Food and Drug Administration. Applying human factors and usability engineering to medical devices. (2016). https://www.fda.gov/downloads/MedicalDevices/.../UCM259760.pdf. (accessed 01.06.17).

[19] US Food and Drug Administration. Is the product a medical device? https://www.fda.gov/MedicalDevices/DeviceRegulationandGuidance/Overview/ClassifyYourDevice/ucm051512.htm. (accessed 01.06.17).

[20] US Food and Drug Administration. Mobile medical applications. guidance for industry and food and drug administration staff. (2015). https://www.fda.gov/downloads/MedicalDevices/.../UCM263366.pdf. (accessed 01.06.17).

[21] US Food and Drug Administration. General wellness: policy for low risk medical devices. Guidance for industry and food and drug administration staff. (2016). https://www.fda.gov/downloads/medicaldevices/deviceregulationandguidance/guidancedocuments/ucm429674.pdf. (accessed 01.06.17).

[22] US Food and Drug Administration. Dissemination of patient-specific information from devices by device manufacturers. Guidance of industry and food and drug administration staff. (2016). https://www.fda.gov/ucm/groups/fdagov-public/@fdagov-meddev-gen/documents/document/ucm505756.pdf. (accessed 01.06.17).

[23] US Food and Drug Administration. Summary of safety and effectiveness data (SSED). (2012). https://www.accessdata.fda.gov/cdrh_docs/pdf11/P110042B.pdf. (accessed 01.06.17).

[24] US Food and Drug Administration. Medical device accessories: Defining accessories and classification pathway for new accessory types. Draft guidance for industry and food and drug administration staff. (2016). https://www.fda.gov/downloads/medicaldevices/deviceregulationandguidance/guidancedocuments/ucm429672.pdf. (accessed 01.06.17).

Index

Note: Page numbers followed by "*f*" and "*t*" refer to figures and tables, respectively.

Printed in the United States
By Bookmasters